For, after all, what is man in nature? A nothing compared to the infinite, a whole compared to the nothing, infinitely remote from an understanding of the extremes; the end of things and their principles are unattainably hidden from him in impenetrable secrecy. Equally incapable of seeing the nothingness from which he emerges and the infinity in which he is engulfed. What else can he do, then, but perceive some semblance of the middle of things, eternally hopeless of knowing either their principles or their end? All things have come out of nothingness and are carried onwards to infinity.

– BLAISE PASCAL, *Pensées*, 1670

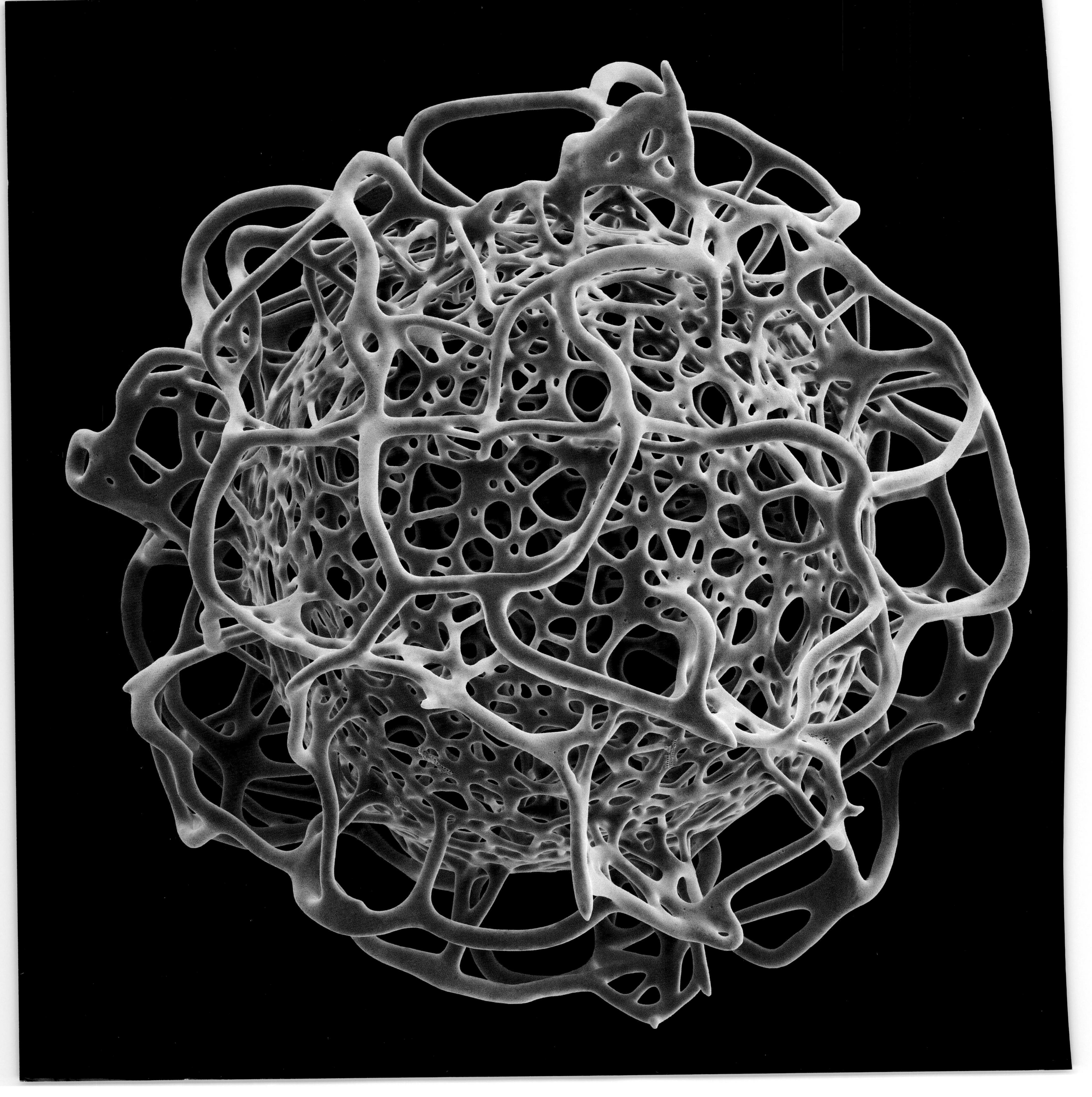

# NANOCOSMOS

## Journeys in Electron Space

Michael Benson

Abrams | New York

In memory of my mom,
Shirley Benson

Sept. 8, 1932—Nov. 7, 2023

The temple bell stops
but the sound keeps coming
out of the flowers.

– Bashō

# Table of Contents

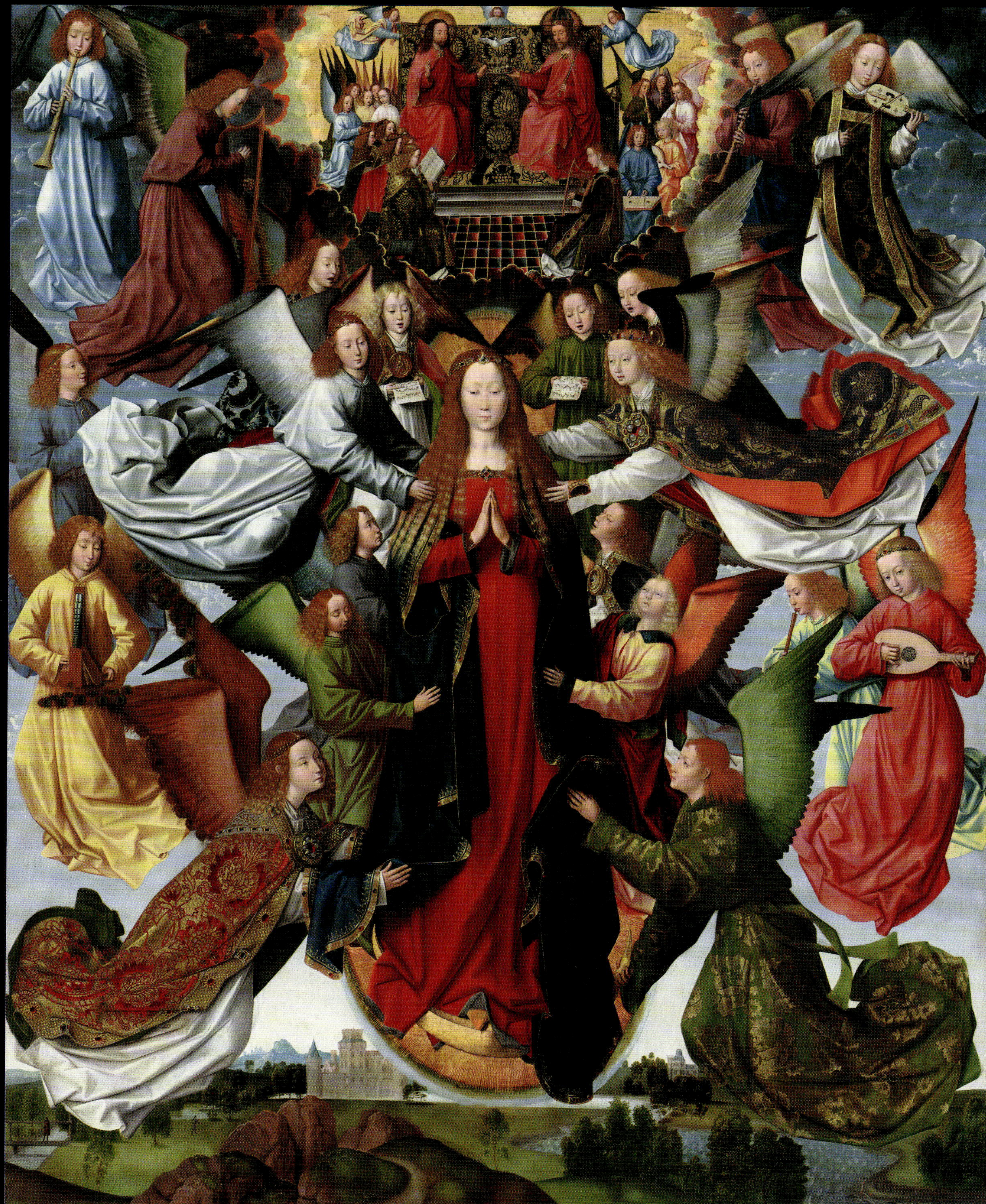

# Introduction



About our place in nature they were never wrong, the Old Masters. How well they understood our human position: how even the most miraculous events—the arrival of a cavalcade of kings to kiss the feet of the squirmy infant; the levitation, decades later, of his mother to heaven, escorted by phalanxes of angels—took place within nature's green frame. Immaculate in billowing silks, dozens of winged forms flutter around the Virgin, a surreal panoply anticipating the weightlessness of spaceflight. Accompanied by an angelic orchestra, Creator and Son flank the brocaded throne. Each holds one side of her crown. At the apex, a white dove: the Holy Ghost. The Queen of Heaven ascends to her coronation.

This glorious assemblage occupies a good nine-tenths of the enormous panel, with only a horizontal band of green at the bottom signifying her point of origin. But the flatlands visible there—a classic Low Countries scene, albeit with a range of distant blue mountains—couldn't be more telling. As base and launchpad to all the metaphysical frippery, it plays a relatively small role. But if you look closely, you see that actually, it's the only section of the painting where *physics* reigns. Indicative not just of nature's presence, but also of what used to be called natural philosophy.

The unnamed artist, identified only as the "Master of the Saint Lucy Legend" and active in Flanders in the late fifteenth century, effectively portrays this narrow zone of damp earth and winding waterways as the only place where he—and it was almost certainly a man—allowed himself the freedom, within the commercial imperatives of the theological construct, to exercise an observational sensibility focused on the phenomenal world.

So, what do we see there, below all that floating Flemish silk? A man on horseback, accompanied by a dog. Trees. Wind-tossed water. A shore. A well-defined, almost three-dimensional cockle

**Left:** Master of the Saint Lucy Legend, *Mary, Queen of Heaven*, circa 1485, oil on panel

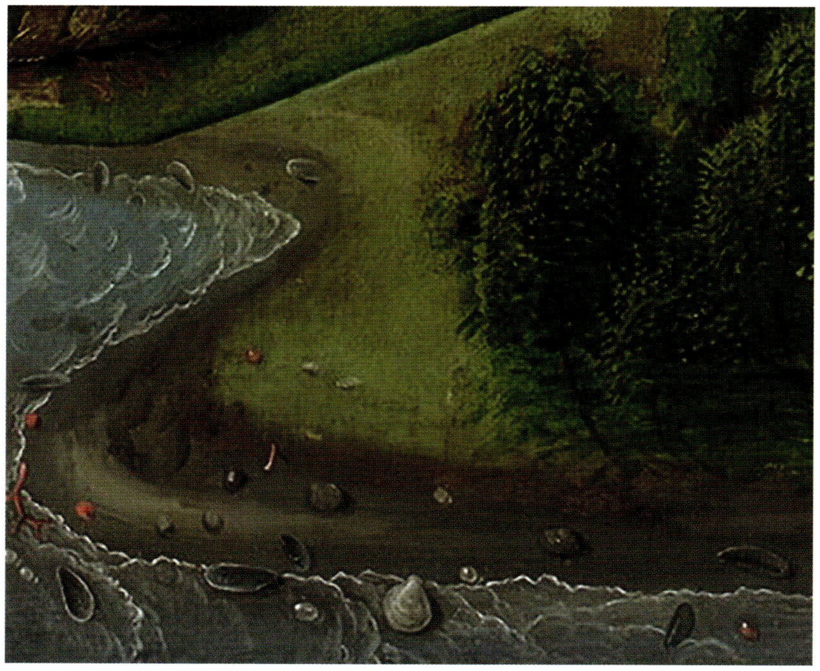

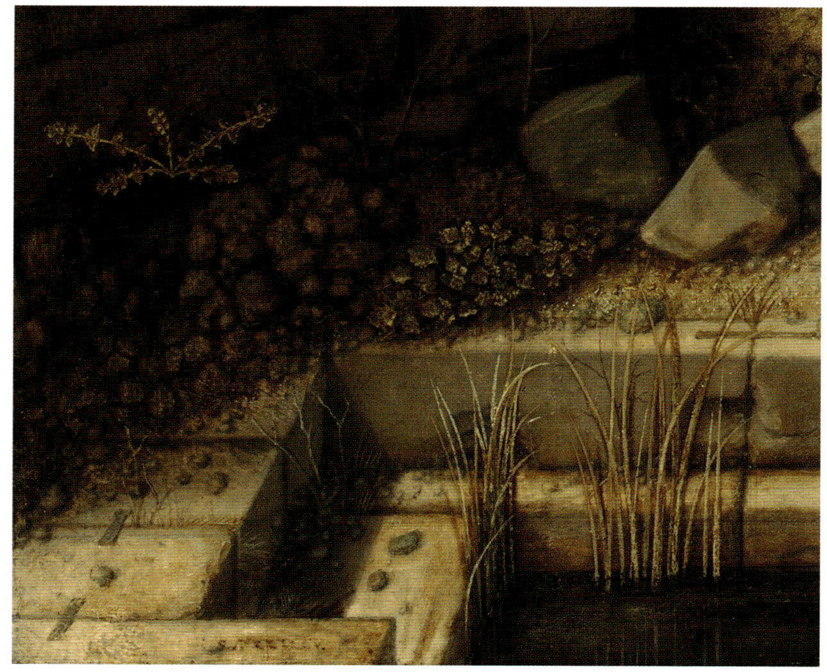



shell. Two dozen mussels, some submerged, some on the shore, some churning at the border of waves and sand. Eight fragments of pink coral. A hint of submerged seaweed. A number of glistening white orbs, on and near the shore. The latter are a bit puzzling. Are they pearls? Jellyfish?

In any case, what we're seeing is clear. Life, emerging from the water.

Let's take another example, this one by Italian virtuoso Giovanni Bellini. Saint Jerome in the desert. Bearded and elderly, clad only in a pale blue loincloth, Jerome reads. His bare feet are splayed comfortably on a pebbly stone floor—one can sense the mineral grains, the warmth emanating. His book stands on a natural rock pedestal. He's in a cave of sorts, but with plenty of foreground light.

The cave's oval mouth provides a frame within a frame. It opens onto the clean deep blues and yellow-brown siennas of a southern landscape seemingly too large for the painting. Distant trees, towers, a bridge. No other human beings, though—as the first to translate classical scripture into Latin, Jerome needed his peace and quiet. There's water here too, this time greenish (a river) and cerulean (the deep blue Mediterranean). Directly behind our recluse, a stratified rock wall displays shrubs, ivy, grass, tiny white flowers. A spindly tree is rooted between stone layers above his head. In its horizontal geometry, this natural bulwark displays clear evidence of geological time scales at variance with biblical chronologies.

In the foreground, dozens of closely observed weeds, reeds, and yet more grass and ivy protrude from flagstones and the reflective water of a cistern. A lizard dozes peacefully. Other animals are present too, including Jerome's pet lion (only the head is visible, on the lower right), a squirrel, and a bird of prey. The latter is perched in the tree, observing the scene. Below, a pair of rabbits frolic, nose to nose on the grass. They are oblivious of the raptor.

Cue *Peter and the Wolf* theme.

A wall label at the National Gallery references the rabbits. Evidently they're symbolic of "lust, but also Christian humility." But which is which? Is that what the hawk's trying to decide?

*Saint Jerome Reading* is actually quite different from *Mary, Queen of Heaven*. Its subject isn't rising above the material world, for one—at least, not physically. Instead, Jerome is entirely enveloped

**Top, Left:** *Mary, Queen of Heaven* detail view. **Top, right:** *Saint Jerome Reading* detail view.
**Right:** Giovanni Bellini, *Saint Jerome Reading*, 1505, oil on panel

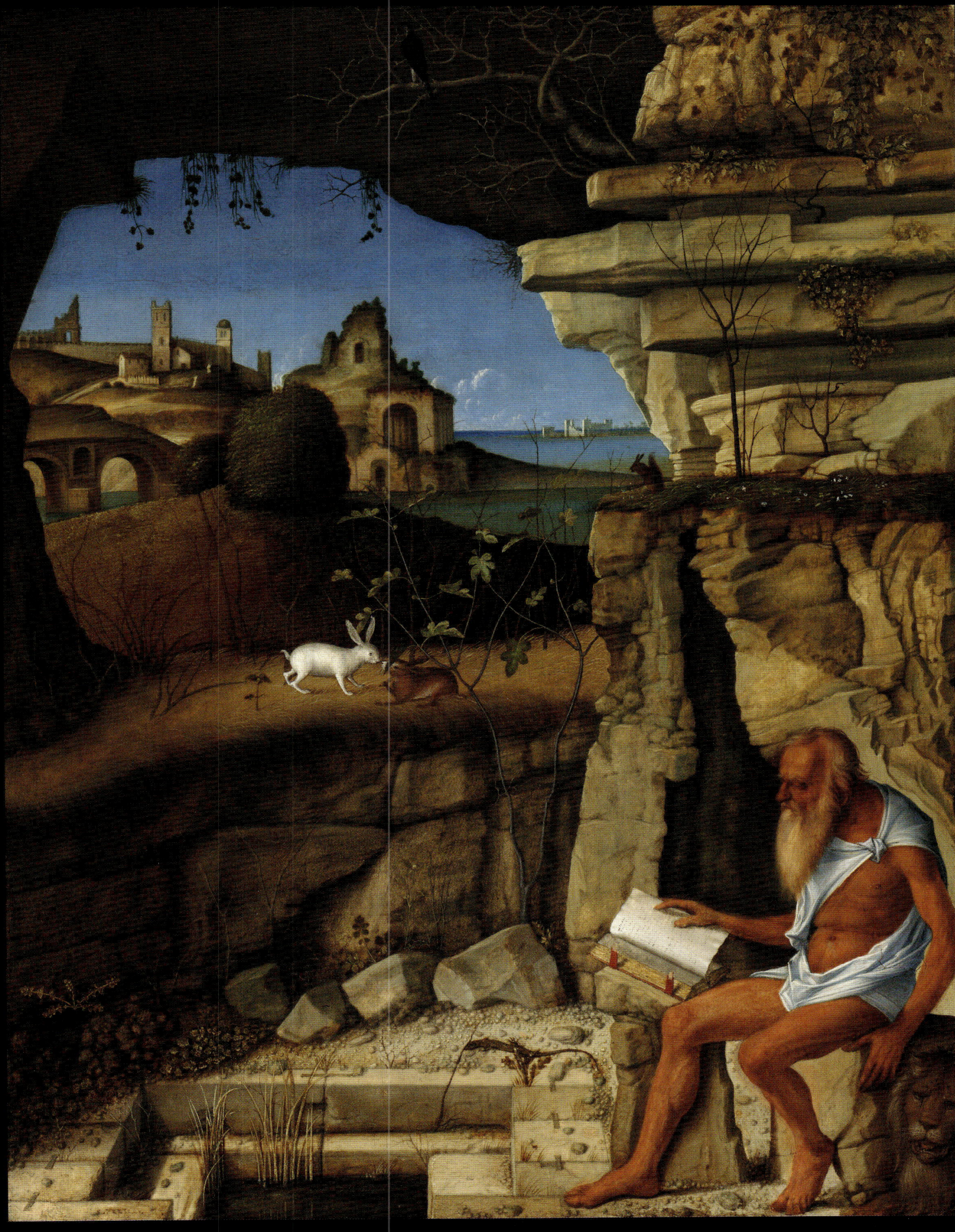

by nature. He's comfortably part of the world that surrounds him. His mind may be elsewhere, but he's bodily present, solidly planted.[1]

But regardless of what their *subjects* are up to, the vivid, almost tactile depiction of nature in each of these paintings marks an emergent new way of examining phenomenal reality.

THE FIRST TIME I USED A SCANNING ELECTRON microscope, or SEM, was in 2013, in the basement lab complex of the Center for Bits and Atoms in MIT's Media Lab in Cambridge, Massachusetts. At the time, its SEM was a fairly simple, even dated piece of equipment—almost an afterthought amid the gleaming new 3D printers and CT scanners—making it an excellent machine to learn on.

My project, provisionally titled *Nanocosmos*, was originally conceived as an investigation of natural design at submillimeter scales. A rather broad subject, as I would soon discover. In any case, the rubric was something of an excuse. I simply wanted to land on the new planet and see what happened. Not to mention bring back evidence, in the form of images.



Your basic exploratory mission, in other words.

Ideally those images would be compelling, absorbing, alien yet somehow also recognizable. Actually, that was the nonnegotiable part. As the latest of several image-based projects, *Nanocosmos* was also intended to be something like the new wing of an ongoing construction project. A building-out from previous work.

So what was this structure I was intent on building? A kind of framework, encompassing aspects of the natural world at microscopic scales. Something capable of orienting its viewers within an unfamiliar realm. Comprised of documents from the frontier, with the latter defined as the border between what we know or think we know, and what remains mysterious. Assembled from evidentiary, possibly even oracular images,[2] while also—and this was important—being worthy of aesthetic contemplation.

Regarding aesthetics, I made no apologies for pursuing what has long been seen as a suspect quality by the gatekeepers who police what's considered worthy: beauty. As a personal driver, it has always kept me focused. To my mind, rejecting such an attribute, however defined, isn't just dismissing most of art history as "retinal," as Marcel Duchamp did early in the last century. It's failing to understand the optic nerve's centrality to the human brain.[3] In any case, although it would be reliant on technologies designed for scientific research, that aesthetic driver should, I thought, provide a clear differentiation between *Nanocosmos* and science.

But I was also hoping the resulting images would provide at least some evidence of 4.3 billion years of terrestrial evolution—all those geological epochs, stacked up through time like the layers in Jerome's cave. Ideally, *Nanocosmos's* micrographs would hint at both the idiosyncratic pathways taken through the world by its subjects, and also the longer arc of the species or type they represent.

So, while the results would inevitably be stills—static images, like all photographs and micrographs—in principle they should hint not only at individual trajectories but also extended Darwinian timescales.

Anyway, that was the plan.

1 Jerome wrote "To me a town is a prison, and the desert loneliness a paradise." Quoted in Rebecca Solnit, "Scapeland," in *Crimes and Splendors: The Desert Cantos of Richard Misrach* (Bullfinch Press / Little Brown and Co., 1996), 37, footnoted to Yi Fu Tuan, "Desert and Ice: Ambivalent Aesthetics," in *Landscapes, Beauty and the Arts* (Cambridge University Press, 1993), 244.

2 Not in the literary sense of grand pronouncements, but rather as a way in which a fundamentally enigmatic material world can communicate some of its qualities via the micrographic or photographic medium. I owe my ideas concerning photography's sometimes-oracular role in part to conversations between 2014 and 2016 with Robert Polidori, who once observed that we both use the medium as an exploratory device, rather than to express subjective internal states.

3 Duchamp said he wanted to "put art back in the service of the mind." At the time he was busy shifting away from painting as his medium of choice and presumably intent on justifying it.

Through the generosity of Dr. O. Roger Anderson, professor emeritus of natural science at Columbia University, I had my initial samples. Five containers filled with a vaguely illicit-looking white powder, constituting the innumerable jumbled silica skeletons of a particularly charismatic class of unicellular marine organisms called radiolarians.

German marine biologist Ernst Haeckel's renderings of these architectonic protozoans, drawn from life using an optical microscope, captured the public imagination at the turn of the twentieth century. Converted into colorful lithographs and published in lavishly illustrated books, his pictorial investigations of these and other diminutive sea creatures impacted design, architecture, and art during the art nouveau period.

The samples I was now carefully transferring onto a dime-sized sample stub—a type of aluminum disc ubiquitous in electron microscopy—had been collected by Dr. Anderson on research trips to the equatorial Pacific and Indian Oceans in the late 1960s. Work that eventually resulted in his magisterial 1983 book on the subject.[4]

Using a fine-point paintbrush, I scattered dozens of white specks across the sticky black carbon tape affixed to the stub, then carefully placed it in the lab's sputter coater—a retro-looking piece of equipment designed to ameliorate the tendency of SEM samples to charge, and therefore distort, in the microscope's electron beam. Switching it on, I covered the samples with a molecule-thin layer of platinum. Then I slid the stub, its surface now gleaming, into a slot on the flat stage at the center of the microscope's vacuum chamber, closed its heavy door, and set the controls to pump out the air inside.

After some trouble focusing the beam, I watched the first radiolarian shimmer into view. A complex wall of polygonal forms loomed out of the murk. Behind, linked to it by fragile-looking tendrils of silica, cryptic internal structures hinted at obscure biomechanical workings. The effect was a bit like Piranesi's labyrinths: inscrutable architectonic structures receding into subterranean darkness. It seemed monumental on the SEM's screen.

Playing with magnification levels, I zoomed in on levels of complexity that I had known must be there but had never before seen myself. I was in a submersible, hovering above the deep seafloor. A landscape of nested forms towered in the darkness around me, visibly organic yet orderly and intricate. An almost machined coherence, producing a clear sense of design. They looked like crashed alien technology.

Like any self-respecting submersible, I had brought my own source of illumination in the form of the SEM's electron beam. Navigating around individual examples, I enlarged structural elements while gradually tilting the stage to change my viewing angle. Many of the fragile radiolarians on view were damaged from decades of knocking against each other in their glass-walled containers. In fact, shards outnumbered intact specimens by about five to one. But even the shards were fascinating. I risked tilting the stub a bit further, creating a kind of horizon line. There was little to indicate that I was observing a microscopic environment. Rather, I had entered their world—a subtly undulating surface of low ridges and convoluted geometries.

I'd never seen anything like it, and by the time I looked up again, over a week had passed. True, I went off to sleep in my Somerville garret every night. But for days and days, I remained transfixed by that single stub. I had absolutely no sense that the topography on view was finite, or anywhere near its limits. And yet I was navigating within an area that to the naked eye was only about 12 millimeters wide. Half an inch.

This sensation was augmented by the superpower of sliding magnification. Manipulating the SEM's controls, I moved from wide to medium shot to close up, until my field of view spanned only 5 microns, or 0.005th of a millimeter, or approximately 2% the width of the specimen. *Sixty times smaller than the width of a grain of table salt.*

My small subset of O. Roger Anderson's samples, collected at the dawn of his career, appeared inexhaustible—a cornucopia of

4 Magisterial, packed with tantalizing information, and yet filled with honest appraisals of how little we actually know about these weirdly geometrical creatures.

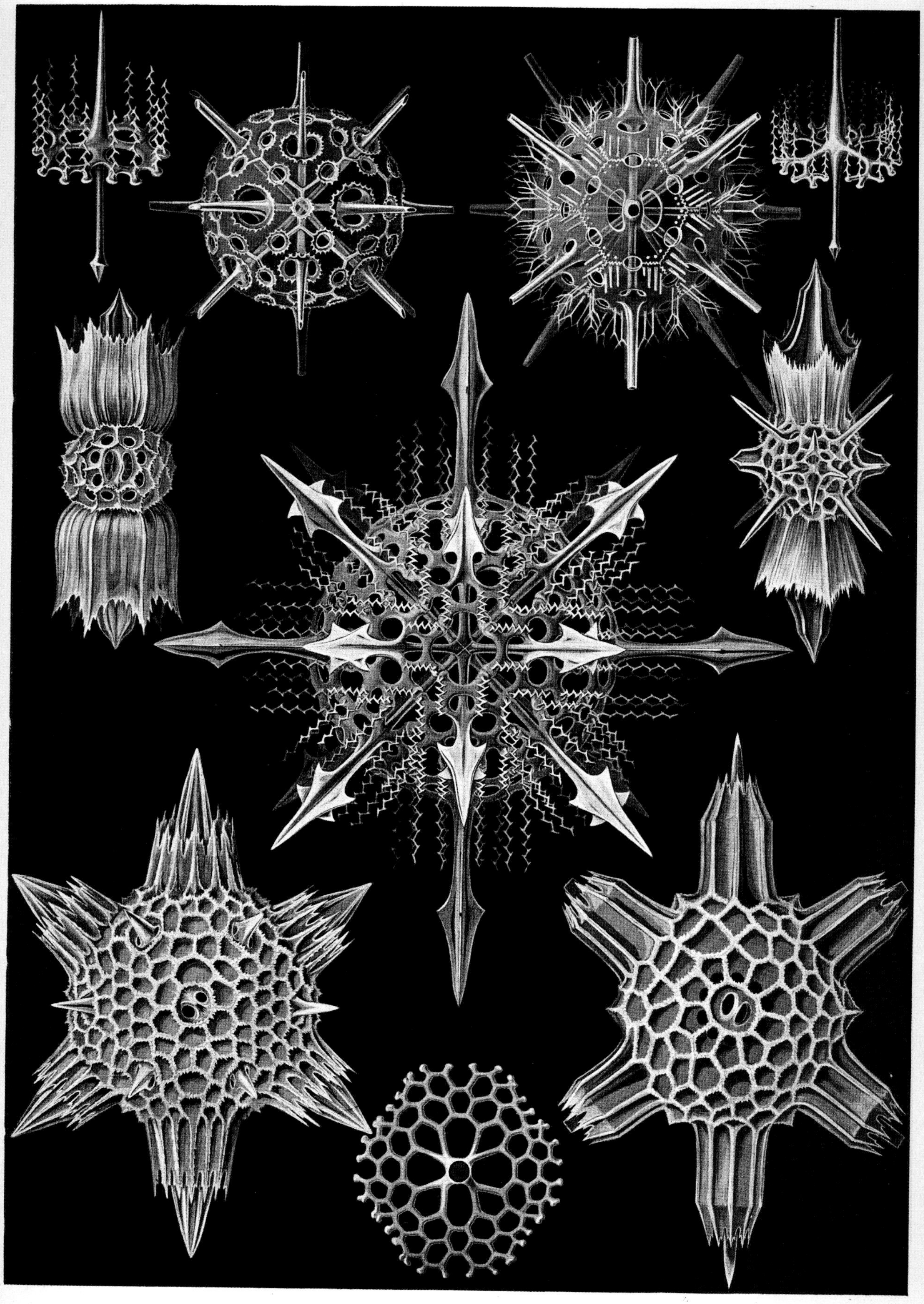

Acanthophracta. — Wunderstrahlinge.

natural geometries. Illuminated by high-velocity energies within the chamber, sculpted by evolution, time, and the circumstances of a global ocean full of competing life forms, those intricate radiolarians gave my sample stub a fractal quality of endless recursion. They were a Mandelbrot place setting, on a holographic dinner table, on a stage set illuminated by subatomic particles—shimmering fragments of universal energy dating back to the colossal bang that started it all—a setting in which the parts somehow encompassed the whole.

Infinity, in the palm of my hand.

WITH THEIR CLEAR BLUE SKIES, SCULPTED stone cities, and grassy landscapes, the Old Masters reflected the circumstances of a planet long before 11 gigatons of concrete were poured across it during the twentieth and twenty-first centuries.[5] Much of what they depicted has been paved over. The global population in the late fifteenth century was somewhere between 400 to 500 million people. As of 2025, we're well over 8.2 billion—16 times as many resource-hungry human beings.

About our position in nature, they weren't wrong—but nature is everywhere in retreat. It's supporting the insupportable.

Clues as to how this happened are already visible in those paintings. Jerome may be physically present, but he's mentally absent. He's contemplating empyrean realms copied out by monkish hands. Nine-tenths of *Mary, Queen of Heaven* depicts just that—a higher heaven where true power resides. For all its beauty, the world below was understood to be corrupted, sinful, degenerate. The empyrean, by contrast, was incorruptible, immaterial, virtuous, and clean—an etherial realm of light, perfection, and perfect timelessness.

Supposedly. In essence, of course, this ideological substructure privileges the invisible over the visible, the immaterial over the material, the notional over the sensorial. Such views lie at the root of Western attitudes toward nature that gradually encompassed most of the world. Although we were considered veritable tools of that corruption, nevertheless the planet had been granted to us—to our species, exclusively—for purposes of exploitation. Collateral damage wasn't important. If we'd already been expelled, how could we tend to the Garden? The gates were locked. Reentry was only possible posthumously, for God's sake.

Still, dialectical elements crept into the flawed construct. Consider the acutely observational quality of those paintings again, with their focus on the visible, the material, and the sensorial. The technologies used to realize the images in this book—the technologies used to realize any photograph, for that matter—resulted from the gradual refinement of ways of seeing already discernible here. An incremental process that governed the rise of ever more sophisticated methods, which were then deployed to further amplify, enlarge, and fix those ways.[6]

The roots of this pre-photographic systemization of vision extend back long before the Renaissance to the ancient Greeks, the Egyptian High Kingdoms, and even further. Their deepest soil lies in seven centuries of methodical data acquisition by Mesopotamian astronomer-priests at the dawn of recorded history. Conducted by thirty-odd generations of anonymous observers, it was the longest continuous body of scientific research in history.[7] It represents among the lowest fertile layers of a mathematicised understanding of space and time now best represented by contemporary physics.

Much of it is still preserved in cuneiform tablets, the oldest of which date to the early second millennium BCE.

5 A gigaton is one billion metric tons, and a metric ton is 2,205 pounds. A gigaton is equivalent to about 20,000 RMS *Titanics*.

6 Rapidly fading from collective memory, for example, is that chemical innovation in analog photography, the fixer, aka sodium thiosulfate, used to remove unexposed silver halide particles from film or photographic paper and intended to make the image permanent and light resistant.

7 Actually, some names are known. For example, Nabu-rimanni, among the first Babylonian astronomers, wrote lunar eclipse computation tables and other mathematical calculations related to celestial bodies sometime between the 6th and 3rd century BCE.

**Left:** Ernst Haeckel and Adolf Giltsch, *Radiolarians*, 1899, lithograph

ARISTOTLE PLACED EARTH AT THE CENTER of creation, surrounded by an unchangeable aether—the so-called fifth element, now replaced in our cosmological models by a vacuum teeming with gravity waves and relativistic particles. He studied animals and plants with considerable attention and described them in great detail. Aristotle grasped that their forms were well matched to their functions. For him, nature's operations required study, explanation, and the shedding of as much anthropocentric and mythopoeic baggage as possible. His examination of biological processes was a first in the history of science. Two millennia later, Charles Darwin called him "one of the greatest, if not the greatest observer, that ever lived."[8]

While Aristotle rejected the assertion, by fifth century BCE philosopher Empedocles, that those forms were likely the result of chance-driven processes—a clear premonition of Darwinian natural selection—he didn't dispute Empedocles's position that life originated due to a confluence of natural forces, and not divine intervention.

Aristotle laid the groundwork for what would eventually be called science. He did this through an emphasis on empirical observation, the classification of living things, the study of patterns within nature and attempts to infer causal explanations for them, and the deployment of the concept of *techne*—ancient Greek for "craft, art"—as a set of *tech*niques aimed at achieving human ends within nature's framework. His construction of a camera obscura to study the Sun foreshadowed technology, or the engineering resulting from applying science to such applications as investigations of the phenomenal world. (The word compounds *techne* and *logia*, "study, knowledge." Aristotle's definition of *techne* further refined this: "a state involving true reason concerned with production.")

It's telling that this definition, which comes down to us across twenty-seven hundred years, could just as easily serve as a description of contemporary capitalism. Ideas like these permitted both an evolving understanding of nature's workings and our current domination of the planet. Ironically, in some ways the story of the twentieth and early twenty-first centuries is a narrative in which the empiricist, rationalist concepts fostered by Aristotle and his successors, lost for centuries and revived during the Renaissance, were increasingly deployed at the service of retrograde modes of thought stressing human centrality and ownership of the world—which in turn led directly to our current environmental impasse. It's the template of an economic model designed to produce limitless growth within a sphere only 7,900 miles wide, or just 0.0003 percent of the solar system's mass.

How can you have limitless growth within such a limited physical body? We have a term for such thing in medicine. Frequently it doesn't end well.

In any case, by examining microscopic worlds using advanced imaging technologies, *Nanocosmos* drew from this complex legacy in part to proffer new ways of appreciating nature's beauty, fragility, and vulnerability.

MY YEAR AT MIT HURTLED BY. IN THE END I produced a considerable pile of SEM scans of varying quality, but the net result was that this period ended up as *Nanocosmos*'s proof-of-concept phase only, and for various reasons none of my MIT work made it into this volume.

Fast forward to Ottawa, Ontario, in the fall of 2018. Like Washington, Canada's capital is home to a formidable array of national museums, including the Canadian Museum of Nature. In my proposal to them I cited Ernst Haeckel's late nineteenth-century radiolarian lithographs, and also the macro plant studies of German photographer Karl Blossfeldt, the latter roughly contemporaneous with Haeckel but producing work into the early 1930s.

I pointed out that both had revealed aspects of nature in startlingly new ways for the time. And I quoted German-Jewish philosopher and cultural critic Walter Benjamin, who wrote that

8 Charles Darwin, letter to J. A. Crawley, February 12, 1879, darwinproject.ac.uk/letter/?docId=letters/DCP-LETT-11875.xml.

Blossfeldt had "played his part in that great examination of the inventory of perception, which will have an unforeseeable effect on our conception of the world." Benjamin noted that one effect of Blossfeldt's detailed examination of vegetable structures was that "we, the observers, walk among these gigantic plants like Lilliputians."[9] In my proposal, I suggested that if properly used, the Canadian Museum of Nature's SEM might produce a similarly unforeseeable effect.

The Museum of Nature is divided into two parts: a large Scottish baronial-style edifice in downtown Ottawa that's open to the public, and a modern research and collections complex sprawling across the countryside about a half hour's drive away in Quebec. The Natural Heritage Campus is only open to the public once a year, and houses a national treasury of approximately 15 million samples taken from across the world's second-largest country. It also contains offices for the museum's fifty or so research scientists, and many well-equipped laboratories—including a windowless cubicle housing the CMN's powerful late-model scanning electron microscope.

What followed was a highly productive period, albeit interrupted by two years of global pandemic, during which the labs were closed. I spent so many hours in that SEM room that I could undoubtedly reconstruct the place from memory with forensic accuracy—something I vaguely remember actually attempting, while deep in REM sleep one morning.

Which brings me to a concept that Walter Benjamin came up with in another essay, "A Short History of Photography": the optical unconscious.

> It is indeed a different nature that speaks to the camera from the one that addresses the eye; different above all in the sense that instead of space worked through by human consciousness there appears one which is affected unconsciously . . . Photography makes one aware for the first time of the optical unconscious, just as psychoanalysis discloses the instinctual unconscious... At the same time photography uncovers in this material physiognomic aspects of pictorial worlds which live in the smallest details, clear and yet hidden enough to find shelter in daydreams, but which, once enlarged and capable of formulation, show the difference between technology and magic to be entirely a matter of historical variables.[10]

In Benjamin's formulation, it seems that our optical unconscious is transformed by photography, which brings unsuspected worlds into the mind's foreground. And that thought again underlines how integral the "retinal" is to the brain's functioning, particularly as concerns the mechanically produced image—itself inconceivable without that image-hungry organ, which in devising ways of recording the phenomenal world photographically may have been intent on "putting art back in the service of the mind," in Duchamp's phrase.

Not to mention its role servicing discovery. We are speaking about images as investigative tools. Earlier in the essay, Benjamin quoted nineteenth-century French physicist François Arago's speech heralding the new medium's revelatory qualities: "When inventors of a new instrument apply it to the observation of nature, the hopes that they place in it are always insignificant compared with the number of subsequent discoveries of which the instrument was the origin."

Nowhere is this more true than in the long history of the microscope.

IN THE SUMMER OF 1674, DUTCH LENS MAKER Antonie van Leeuwenhoek collected a vial of water from a boggy pair of lakes near

9 Benjamin's essay, published in the Berlin journal *Die literarische Welt* under the headline "News from Flowers," reviewed Blossfeldt's first book, *Urformen der Kunst* (Ernst Wasmuth,1928), which came out when Blossfeldt was already in his sixties. Literally translated, its title means "Original Artforms," but the English edition was titled *Art Forms in Nature*—the same as the English-language editions of Haeckel's book.

10 "A Short History of Photography" was originally published in *Die literarische Welt* of September 18, September 25, and October 2, 1931.

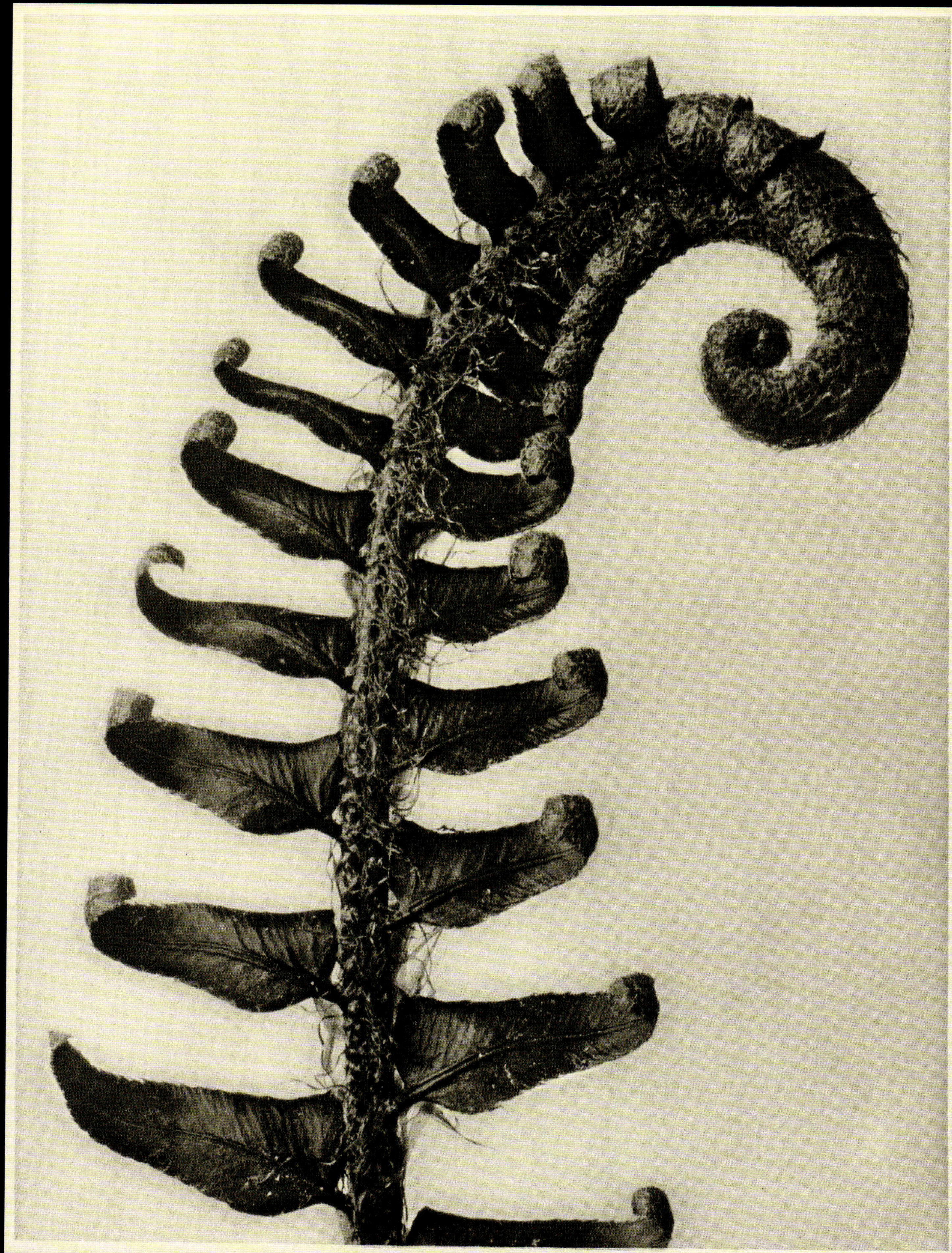

Delft called the Berkelse Meer and became one of the first observers of microbial life.[11] He did so through a single-lens optical microscope of his own design. Although van Leeuwenhoek didn't invent the microscope, he built deceptively simple instruments of unprecedented magnifying power.

He summarized his findings on September 7 of that year in a letter to the Royal Society in London. Among clumps of green algae, van Leeuwenhoek wrote, he had observed:

> . . . very many little animacules, whereof some were roundish, while others, a bit bigger, consisted of an oval. On these last I saw two little legs near the head, and two little fins at the hindmost end of the body. Others were somewhat longer than an oval, and these were very slow a-moving, and few in number . . . And the motion of most of these animalcules in the water was so swift, and so various, upwards, downwards, and round about, that 'twas wonderful to see: and I judge them some of these little creatures were above a thousand times smaller than the smallest ones I have ever yet seen.

Van Leeuwenhoek had discovered what amounted to a new dimension of life. Aqueous and teeming, it was packed with "creatures . . . a thousand times smaller" than any previously witnessed. Apart from the paramecium, rotifers, and single-celled microorganisms usually found in lake water, it is now thought that he had probably observed a particularly wriggly type of cyanobacteria called Anabaena. As a result, and standing apart from his role as pioneering microscopist, he is now considered the founder of an important scientific discipline, microbiology.

The genealogy of the device van Leeuwenhoek used to accomplish this is surprisingly opaque. The first known mention of lenses of any kind dates back to about 2000 BCE and the Babylonian *Epic of Ishtar and Izdubar*, a poetic retelling of the *Epic of Gilgamesh* in which a "sacred glass" was used to start fires by focusing the Sun's rays. But it took three millennia for simple magnifying devices to start appearing in China and Europe. Sometimes called "reading stones," they were made from fragments of quartz crystal or glass polished into oval shapes and placed directly on top of manuscripts to enlarge text.

By the late thirteenth century, a new innovation was developed in northern Italy: paired eyeglasses. The high-quality glass made in Murano near Venice was particularly prized for the purpose. The devices quickly spread, and by the mid-fourteenth century they could be found perched on noses throughout Europe.

Within a couple hundred years of *that*, the center of lens research and development had shifted decisively to the Netherlands. In the late sixteenth century, a vigorous competition between opticians spurred many innovations there. One of them, the compound microscope—a device featuring multiple lenses encased in a tube—was probably invented by Zacharias Janssen, a spectacle maker based in Middelburg, Zeeland's capital. Janssen's microscope appeared on the scene around 1590 and allowed magnifications of about nine times. This was an improvement over the simple hand-held magnifying glass, but not by much.

Working in Delft late the following century, van Leeuwenhoek, who is known to have made upwards of 500 lenses during his lifetime, set about increasing the device's resolving power. He did so by simplifying it. Actually, his experiments with microscopes came after reading British scientist Robert Hooke's book *Micrographia*, which had caused a sensation in London in 1665. Apart from its attention-grabbing gatefold etching of a flea seen in exceptionally hairy detail, the book contained detailed descriptions of how Hooke's microscopes were made. In one case this included the grinding of a single lens mounted on a metal plate. This was the design that van Leeuwenhoek imitated and improved on.[12]

11 German Jesuit polymath Athanasius Kircher preceded van Leeuwenhoek in observing microscopic "animalcules," which he described in his 1658 book on contagious disease. Although English scientist Robert Hooke didn't describe single-celled organisms prior to van Leeuwenhoek, in 1665 he coined the term *cells* to describe the structure of a piece of cork he had studied through his microscope—this after the Latin term for the small rooms where monks live.

12 Brian J. Ford, "Breaking the Myths of Microscopy," *The Microscope* 62, no. 2 (2014):

**Left:** Karl Blossfeldt, *Polystichum munitum*, 1928, photogravure

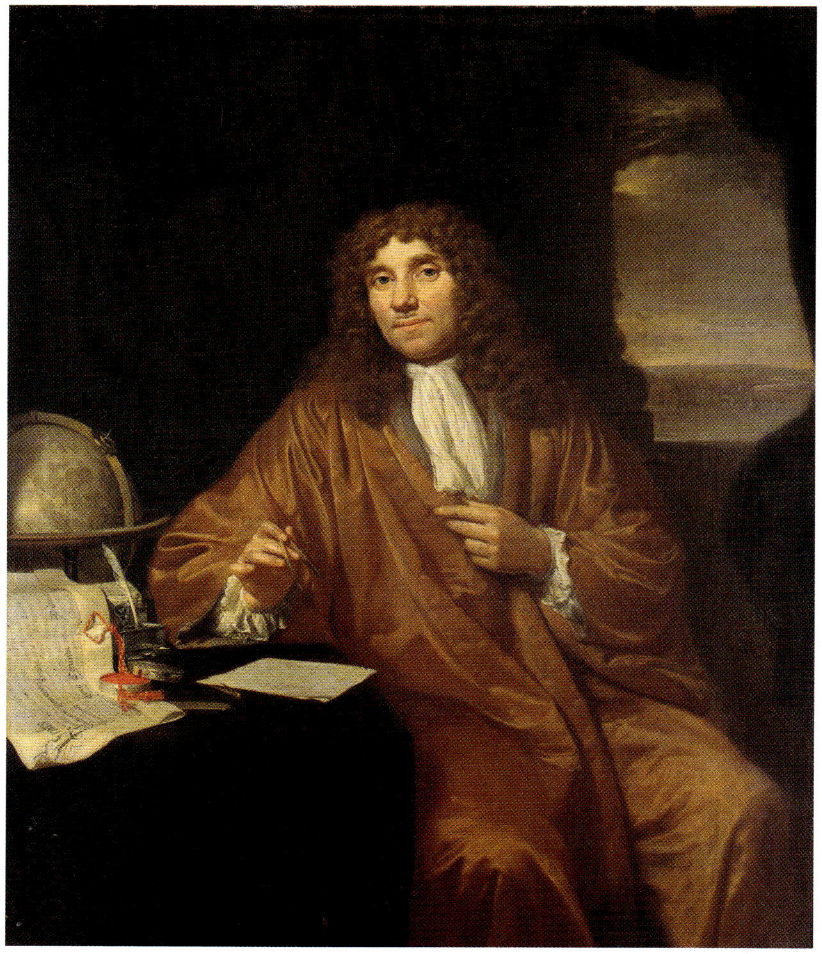



His techniques evolved over time, and finally included a glass-blowing technique in which glass tubes were formed into bulbs, which in turn permitted the forging of nonspherical lenses with superb optical qualities. This enabled van Leeuwenhoek's single-lens instruments to achieve between 265 to 275 times magnification—a resolution approaching a single micron, or ten thousandth of a centimeter. By contrast Hooke mostly used compound microscopes built by London instrument maker Christopher Cock, the best of which only magnified about fifty times. While this was substantially better than Janssen's original device, they suffered from chromatic aberrations, meaning they didn't focus color wavelengths to the same point, producing a distorted image, particularly at higher magnifications.

In increasing the power of an existing more simple design and proceeding to make major discoveries with it, van Leeuwenhoek is comparable to the best-known natural philosopher of the time, Florentine astronomer Galileo Galilei. After his conviction for heresy by the Roman Inquisition and nine years of house arrest, Galileo had passed away in 1642—just three decades before van Leeuwenhoek's Berkelse Meer revelations. In 1609 Galileo had constructed a telescope based on a description of the earliest known such instrument. The "Dutch perspective glass" he modeled his device on was invented by Middelberg lens maker Hans Lipperhey, who had applied for a patent the previous year—about twenty years after Zacharias Janssen built the first compound microscope in the same city. It's notable that the technologies enabling the observation of both distant celestial objects and infinitesimal terrestrial ones had been conceived and built in the same provincial capital within a quarter century of each other.

Galileo ultimately improved Lipperhey's telescope by a factor of ten, enabling his discovery of mountains on the Moon, Jupiter's satellites, and the phases of Venus. He made another notable finding as well: that when inverted, his telescope could also magnify small things—albeit in somewhat blurry fashion. As a result, Galileo is himself sometimes erroneously credited with the invention of the microscope. But by the time he noticed this capability, Janssen's microscope was already over a decade old. It was also a better instrument— at least when it came to enlarging small subjects.

That wasn't the end of the story, however. In 1624 Galileo, by then increasingly embroiled in his conflict with the Catholic Church, risked traveling to Rome to examine a more sophisticated device than Janssen's—a compound microscope produced by Dutch inventor Cornelis Drebbel. True to form, he soon built an improved version.[13]

---

63–73. Ford states that "Hooke was the source and original inspiration for Leeuwenhoek's lifetime of devotion to microscopy."

13 In fact, the term *microscope* was coined by Drebbel in response to Galileo's innovative 1624 improvement of his design; Galileo himself called it *occhiolino*, or "small eye." Stephen Jay Gould, Chapter 2 in *The Lying Stones of Marrakech* (Harmony Books, 2000).

Jan Verkolje, *Portrait of Antonie van Leeuwenhoek*, circa 1680, oil on canvas

As for van Leeuwenhoek, who had the benefit of operating without fear of papal censure, his lake-water "animalcules" were soon followed by the first descriptions of red blood cells, the discovery of spermatozoa, and many other findings. By the turn of the eighteenth century, his reputation was such that it was virtually impossible to mention microscopy without citing his name.

He never wrote a true scientific paper, instead transmitting his findings in over 500 detailed letters to the Royal Society in London. They were translated into English by Society secretary Henry Oldenburg, who had learned Dutch specifically for the purpose. All of them were published in *The Philosophical Transactions of the Royal Society*, founded by Oldenburg in 1665 and the first real scientific journal with peer review.

None of Oldenburg's letters in response survive, but we know he encouraged his correspondent and suggested various other subjects for microscopic study. When Antonie van Leeuwenhoek was finally elected as a member of the Royal Society in 1680, three years after Oldenburg's death, he was reportedly "taken aback" by the honor.[14]

DESPITE THE CANADIAN MUSEUM OF NATURE'S 15 million samples, my work on *Nanocosmos* involved much prowling across various landscapes with a collector's toolkit attuned to the needs of electron microscopy. A butterfly net. Several tweezers of various shapes and sizes. A drawstring bag stuffed with dozens of small plastic vials filled with 80 percent ethanol—the best preservative for SEM samples, doing the least damage to flora and fauna.

Apart from feeling faintly ridiculous—a middle-aged naturalist blundering through the countryside while everyone else cavorted in the waves—I felt a bit like a small-game hunter stalking specimens along the Croatian coast, through Caribbean rain forests, or across the Canadian outback.

14 Clifford Dobell, "A Protozoological Bicentenary: Antony van Leeuwenhoek (1632–1723) and Louis Joblot (1645–1723)," *Parasitology* 15, no. 3 (September 1923): 46–50.

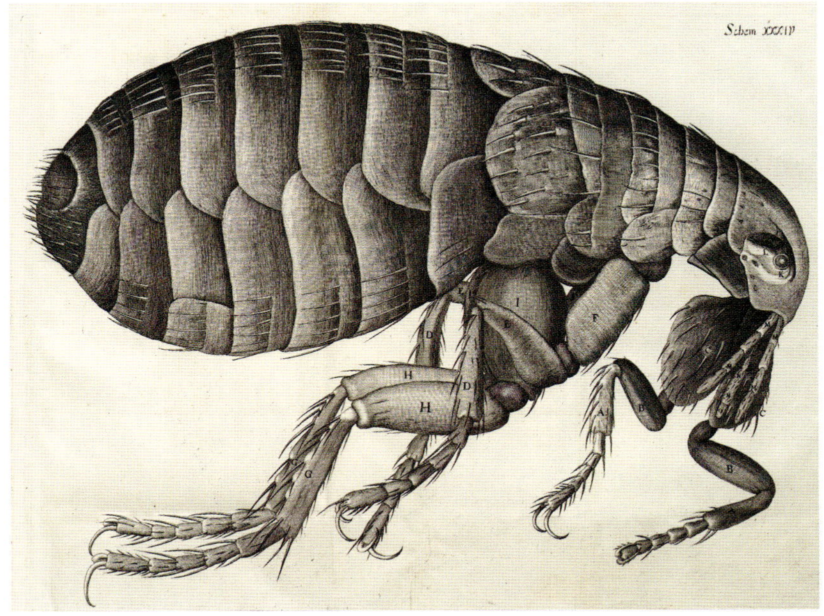

Robert Hooke, *Flea*, 1665, engraving

Not all of my samples were collected by me, though. Many were indeed taken from either the CMN's holdings or those of its researchers, notably Paul Hamilton's diatoms and the many insects provided by entomologists Robert Anderson and Thomas Onuferko. I was also privileged to borrow material from such other leading scientists as the late Marie Faust of the Smithsonian's National Museum of Natural History (the dinoflagellates between pages 164–207) and O. Roger Anderson of Columbia (the radiolarians on pages 62–119). But all of the subjects visible here were in effect sacrificed so that they could be studied, either for scientific purposes or for this project—a fact I acknowledge with respect.

While collecting samples I had a series of memorable encounters with insects that couldn't help but modify my previously inchoate sense of what these sophisticated creatures were all about. They prompted side trips through the writings of those who have attempted to grapple with subjects such as animal consciousness and larger questions concerning our dominant position within nature and how it impacts other living things.

By contrast, my encounters in the field were entirely wordless. In one, I came across a dragonfly resting on a leaf in the flowering garden of a vacation house on Fire Island's north shore, and slowly approached it.[15] To my surprise, it didn't immediately launch itself into the air, but simply looked back at me. In this case, I had no net. Like the dragonfly, I wasn't in hunting mode.

The first thing that struck me was those huge orb-like eyes, mounted side by side in its head to facilitate stereoscopic vision. As with wolves, cats, and raptors, it's the sure sign of a hunter. Its wings were swept forward and down, with the rear pair slightly larger than the front. All four wings were sleekly aerodynamic and thinly veined by dark lines. Its thin, fuselage-like body extended behind a streamlined central thorax anchoring its wings.[16] It was a beautiful creature.

The second thing that dawned on me—rapidly and with force—was that this was a conscious entity. The twin domes of its eyes tracked me as I approached, leaving no doubt whatsoever that it was aware of my presence. As I edged forward, rather than simply flitting away—an easy action for an insect that has been called a flying shark—it let me get quite close.



Gradually it tilted its head as I rose in its field of vision, and I became aware of an optical transparency to its compound eyes. When I moved sideways, its head pivoted smoothly to track me. There was a glinting depth to its thousands of ommatidia, the clustered hexagonal photoreceptors of which some dragonfly species have up to 30,000 per eye. I had already marveled at the geometric regularity of insect eyes seen in the SEM. Now I observed a turquoise transparency and three-dimensional shimmer within them.

I knew that each tiny eye element had its own photoreceptors, overlaid by a clear cornea. And I knew that each of those circa 60,000 ommatidia—the maximum for any insect type—was feeding one small picture element down the cable of its axon nerve bundle to the creature's brain, which assembled those "pixels" into an image. Live, in real time, and with stereoscopic depth.

An image of myself.

And I was doing much the same, looking back.

There is no doubt that this locking of the eyes—that sense of looking at a creature that's returning one's gaze—is integral, in human cultures, to identifying consciousness. My perception of awareness on the part of this stiletto-shaped insect is probably easy to dismiss as anthropocentric. Yet that Fire Island dragonfly undeniably overrode an instinct, rooted in some 325 million years of evolution, urging it to fly away. [17] And instead calmly looked me in the eye.

Wasn't that decision itself evidence of agency? Obviously the impulse to self-preservation common to all animals lies at the root of a sense of self to begin with. And that's probably true not only among complex multicellular creatures such as insects and humans, but also—at least arguably—even among unicellular creatures like van Leeuwenhoek's "animalcules." After all, many have hunted each other with carnivorous intent for millennia, and to this day can sometimes be seen diving desperately to avoid being vacuumed up by the vast maw of a glass pipette within the liquid suspension of a laboratory petri dish.

Because how can one act to preserve one's continued existence without first differentiating oneself from the rest of nature?

One. Self.

Surely such individuation underpins consciousness, which is not the same thing as intelligence but a necessary precursor. Rather than "I think, therefore I am," shouldn't it be "I act to preserve myself, therefore I am"? Or even, "I ignore the impulse to preserve myself because I assess that this curious bipedal creature in the funny hat is no threat—therefore I am"?

In other words, couldn't it just as easily be argued that a denial of agency on the part of other creatures is the most anthropocen-

15 Specifically, the garden of Eric Himmel, the editor in chief of Abrams, the publisher of this book, and his wife, Caroline Miller, a great gardener and wit.

16 Leonardo da Vinci was fascinated by dragonflies, and his notebooks contain a number of renderings of them, in particular their wings. As part of his broader interest in flight, he even sketched a mechanical dragonfly designed to mimic the movements of the actual insect.

17 The earliest insects exhibiting a shape practically indistinguishable from contemporary dragonflies were the giant griffinflies of the Carboniferous period. By the early Permian, about 300 million years ago, they had produced the largest flying insect ever, *Meganeuropsis permiana*, which had a wingspan of some 71 cm, or 28 inches.

tric characteristic of all? As with Descartes, who famously dismissed other animals as mere mechanical automata? *We* were the only ones who thought and were. *They* just reacted instinctively.

By contrast, Aristotle argued that all animals possessed souls.

THE ELECTRON MICROSCOPE (EM) WAS FIRST developed at the Technical University of Berlin in 1931 by researchers Max Knoll and Ernst Ruska. Rather than the visible light used by optical microscopes, the new device deployed electrons in a focused beam to magnify a subject. Because electrons operate at far shorter wavelengths than visible light, superior imaging resolutions were thought possible.

Working in the dying days of the Weimar Republic, Knoll and Ruska figured out how to focus their electron beam using magnetic lenses inside a vacuum chamber. Within a couple of years, they had indeed exceeded the power of the best optical microscopes, and by 1939 a team of researchers including Ruska's brother, physician and biologist Helmut Ruska, used the new instrument to make the first detailed images of a virus. Their study of the helical structure of the TMV, or tobacco mosaic virus, constituted a major advancement in virology—a new division within microbiology, the field Antonie van Leeuwenhoek had founded in 1674.

By the 1960s the first commercial versions of a variant using an electron beam scanner—an SEM, or scanning electron microscope—was released in the United States. The SEM uses a raster technique in which the tightly focused beam moves across the sample in a grid pattern, assembling the final image line by line. At each point of the scan, the microscope's detectors gather data about the interactions of the electrons with the subject—information then mapped back onto the scanned grid and converted into readable visual qualities such as brightness and contrast. While each given point is usually exceedingly tiny—typically just a few microns wide[18]— when scaled up it is the basis for an enlargement, much as the individual dots in an ink-jet print collectively comprise a final readable image.

By the time Ernst Ruska received a Nobel Prize in Physics in 1986 for his work designing the first electron microscope in the early 1930s, the Nobel Committee described it as among the most important inventions of the twentieth century. Despite their significant contributions, neither Max Knoll nor a researcher at German industrial giant Siemens, Reinhold Rüdenberg—who had obtained the first EM patent in 1931 and oversaw the release of the initial commercial production model in 1939—were included. Longevity maximizes eligibility for the Nobel, which is never granted posthumously, and both Knoll and Rüdenberg had passed away in the 1960s.

In the century since its introduction, the electron microscope has been pivotal to the progress of science. It contributed significantly to discoveries in biomedical research, drug development, virology, nanotechnology, and many other fields. Recently it also became an important tool in the new discipline of quantum information science, where it is used to study materials exhibiting quantum behaviors—substances like superconductors, nanoparticles, and quasicrystals.

As of this writing, some electron microscopes are capable of two million times magnification—a percentage increase of 22,222 percent over Zacharias Janssen's compound microscope, devised and built during the Dutch Golden Age.

MY USE OF TECHNOLOGIES DESIGNED to expand our senses into realms inaccessible without them inevitably caused me to meditate, at least occasionally, on my own motivations in doing so and to what extent my activities could be compared to others'. I wasn't making scientific discoveries, let alone scribbling descriptions of them with a quill pen and mailing them off to the Royal Society—though I allowed myself the hope that other kinds of revelations might arise. Certainly, what I was doing involved a

18 A micron, sometimes called a micrometer, is one millionth of a meter, one thousandth of a millimeter, or one ten-thousandth of a centimeter.

type of nonverbal communication, perhaps a bit one-sided, in which the microscope's resolving power and my image processing efforts allowed aspects of natural design visible only at small scales to find their expression.

But there was more to it than that. My Fire Island dragonfly and other suggestive experiences in the field led me to a whole genre of writing about consciousness within nonhuman species and the nature of animal experience. This included the almost forgotten work of the brilliant Afrikaner naturalist and poet Eugène Marais, who in 1937 published a groundbreaking book, *The Soul of the White Ant*, which explored the complex social structure and collective behavior of termites.

After studying colonies of the insect for decades in the countryside north of Pretoria, Marais wrote that "the soul of the termite—the psyche, we should say—is the property of the entire society." He suggested that in aggregate, the behavior of his "white ants" wasn't merely instinctual but rather exhibited signs of consciousness—a collective awareness within the "superorganism" of the termite colony. While the multiple lines of evidence he marshalled in support of his thesis are beyond the reach of this Introduction, recent field work focusing on a related termite species on South Africa's western coast gives intriguing grounds for how such a collective awareness might have had time to emerge.



The Namaqualand region is so dominated by large termite mounds (up to 130 feet in diameter, or 40 meters) that they are estimated to cover about 20 percent of the surface of this arid coastal province. Each low mound is actually the dome-like roof of an intricately reinforced subsurface network of chambers, tunnels, and nests extending as much as eleven feet underground—over three meters. The mounds feature complex ventilation ducts capable of regulating both cooling and heating, and a pheromone-based security system allowing worker termites to know exactly when and where to repair the colony when it is compromised, even as soldier termites swarm in to guard them.

According to a staggering 2024 finding by a research team from nearby Stellenbosch University, one of the three mounds chosen for radiocarbon dating in Namaqualand had been continuously occupied for 34,000 years. This means it was constructed at a time when Neanderthals still walked the earth.[19] That's three times longer than the oldest known continuously inhabited human settlement, Damascus, where evidence of settlement dates back about 12,000 years.

In his book, Marais proposed the concept of "two-fold thought," a term suggesting that termites possess both individual and collective modes of thinking. The complex behaviors and aggregate intelligence exhibited by such colonies, he wrote, was the net result of an interplay between these two modes as well as between termite castes. He also documented wars of territorial conquest between competing colonies as they attempted to secure resources and assert dominance.

On reading this I found myself asking: Is there anything here that doesn't resemble the human race as seen from 30,000 feet? What meaningful differences exist between Eugène Marais's vision of a termite colony's organizational principles and behavior and how we structure our own societies?

During a subsequent discussion with a friend, Mikkel Eskjaer of Copenhagen's Aalborg University, I mentioned Marais's work and he advised me to check out German biologist Jakob Johann von Uexküll's 1934 book *A Foray into the Worlds of Animals and Humans (Streifzüge durch die Umwelten von Tieren und Menschen)*. In 1909, von Uexküll adopted the neologism "Umwelt" to describe the sensory perceptual environment surrounding nonhuman organisms, thereby laying the foundations for what today is known as biosemiotics—a field in which prelinguistic meaning formation and communications processes within the biological realm are studied. His book is now considered an early landmark of posthumanist thought.

As Mikkel pointed out, by viewing living nature as a kind of matrix of overlapping sensory modalities—a complex environ-

19 "Calcareous Termite Mounds in South Africa Are Ancient Carbon Reservoirs," *Science of the Total Environment* 926 (May 2024). See also Frank Lidz, "The World's Oldest Termite Mound is 34,000 Years and Counting," *New York Times*, October 4, 2024. (Neanderthal link originally by Mr. Lidz and used here with thanks.)

ment of interlinked organisms—von Uexküll's Umwelt concept helped bring about what we now call environmentalism. And his claim that creatures actively shape their environments, in effect creating their own niches depending on their needs, was very much in keeping with Marais's observations of how termites construct their surroundings—something we now know can provide a stability lasting for thousands of years.

After reading von Uexküll and Marais, I was reminded that perhaps the most outstanding example of organisms modifying their environment is the outsized impact that diatoms have on the habitability of our world—not just for themselves but for all animal species. Despite their microscopic size, these tiny organisms produce upwards of 50 percent of all the oxygen in the atmosphere. As I touch on later in the book, in this they provide empirical evidence in support of the Gaia hypothesis, which posits that the entire biosphere is a kind of self-regulating superorganism.

Of course, few species can compete with human beings when it comes to actively shaping their environment. It's just that stability increasingly appears to be a casualty and not a consequence.

~

While using the SEM to examine the quasi-calligraphic scars on the wings of a Quebec dragonfly, I started to think about how an insect that I had brought to absolute stasis, and then further fixed in still images, nevertheless gave compelling evidence, in its wing abrasions, of high-speed, free-flying aerial trajectories while alive.[20]

This in turn led me through gnomic fragments by Heraclitus ("Everything flows" "The way up and the way down are one and the same") to Alfred North Whitehead's process philosophy, with its emphasis on events—change and flux—and its rejection of what he regarded as a misplaced materialist focus on concrete objects within the physical universe, which he argued tended to ignore the continuous flow of process, transformation, and relationships in nature. A very Heraclitean thought, and resonant with his most famous fragment, the one about never stepping twice into the same river.

After zooming in on the sensory apparatuses of butterflies, moths, and bees, with their complex, sometimes ziggurat-like antennae tips, I started to wonder how we could even begin trying to grasp the Umwelt of such creatures. Can we ever truly conceive of what their ways of perceiving the world are like? This led me to American philosopher Thomas Nagel's landmark 1974 essay "What Is It Like to Be a Bat."[21] With his statement, "An organism has conscious mental states if and only if there is something that it is like to *be* that organism—something it is like *for* the organism," it became a foundational text in consciousness studies and the philosophy of mind.

Nagel's thought experiment highlighted the subjective quality of lived experience, using bats as examples of creatures whose experience will forever remain out of reach of our understanding. While he appears to have been unaware of von Uexküll, their ideas were very much in alignment.

As it happened, by the time I read Nagel's essay I was in a good position to evaluate one of his key assertions:

> I have said that the essence of the belief that bats have experience is that there is something that it is like to be a bat. Now we know that most bats . . . perceive the external world primarily by sonar, or echolocation, detecting the reflections, from objects within range, of their own rapid, subtly modulated, high-frequency shrieks. Their brains are designed to correlate the outgoing impulses with the subsequent echoes, and the information thus acquired enables bats to make precise discriminations of distance, size, shape, motion, and texture

20 In my last trip out of Canada before the COVID-19 pandemic, I described these markings to Shanghai artist Wenda Gu, who has produced many intriguing calligraphic works written in pseudo-languages. In fact I still owe him some examples, and at this rate will have to simply send him this book.

21 *The Philosophical Review* 83 no. 4 (October 1974): 435–450. Cornell University.

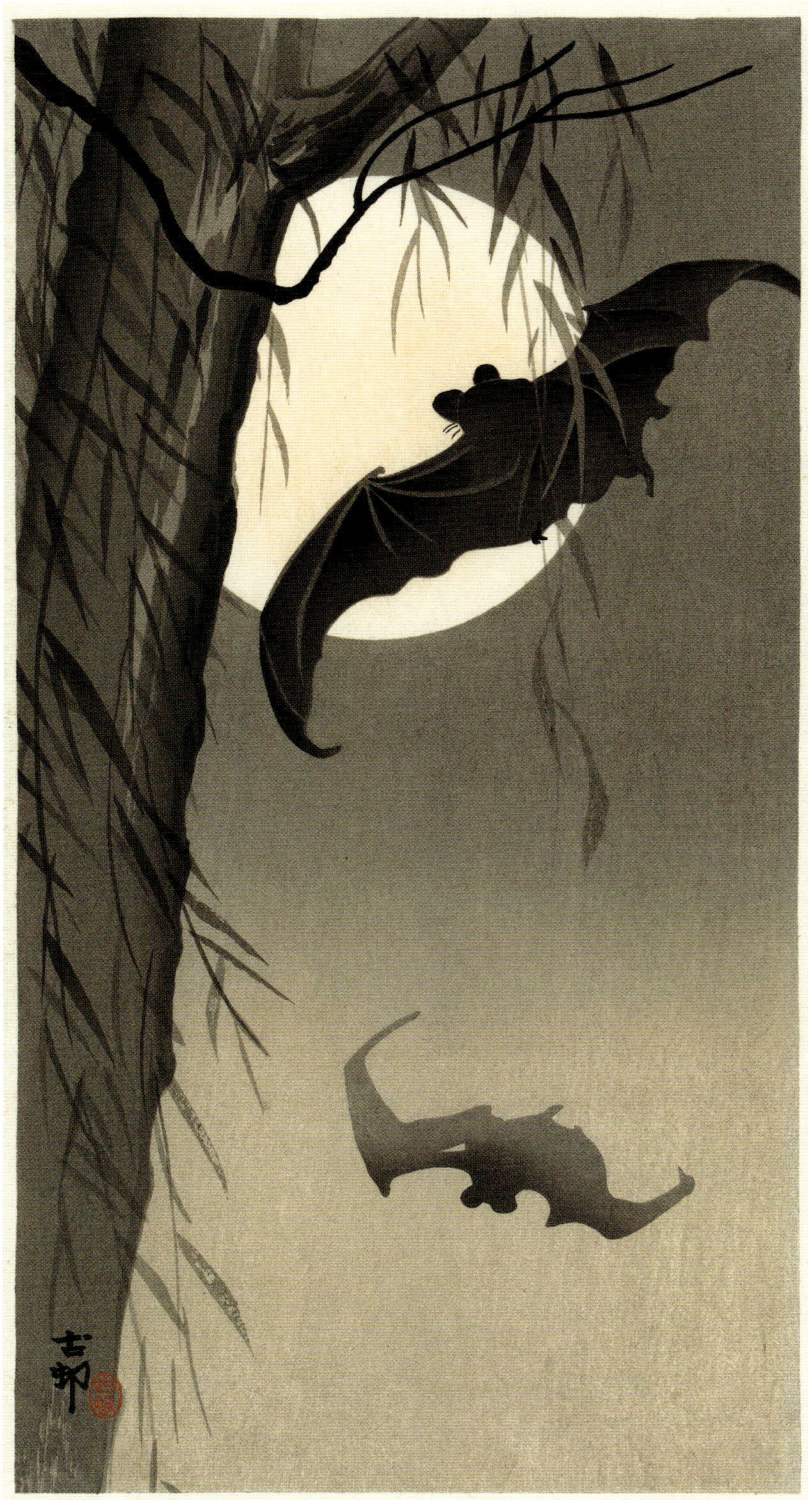

> comparable to those we make by vision. But bat sonar, though clearly a form of perception, is not similar in its operation to any sense that we possess, and there is no reason to suppose that it is subjectively like anything we can experience or imagine. This appears to create difficulties for the notion of what it is like to be a bat. We must consider whether any method will permit us to extrapolate to the inner life of the bat from our own case, and if not, what alternative methods there may be for understanding the notion.

While Nagel's statements concerning the difficulty he describes are largely irrefutable, on reading this I immediately thought of how similar the workings of the scanning electron microscope are to bat echolocation, up to and including the "subtly modulated, high-frequency . . . outgoing impulses." Could it be that the SEM provides a path to understanding at least some of what it is like to be a bat, however technologically mediated?

As with bats, but not with human vision, the source of the "illumination" of the external world (for bats) and the internal world of the vacuum chamber (for the SEM) lies within the projective apparatus of the perceiver. And this is true in both cases, that of the bat (which does so via echolocation) and of the SEM operator (who does so using tightly focused electron beams). The microscopic realms accessible to our understanding via SEM are reached by deploying electrons at wavelengths far shorter than those of visible light (they are in fact high frequency impulses), thereby revealing images of subjects we cannot immediately see, illuminated in ways we're familiar with. They're a form of translation, in other words, into a visual language we already know, namely the black and white image. (Of course, the latter also constitutes a kind of translation: from phenomenal reality into the type of two-dimensional monochrome picture that has been with us since the invention of photography, two hundred years ago.)

This doesn't really solve the difficult problem Nagel poses, namely how to get inside the subjective world of a bat. But I be-

Ohara Koson, *Bats Against Full Moon*, early twentieth century, color woodcut on paper lieve it undermines his assertion that bat sonar is so dissimilar to any sense we possess as to defeat such an attempt in advance.[22]

As to the production chain of my own SEM image-making—which in effect correlated "outgoing impulses with the subsequent echoes"—the procedure was intended to yield high-resolution, large format prints, and also the micrographs reproduced in this volume. Ultimately it required me to follow a process I was familiar with from years of working with raw image data from interplanetary spacecraft, namely the construction of final composite images from multiple mosaic "tiles." But it didn't even occur to me—at least not at first—to consider the role my procedure might play in attempting to discern the subjective experience of any of the creatures I was studying.

Inevitably, however, I spent extended periods of time in close proximity to various small organisms, greatly enlarged. Some of them I might initially have viewed with a degree of incomprehension, even alienation, in particular the insects. But after long exposure, at least a few became rather familiar. And finally, it was almost like seeing an old acquaintance again as I resumed work every morning. While this didn't necessarily allow me to extrapolate what their inner lives may have been like, perhaps the procedure represented the very beginnings of taking up that creature's point of view—if only "roughly, or partially," as Nagel put it.

For example, the bee visible on pages 264–269 started out as very much an alien presence on the screen. But after days of grappling with multiple SEM scans recording its shaggy front end at various magnifications, what at first may have seemed bizarre or even repugnant began to acquire a real individuality, even a personality, in my eyes—an almost benevolent aspect. And much the same held true for the weevils and scarabs seen between pages 232–245. Despite their bizarre exteriors—in the case of the scarabs, their armor plating and masked Samurai faces—I concluded that these industrious creatures simply meant business, and certainly not harm. However strange to us, their forms followed their function, just as with the much larger creatures we're more used to seeing. They were all just trying to earn an honest living, in other words, before their unfortunate final encounters with an entomologist doing much the same.

And I remembered the response of the sentient HAL-9000 supercomputer in Stanley Kubrick's 1968 film *2001: A Space Odyssey* to a BBC journalist's probing questions about its sense of self: "I am constantly occupied. I am putting myself to the fullest possible use, which is all I think any conscious entity can ever hope to do."[23]

CONCERNING FORM, FUNCTION, AND DESIGN, FEW THINGS in nature are more visibly utilitarian than representatives of the class Insecta, by far the most diverse group of creatures on earth. They tend to show up with all their tools and sensors—their mandibles, antennae, and tarsal appendages—external to their exoskeletons, themselves sculpted marvels of evolutionary refinement. For example, the Ontario carpenter ant on page 236, which I picked off from a posse industriously reconnoitering our kitchen in the spring of 2024, started to possess the sleek quality of a Pininfarina sports car—at least to me, after sustained observation. A design masterpiece optimized for speed on six legs.

This shift in perception was very much on my mind when I became intrigued by eco-philosopher David Abram's 1996 book *The Spell of the Sensuous: Perception and Language in a More-Than-Human World,* which argues for the importance of direct sensory

22 Actually, Nagel has an intriguing footnote in which he seems to doubt elements of his own thesis, namely that such echolocation is impossible for us to really understand. In it he cites the use, by visually impaired people, of canes and even "vocal clicks" to map out their surroundings. He concedes that "it may be easier than I suppose to transcend inter-species barriers with the aid of the imagination," concluding on p. 442: "My point, however, is not that we cannot *know* what it is like to be a bat. I am not raising that epistemological problem. My point is rather that even to form a *conception* of what it is like to be a bat (and a fortiori to know what it is like to be a bat) one must take up the bat's point of view. If one can take it up roughly, or partially, then one's conception will also be rough or partial. Or so it seems in our present state of understanding."

23 It will be interesting to see how the moral and ethical implications of our interactions with AGI, or artificial general intelligence, when it finally arrives, may resonate with related questions concerning animal consciousness and our relationships with other living creatures.

experience of nature.[24] One of its more thought-provoking sections centered on his realization, while embedded in South Asian rural communities in the early 1980s, that western interpreters of local customs had gotten the story all wrong concerning the activities of the local sorcerers, or *dukuns*, of Indonesia, and the *dzankris*, or shamans, of Nepal. Abram was assisted in his communications with representatives of both communities by the fact that he arrived not as a researcher or anthropologist but as an accomplished sleight-of-hand magician—something that allowed him to be taken as a kindred spirit, even a peer, and not an outsider, at least not entirely.

As with Eugène Marais's *The Soul of the White Ant*, there really isn't room here to do justice to Abram's absorbing book. But in its introduction, the author recounted how he gradually realized why the *dukuns* and *dzankris* always seemed to live at the edge of their villages. It was because their position was designed to expedite communication "between the human and more-than-human worlds." And that communication went both ways: "The traditional or tribal shaman, I came to discern, acts as an intermediary between the human community and the larger ecological field, ensuring that there is an appropriate flow of nourishment, not just from the landscape to the human inhabitants, but from the human community back to the local earth."



The shaman or sorcerer, Abrams concluded, "is the exemplary voyager in the intermediate realm between the human and the more-than-human worlds, the primary strategist and negotiator in any dealings with the Others." In his account, such a role had been entirely missed by early Western observers, many of them Christian missionaries, and even the trained anthropologists who followed them. Due to their cultural backgrounds, he argued, both missionaries and anthropologists simply assumed that the shamans were engaging with "otherworldly" forces:

> The primacy for the magician of nonhuman nature—the centrality of his relations to other species and to the earth—is not always evident to Western researchers. Countless anthropologists have managed to overlook the ecological dimensions of the shaman's craft, while writing at great length of the shaman's rapport with "supernatural" entities. We can attribute much of this oversight to the modern, civilized assumption that the natural world is largely determinate and mechanical, and that what is regarded as mysterious, powerful, and beyond human ken must therefore be of some other, nonphysical realm above nature, "supernatural."

24 David Abram, *The Spell of the Sensuous* (Vintage, 1997).

But in fact—Abram continued—what was regarded "with the greatest awe and wonder" by the shamans and also the communities they served was "the multiple nonhuman sensibilities that animate the local landscape." Not some hazy metaphysical realm, but "none other than what we view as nature itself."

There is plenty of mystery to go around within realms directly accessible to our senses, in other words—albeit sometimes with the help of technological mediation. In any case, such nuanced, thoughtful views helped inform my work on this project.

Another impressive book focusing on the innermost experiences of nonhuman creatures is science writer Ed Yong's recent *An Immense World*, which explicated and updated Jakob von Uexküll's Umwelt concept with contemporary research, greatly expanding on prior attempts to consolidate our efforts to understand how animals perceive and act.[25] Even before reading it, I heard Yong expound on Terry Gross's radio talk show *Fresh Air*, and was quite taken by a point he made:

> One of my favorite things about the relationship between insect vision and flowers is that if you took all the colors in all the flowers that were out there and you asked, "What kind of eye, what kind of color vision is

25 Ed Yong, *An Immense World* (Random House, 2022).

> best at discriminating between these colors," what you'd get is an eye that's basically almost what a bee has. An eye that is maximally sensitive to green, blue and ultraviolet. And you might think then that the bee eye has evolved to see the colors of flowers really well. And that's exactly the opposite of what happened. Because the bee eye came first, the flowers evolved later. And so, the colors of flowers have evolved to ideally tickle the eyes of bees. And I think that's a truly wondrous result. It means that beauty as we know it isn't only in the eye of the beholder, it arises because of that eye. Eyes in viewing nature's palette also affect its paintings[26]

This jogged my memory, and I realized it reminded me of a passage by Italo Calvino from his epic 1965 book *Cosmicomics*: "So our efforts led us to become those perfect objects of a sense whose nature nobody quite knew yet, and which later became perfect precisely through the perfection of its object, which was, in fact, us. I'm talking about sight, the eyes; only I had failed to foresee one thing: the eyes that finally opened to see us didn't belong to us but to others."[27]

Subsequent research revealed that actually Yong had slightly elided the matter. In fact flowering plants appeared 140 million years ago, with bees only evolving about 20 million years later. So while the insect pollination of flowers was already well underway when bees arrived, largely by beetles, so too was the feedback loop he described—the evolution of specific flower traits in response to their environment. For example, wind-pollenated plant flowers are typically rather inconspicuous, but by necessity produce immense quantities of pollen. Whereas flowers that rely on insect pollination are usually attractively showy and inviting, but require far less pollen production. One example of this kind of mutually beneficial symbiosis can be seen on page 263, in which an Ontario honeybee is seen lifting off from a flower with a sticky cargo glued to the tibia of its rear legs—or so-called pollen basket.

In any case, Yong highlighted something truly wondrous in that interview. Although flowers already existed, they did indeed evolve such traits as their bright coloration and bouquet of scents (quite literally) as a direct result of the sensory spectrum of the bees and other insects that pollenated them. And it only adds to the wonder of it to observe that the contrary also holds true: in a co-evolutionary process, the insects further tuned their sensory systems to the flowers. And so on, ad infinitum amen.

Long may it continue.

ELECTROMAGNETIC LENSING—ONE OF THE BREAKTHROUGHS permitting the electron microscope's appearance a century ago—uses magnetic fields generated by coils of copper wire called solenoids to bend the path of the electrons and focus them. This is quite similar to glass's ability, when ground into a convex shape and polished, to bend and magnify rays of visible light. Apart from reducing the diameter of the electron beam, allowing it to be concentrated onto a small spot on the sample, the coils allow the focused electrons to be scanned across the subject in parallel lines—the raster pattern already described, which ultimately produces final images based on the cumulative "illumination."

While it might seem counterintuitive, lensing isn't always something devised by our ingenuity as expressed in our technology. It also occurs in nature. Some examples span most of the observable universe. Others can take thousands, even millions of years of natural selection to unfold. Still others are more or less instantaneous, as when Aristotle constructed his very early camera obscura and observed that no matter what shape he made its aperture, the inverted image projected on the far side stayed true to the Sun's circular form.[28]

26 Terry Gross, *"Fresh Air,"* NPR, June 22, 2022.

27 Italo Calvino, *Cosmicomics* (Macmillan, 1965). Used as a kind of personal mission statement for a lunar time capsule project I'm involved with called Sanctuary. See: sanctuaryonthemoon.com.

28 You could argue that this is actually an early example of technology, rather than a

Although it unfolds over far longer time spans—at least from our perspective[29]—perhaps the closest natural analog to the way the SEM's solenoids bend its electrons is a phenomenon known as gravitational lensing. As it happens, the theoretical framework behind this phenomenon was laid less than a decade before the introduction of the first electron microscopes, when Albert Einstein suggested, in his general theory of relativity of 1915, that massive objects could bend light from more distant sources, thereby creating a lensing effect.

But it wasn't until American space agency NASA lofted the Hubble Space Telescope in 1990, and more precisely after the telescope's faulty 2.4-meter-wide mirror was fixed with a corrective optical package installed during five difficult space walks from space shuttle *Endeavour* in December 1993, that we began seeing clear images of the phenomenon. And since then, the telescope has observed multiple instances in which large mass concentrations such as foreground galaxy clusters have served to magnify the light of far more distant objects.



The most compelling examples are when individual galaxies from the universe's earliest epochs, typically distorted into elongated arcs due to the uneven mass distribution of the foreground cluster, are in effect shunted forward via such gravitational lensing to the orbiting telescope. In one example, considered the most complete "Einstein ring" ever seen, a single, exceedingly distant galaxy appears coiled around the largest mass concentrations at the center of galaxy cluster 0220585—a concatenation of some thirty galaxies in the Fornax constellation, about 4 billion light years away. The magnified background galaxy, distorted and foregrounded by this "funhouse mirror" effect, is thought to be 9.4 billion light years away, putting it near the peak of star formation in the universe.

In *Beyond Good and Evil* (1886), Friedrich Nietzsche warned that "if you gaze for long into an abyss, the abyss will gaze back into you." It's worth keeping his statement in mind when considering such lensing, because one intriguing aspect of it is that even as we observe those starry photons funneling past intermediary mass concentrations—which in turn stream them across much of the visible universe, all the way to Hubble's mirror and our eyes—the reverse may also be happening. Because if we can see that distant magnified spiral curving around that spherical foreground galaxy within cluster 0220585, so too must our own Milky Way be visible, in similarly distorted fashion, on the universe's far side, where it too could be seen by observers in the very galaxy we're looking at. If so, like us they would see a spiral galaxy during its earliest epochs circa 9.4 billion years ago, or 5 billion years before the formation of the Sun.

When contemplating that Hubble image, in other words, we may be looking at something capable of returning our gaze.

Not long ago I encountered a particularly memorable example of another kind of lensing, the kind discernible due to natural selection, which by comparison is thought to have had a total duration of a mere 3.5 to 4 billion years.[30] Fortuitously it came within the frame of the very museum where I was working, as I was admiring the richly detailed moth images produced by a man who in some senses was my predecessor at the Canadian Museum of Nature, Jim des Rivières.[31] Apart from his day job as a software engineer, Jim's life's work involved collecting hundreds

natural phenomenon—indeed, I did so earlier in this essay. And yet Aristotle created his device in response to seeing dozens of crescent suns projected on the forest floor during a solar eclipse—a natural occurrence produced by the pinhole lensing of innumerable leaves, one that he was simply replicating at a time when the Sun was not partially obstructed by the Moon. In any case, it's worth considering that Aristotle's camera obscura was both a literalization of Plato's allegory of the cave and an inversion of it: the projected image actually produced empirical truths rather than grounds for erroneous belief. It therefore resembles a photographic negative.

29 Because an intriguing corollary of Einstein's theory of relativity is that from the point of view of a photon traveling at the speed of light, what to us might seem like billions of years would pass in an instant, and the distance traveled would also appear to be zero.

30 That is, from our so-called Last Universal Common Ancestor (LUCA), thought to be a single-celled organism, to late-stage life of all kinds, including Homo sapiens.

31 The museum doesn't have a formal artist-in-residence program, so in practice those who seek collaborations are considered on a case-by-case basis.

**Right:** Michael Benson, *Last Launch of Endeavour*, 2011, digital chromogenic print

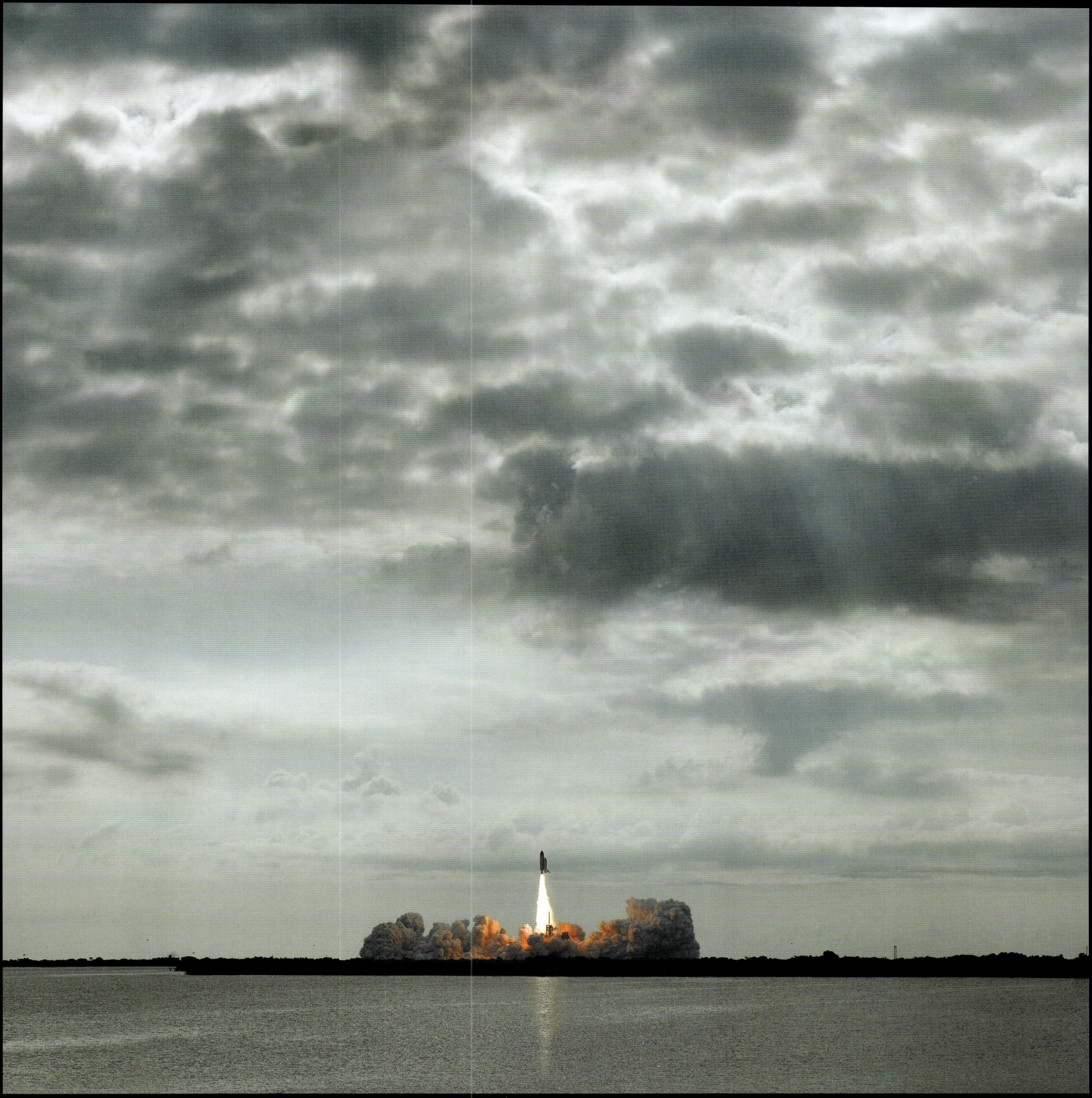

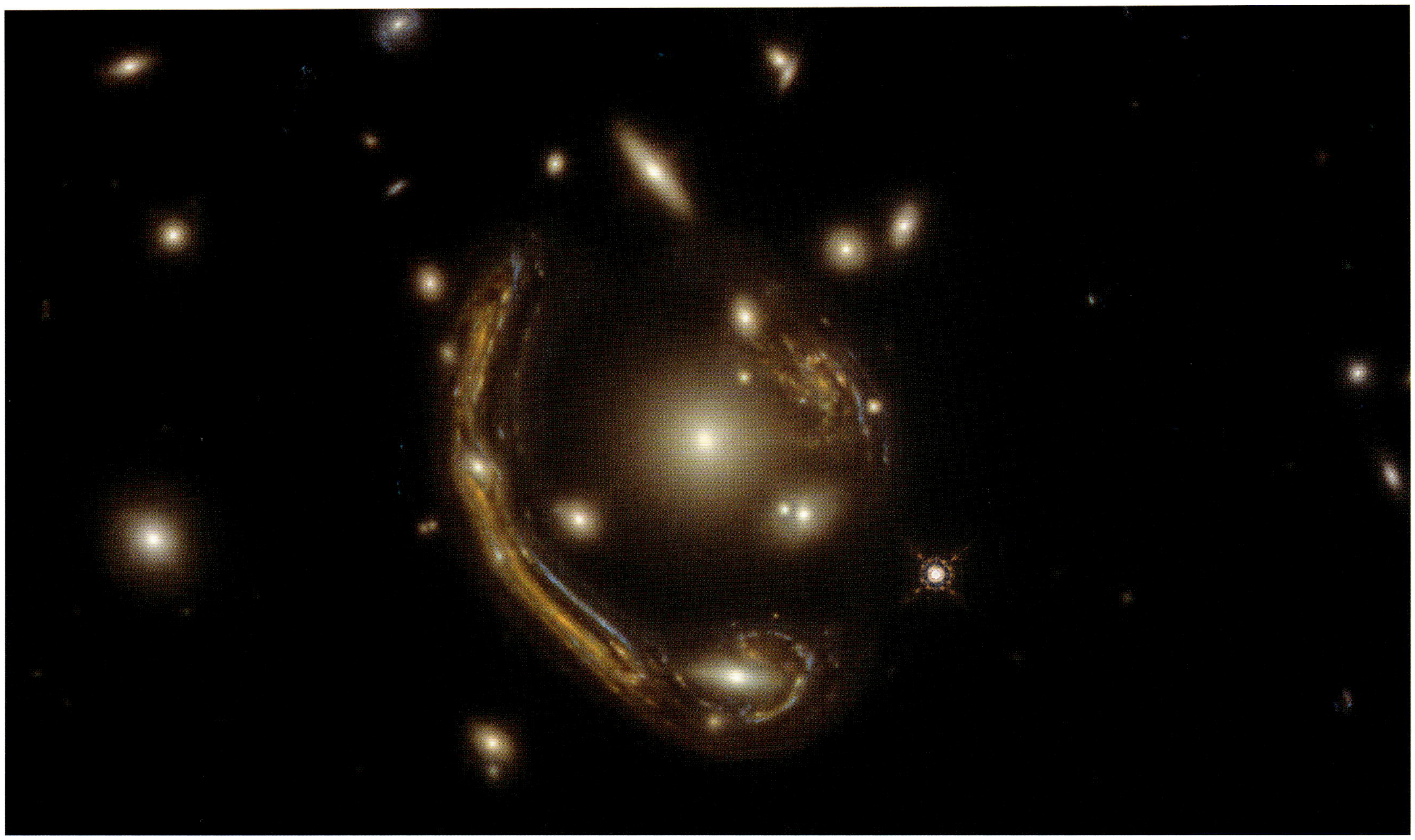



of moths and butterflies, which he imaged at high resolutions using a flatbed scanner. After which he produced vivid large-scale prints, many of them on permanent display in the same Research and Collections building housing the SEM.

I never met Jim, who passed away in 2019. But when visiting an exhibition of his work at the main museum building in downtown Ottawa a couple years ago, his widow Kathryn Finter mentioned that he'd captured an excellent example of evolution's extraordinary ability to inscribe readable images on moths—in this case, a species native to tropical Asia, *Macrocilix maia*, that was not presented in his show. When Kathryn eventually sent me Jim's image, I was amazed.

As visible here, *Macrocilix maia* moths use what amounts to the focal plane of their wings to deter predators. They do so by presenting discernible images of flies feeding on wet excrement, one per wing. And just in case the meaning isn't clear, both "flies" and "excrement" have shiny bright spots on them—a simulacrum of reflected tropical sunlight. To add another note of realism, the bright spot on each red-eyed fly—easily readable as a species commonly called the oriental latrine fly, one well known to enjoy fresh feces[32]—even mimics the blue-white sheen visible on the actual insect's back and wings.

32 *Chrysomya megacephala*.

Hubble Space Telescope, *Gravitational Lensing in the Constellation of Fornax*, 2020, photograph

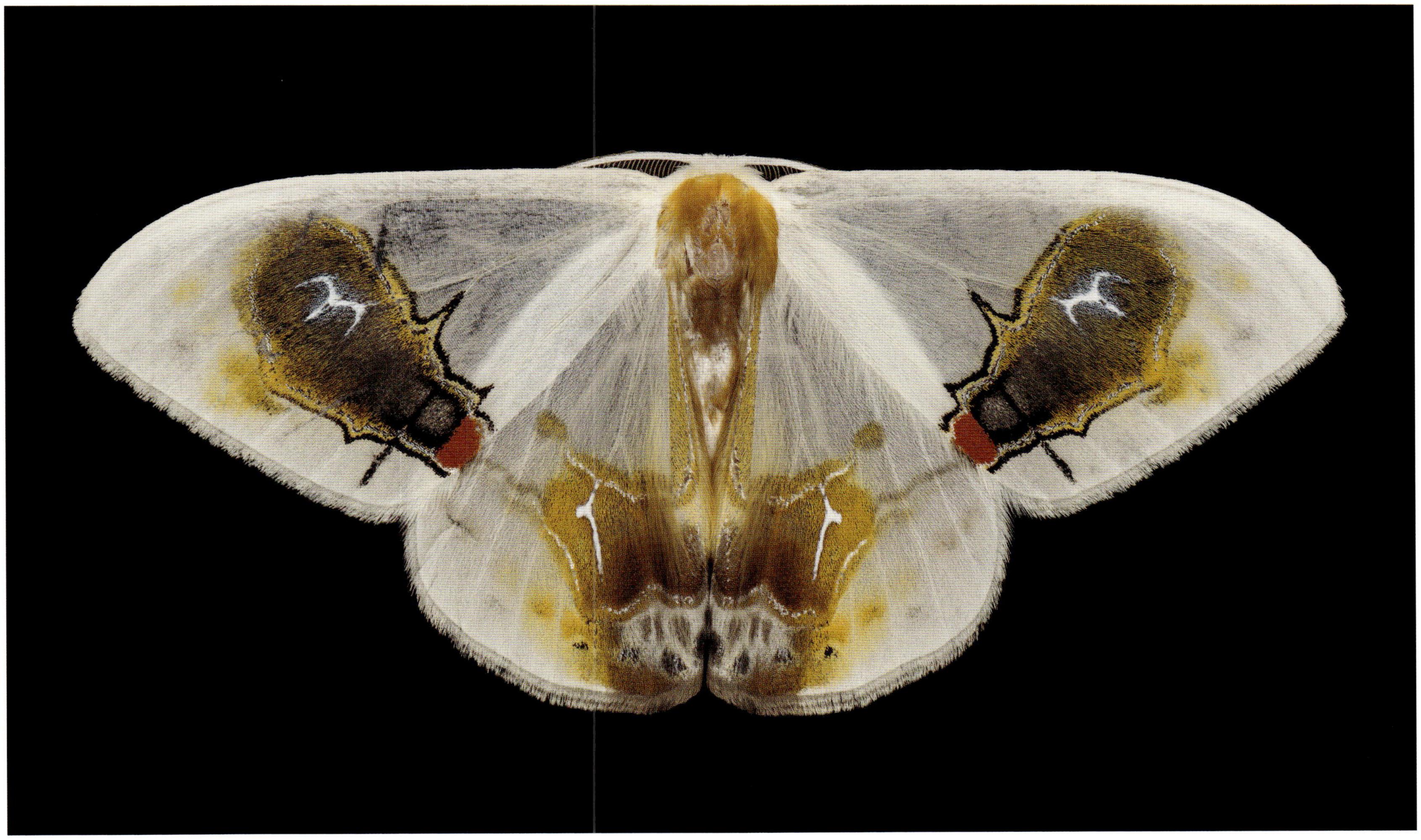

On close inspection, yet another evolutionary innovation is apparent. Because the ersatz "flies" are on the moth's forewings, and the "excrement" on the back ones, with a mere flick of their wings these moths can effectively keep their flies excitably jumpy around their meal. And apart from this tour de force of evolutionary mimicry, as they rest under the protection of their trompe-l'oel camouflage during the day these moths provide a further negative inducement to potential predators by emitting a pungent smell of . . . well, you guessed it.

So how is this lensing, exactly? Isn't my use of the term metaphorical?

On the contrary. Because we have a focal plane (the wings) upon which eons of evolutionary optimization have yielded distinct images, this is in fact lensing according to the dictionary definition of the term, and specifically in physics. Those features gradually acquired their distinctive characteristics across countless generations in which the visual aspects most likely to deter predation were passed down. And as a corollary, whatever details failed to deter had less likelihood of remaining, simply because the moths who possessed them had less time to reproduce.

To sum up, unfavorable picture elements became less distinct, while those capable of increasing longevity became ever more sharply focused. And this is true all the way down to the prismatic white spots on both the jumpy "flies" and the "wet" excrement,

Jim des Rivières, *Macrocilix maia*, circa 2010, digital scan

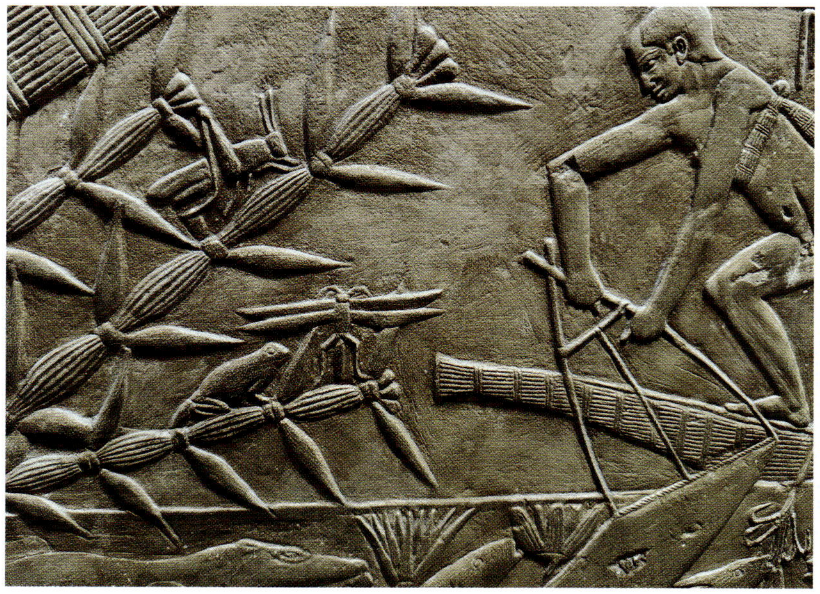

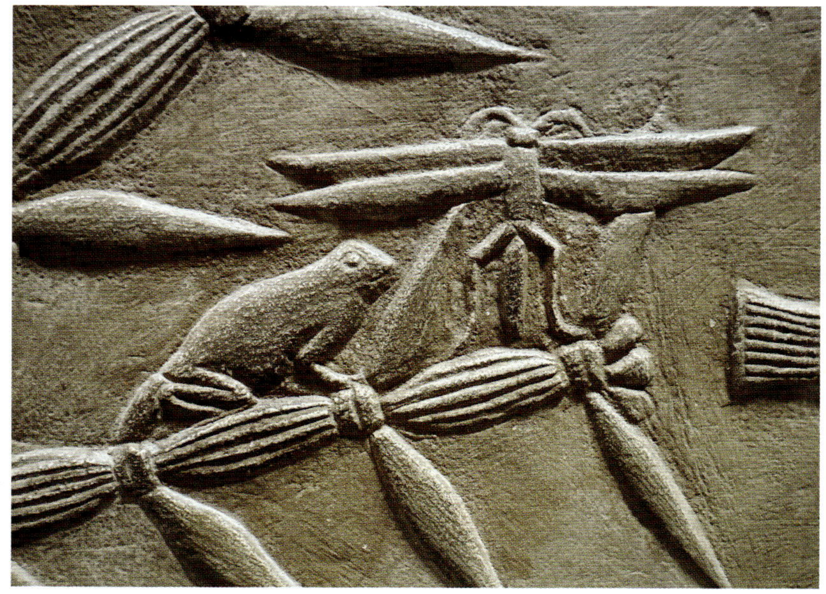



signifying reflected sunlight—ostensibly artifacts of the same bright Milky Way star that ultimately powered the whole evolutionary story.

In short, lensing.

~

On my way back from a trip to Jeddah, Saudi Arabia, in January 2024, I decided to visit Cairo for a few days. After touring the Step Pyramid, the first such giant structure in the history of Egypt and built around 2670 BCE during the Third Dynasty, I descended down a claustrophobic inclined plane into the inner chamber of a nearby subterranean structure, the Tomb of Unas, and saw the Pyramid Texts—among the oldest religious writings in the world. Designed to usher the pharaoh into the afterlife, they're effectively the earliest draft of the Book of the Dead.

Although hieroglyphs are themselves pictorial symbols, in principle the Pyramid Texts are not illustrated. But then I proceeded past the stark squared columns at the entrance to King Djoser's Southern Cemetery and delved into a well-preserved tomb complex in the desert nearby. There a set of reliefs on the limestone walls of the Tomb of Kagemni gave clear evidence of a marshy environment exploding with wildlife during the Sixth Dynasty, or around 2300 BCE. They featured birds, fish, grasshoppers—and a dragonfly perched on a tuft of river reed, its paired wings extended and ready for flight. After forty-seven centuries, I could almost see it looking back at me.

The next day I went to visit the old Egyptian Museum on Tahrir Square, until recently the largest in Africa, and was fortunate to be among the last to see it before most of its holdings were transferred to a shiny new billion-dollar facility out at Giza. Far better to see everything here, I thought, with dust still on its shoulders and the occasional yellowing, hand-typed vitrine label still visible, the product of long-gone Egyptologists.

As with the work of Abram and Marais, there's no sense trying to summarize the manifold impacts of this museum at the end of its long reign. Yet something strange unfolded on the second floor that bears relating. After admiring the gilded splendor of Tutankhamen's treasures, I found myself in a series of less traveled galleries. They featured a number of unfamiliar ancient statues: life-like, life-sized figures carved from sycamore. These skilful representations depicted pharaohs and important court function-

Michael Benson, *Wide and close view of limestone relief, 2340 BCE* (seen in Tomb of Kagemni, Saqqara, Egypt), 2024, photographs

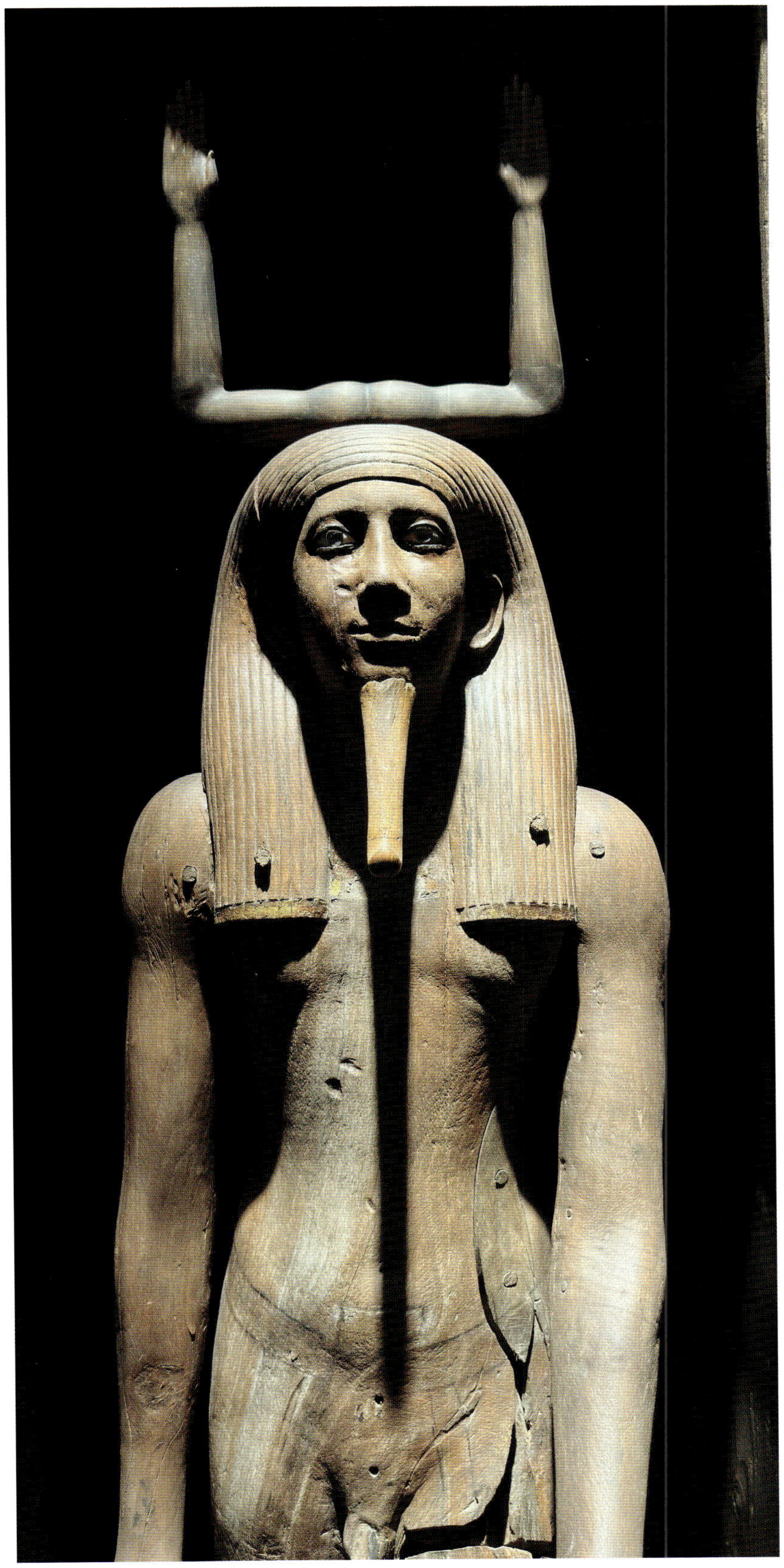

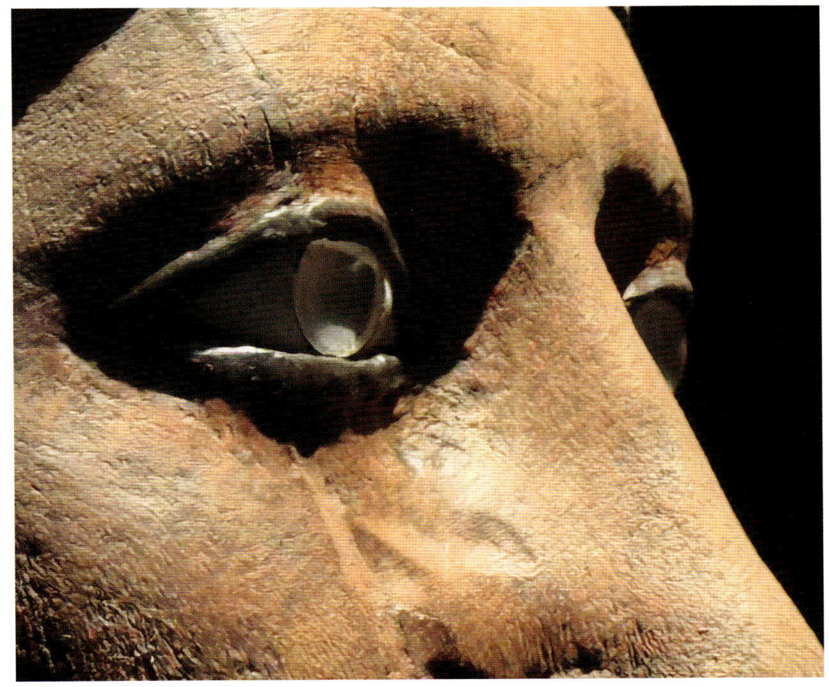



aries from a period extending between the Fifth and Thirteenth Dynasties, four to five thousand years ago.

I've discovered, when attempting to absorb the contents of a museum that has fully captured my attention, that two kinds of vision are at play: a sharply detailed central field of focus, and a second, peripheral field. As any amateur astronomer will tell you, the latter tends to be more light-sensitive, but it is also less capable of registering detail. It was this retinal region that started sending me strange signals to the effect that *one of the statues was looking at me*.

At first these signals were largely subliminal. Being also peripheral, I ignored them. But gradually they overrode everything else—and finally I turned to directly face the standing figure of Hor I, Pharaoh of the Thirteenth Dynasty in the late Middle Kingdom. And discovered that *he was indeed looking back at me*. And not only that, but he had a subtle, unmistakably intelligent gleam in his eyes.

This was a shock. I approached Hor's vitrine with some trepidation—after all, he had effectively come to life. Gradually I saw that the cornea of each eye was transparent, polished rock crystal;

Michael Benson, *Hor I, Pharaoh of the Thirteenth Dynasty* (sculpted by unknown artist in Sycamore, circa 1750 BCE), 2024, photographs

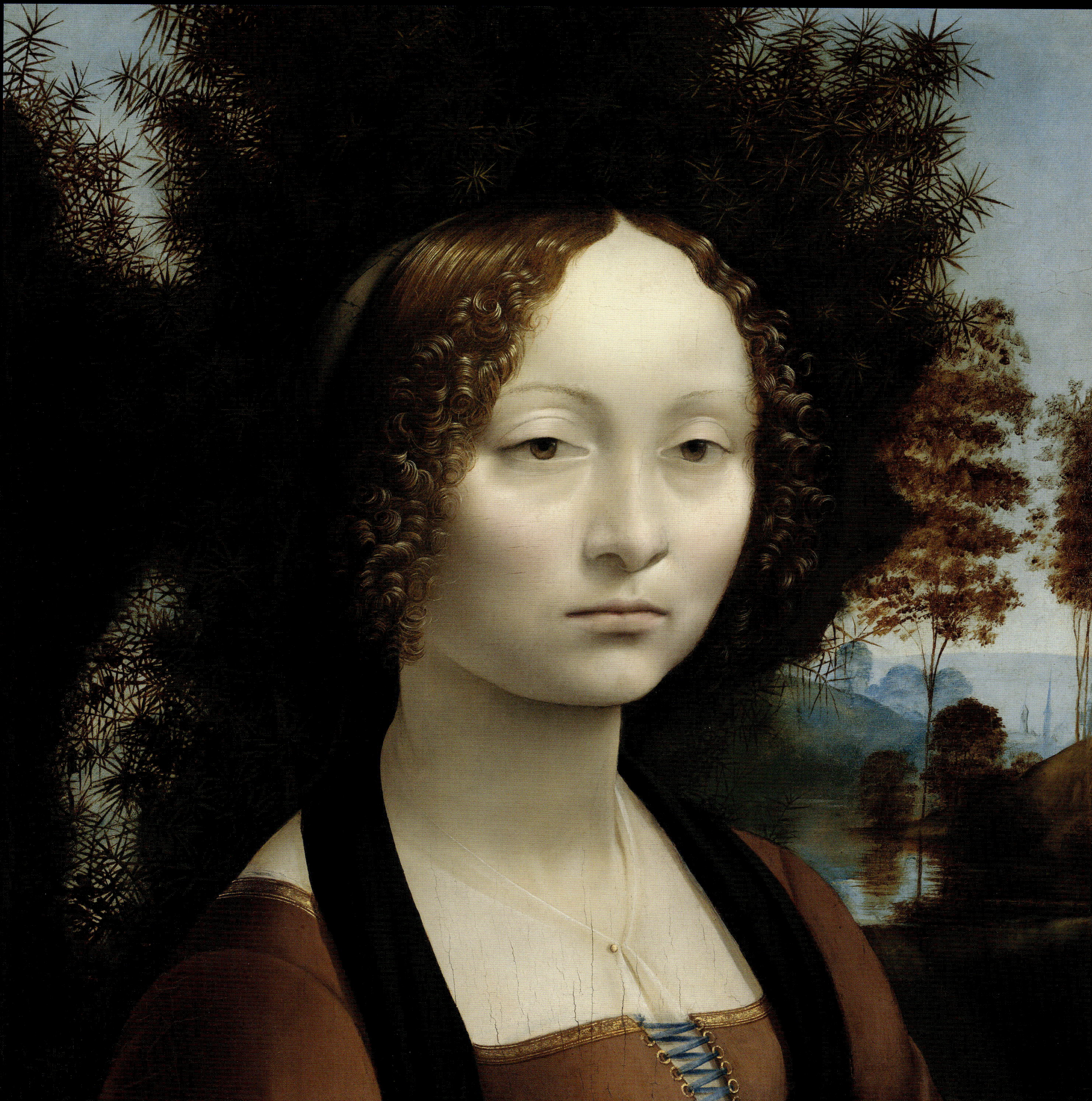

the whites were opaque quartz; his quizzical, questioning aspect was in part due to rims of copper inlay; and the total effect of a probing, living personality was no hallucination, nor was it an accident. The Ancient Egyptians, it seems, had taken clear polished stone of the type used to start fires in the Babylonian *Epic of Ishtar and Izdubar* and adapted them for another purpose. The key to the effect was that shine and depth to the pharaoh's eyes. It occurred to me that effectively I was in the presence of a lens-based realism—one in which both the artist and subject had long since departed, leaving behind the palpable presence, even the living soul, of an individual. Still kicking after four thousand years.

And I imagined French archeologist Jacques de Morgan opening Hor's tomb near the ancient capital of Memphis in 1894, to see the pharaoh silently staring back from the gloom. It must have been quite a shock.

As I absorbed the implications of my experience, I realized that the use of lenses to convey lifelike representations of human beings actually spans most of recorded history. Their deployment in this case to bring a specific pharaoh's personality into sharp focus—his quizzical intelligence and good humor—reminded me of the specular reflections imparting three-dimensionality to the scenes inscribed by natural selection onto the wings of *Macrocilix maia* moths. They also brought to mind the subtle gleam in the eyes of another work at the National Gallery, the only painting by Leonardo da Vinci in North America.

As with Hor I's uncanny ability to look back, clear-eyed, across the centuries at his visitors, Leonardo's portrait of Ginevra de' Benci is capable of meeting your gaze any day of the week in gallery six, and returning it. But unlike our sociable pharaoh, she seems rather reserved. Around her face, a descending torrent of curls meets the spiky leaves of a juniper bush like a waterfall under the stars. Behind her, trees recede along a riverbank into the blue haze of a summer's day.

About our position in nature they were never wrong, the Old Masters.

Every narrative has to come to an end, be it a retelling of the Epic of Gilgamesh in cuneiform or this humble essay printed in Shenzhen, China, at the behest of Abrams in New York.

In June of 2022, I borrowed a small number of pinned insects from the holdings of AgCanada, the governmental department responsible for regulating agriculture across that vast country. Although Owen Lonsdale, the entomologist who approved the loan, was manifestly sympathetic to my project, his requirement was that the insects had to be returned exactly as I received them. This was entirely reasonable, of course, but it meant I couldn't use my laboratory's sputter-coater to lay a molecule-thin layer of platinum on his samples. Which in turn meant I would have to contend with electrostatic charging.

Owen had chosen insects that he thought were particularly unusual and interesting. One of them was a fine example of a stalk-eyed fly. In this family of insects, what at first might be taken for particularly long antennae are actually stalks projecting from an eyeless head, with the eyes to be found at the end of each stalk—something giving them a decidedly weird appearance, even by insect standards. But I knew all that projective sensory equipment might serve to further complicate matters, because it tends to act like a lightning rod under the electron beam—which in turn requires various actions to keep the resulting charging from distorting the image beyond reasonable use.

And that's exactly what happened. But here I just want to note that the passage of a subject from its natural setting to the inner sanctum of an SEM is actually quite extreme. Even without a coating of heavy metals, placing my stalk-eyed fly into the high-tech cube of that vacuum chamber was somehow reminiscent of the days when the USSR was launching a succession of space-suited dogs into orbit, and the USA responded by locking chimpanzees into coffin-shaped pressure chambers and vaulting them on suborbital trajectories. It's one thing to belong to the species that devised the technology and go for a joy ride, in other words, and quite another to be placed in that position involuntarily.

Anyway, such were my thoughts as I tweezered Owen's weirdly



**Left:** Leonardo da Vinci, *Ginevra de' Benci*, circa 1474–1478, oil on panel

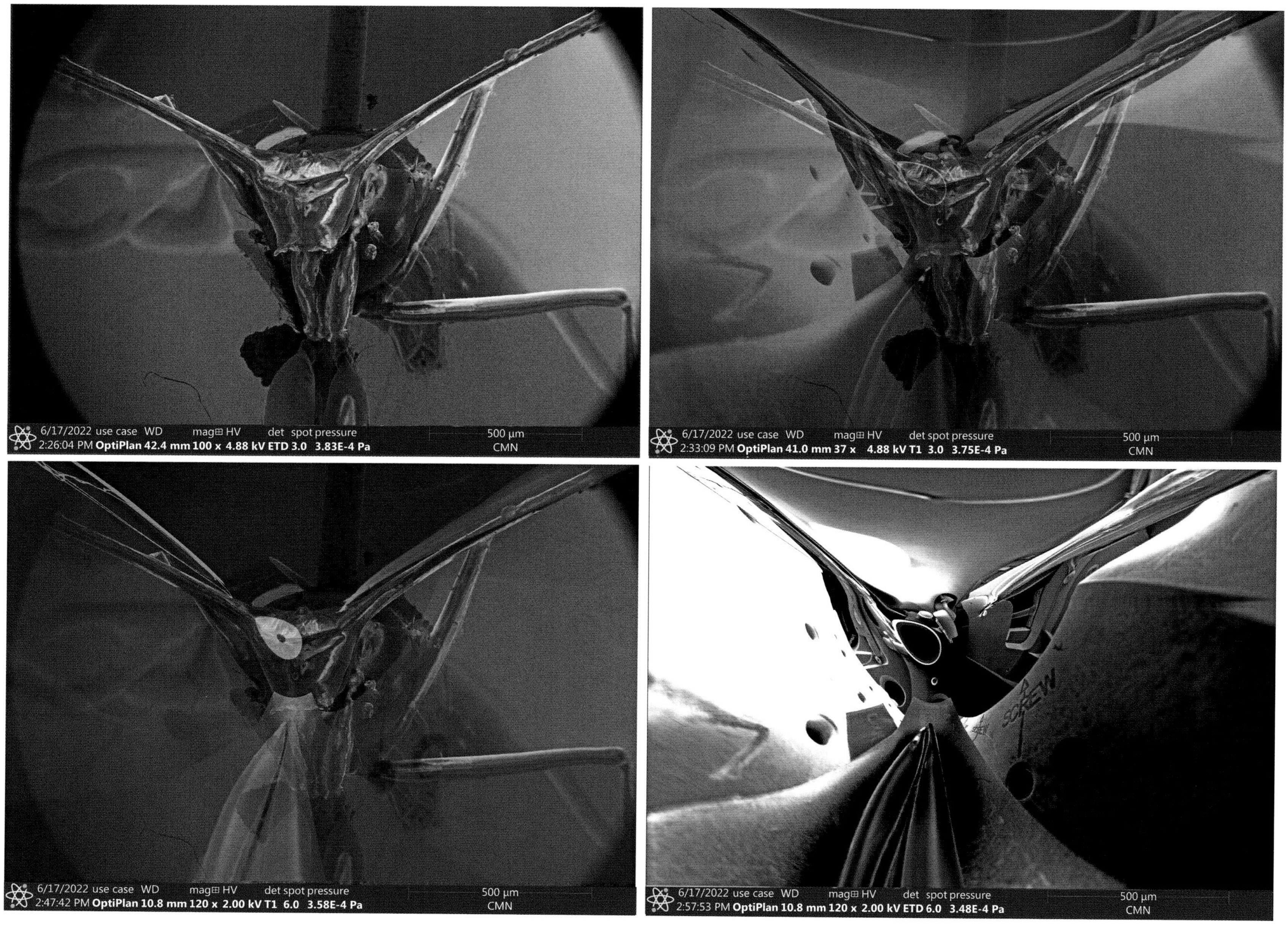




panoptic fly into a circular slot in the SEM's sample stage, closed the heavy chamber door, and pumped out the air. As usual, the image that flickered into view on the LCD screen in front of me contained my subject weirdly foregrounded against a muted gray background. And as I knew would probably happen, my first angle was less than ideal.

To begin with, the morphology of the head, which formed a "Y" shape, was indeed unutterably strange. But in my initial framing, the eyes at the top left and right tips of the "Y" were so far from the center of the screen that they couldn't even be seen, and the creature's paired forelegs extended awkwardly below its central thorax, to be cut off by the bottom of the frame. It wasn't a particularly promising image, in other words, and it was already acquiring a silvery, solarized sheen—which I knew meant that an electrostatic charge was probably building up.

I decided to get a four-minute scan anyway while considering

Michael Benson, *Electrostatic Mirroring of Stalk-Eyed Fly*, 2022, SEM images what to do. I was aware that I could switch to low vacuum mode from high vac, which should reduce the charging.

***Suddenly there came a tapping***

But I wasn't familiar with using the scope in that mode, and I knew that it would come at the expense of image quality. In short, it was a trade-off.

***As of someone gently rapping***

But meanwhile, as I considered my next move, my subject seemingly decided to make one of its own.

***Rapping at my chamber door***

Because even as I watched, the image gradually morphed. A seamless dissolve was taking place. The picture now gradually emerging on the screen remained true to the "Y" shape of the animal's head. But against all sense, *it revealed the point of view of my stalk-eyed fly within the vacuum chamber.*

With a jolt, I realized that the central nexus of the strange fly's head had somehow transformed itself into a view of the circular lens aperture at the top of the chamber. And that warped foreground with crater-like holes in it was the microscope's central stage, perforated to accept sample stubs. Meanwhile those irregular lobster-like projections forming the lower limb of the "Y" were the creature's legs—now as seen from their other end.

*All of it exactly as would have been visible to its eyes.*

Trying to understand what was happening, I flipped rapidly through various possibilities. Finally I realized that a dome of charged electrons had built up over the insect during the first scan. It was acting as an electrostatic mirror. But because of the irregular shape of my subject, that mirroring was distorted. Not unlike that distant elongated spiral galaxy, warped forward by the irregular mass concentration of cluster 0220588, all the way to the Hubble Space Telescope.

A kind of transubstantiation was happening—live and in real time. Because like a patient waking up on the operating table, Owen Lonsdale's fly was looking back at me.

And I could see that clearly, through its eyes.

***In there stepped a stately Raven of the saintly days of yore***

Scanning Electron Micrographs

# PLATES

LUNA

# THE MOON

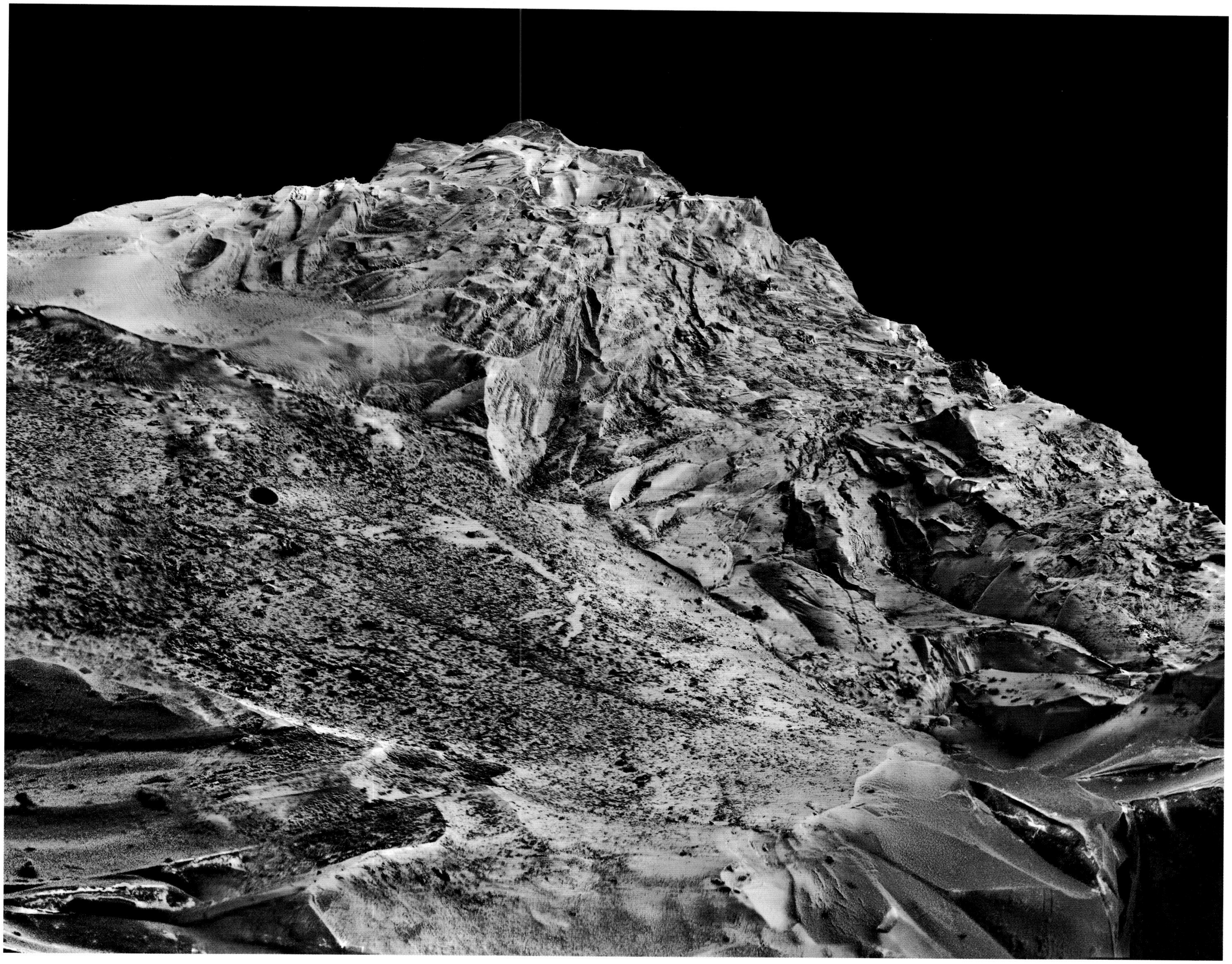

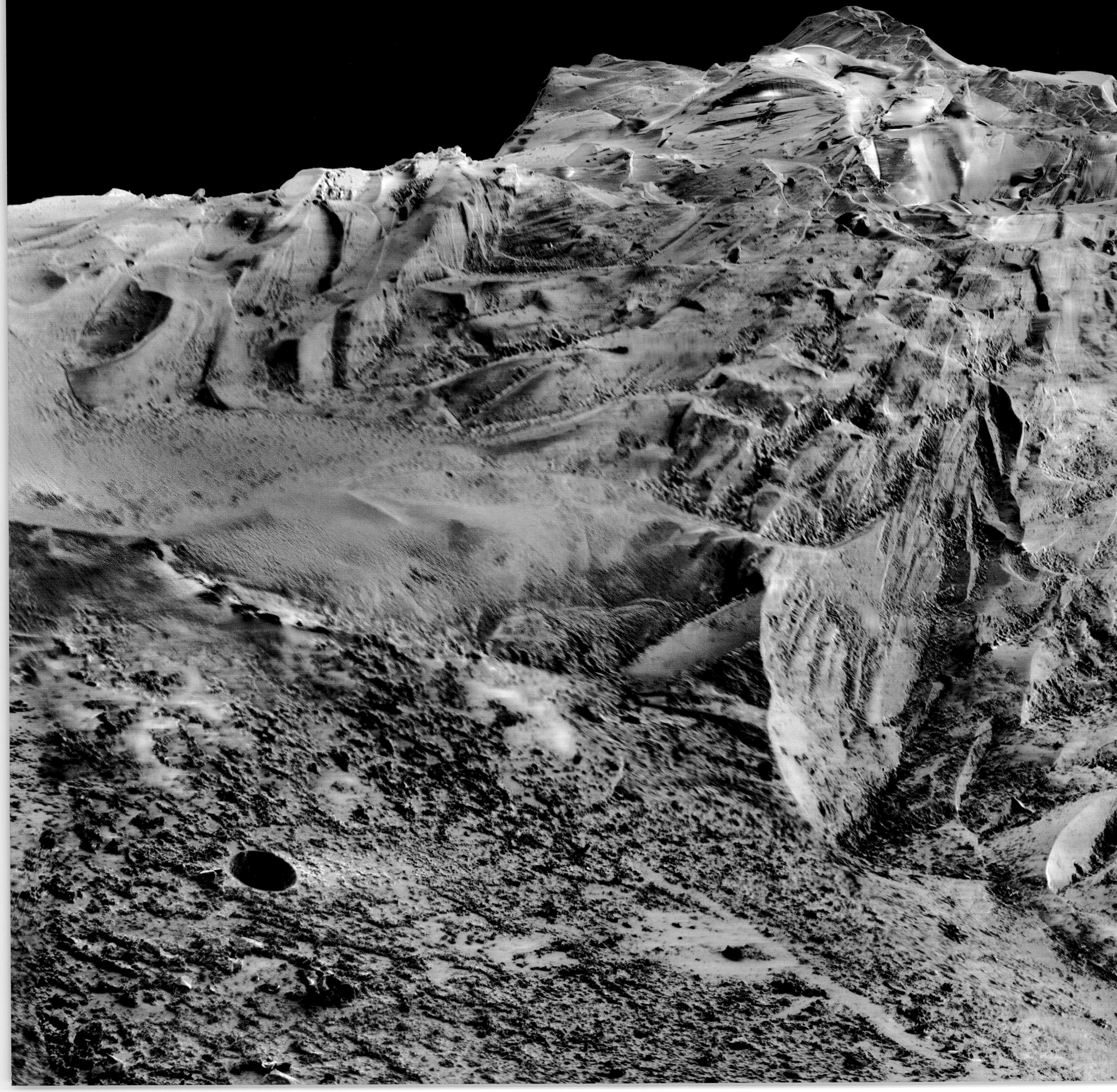

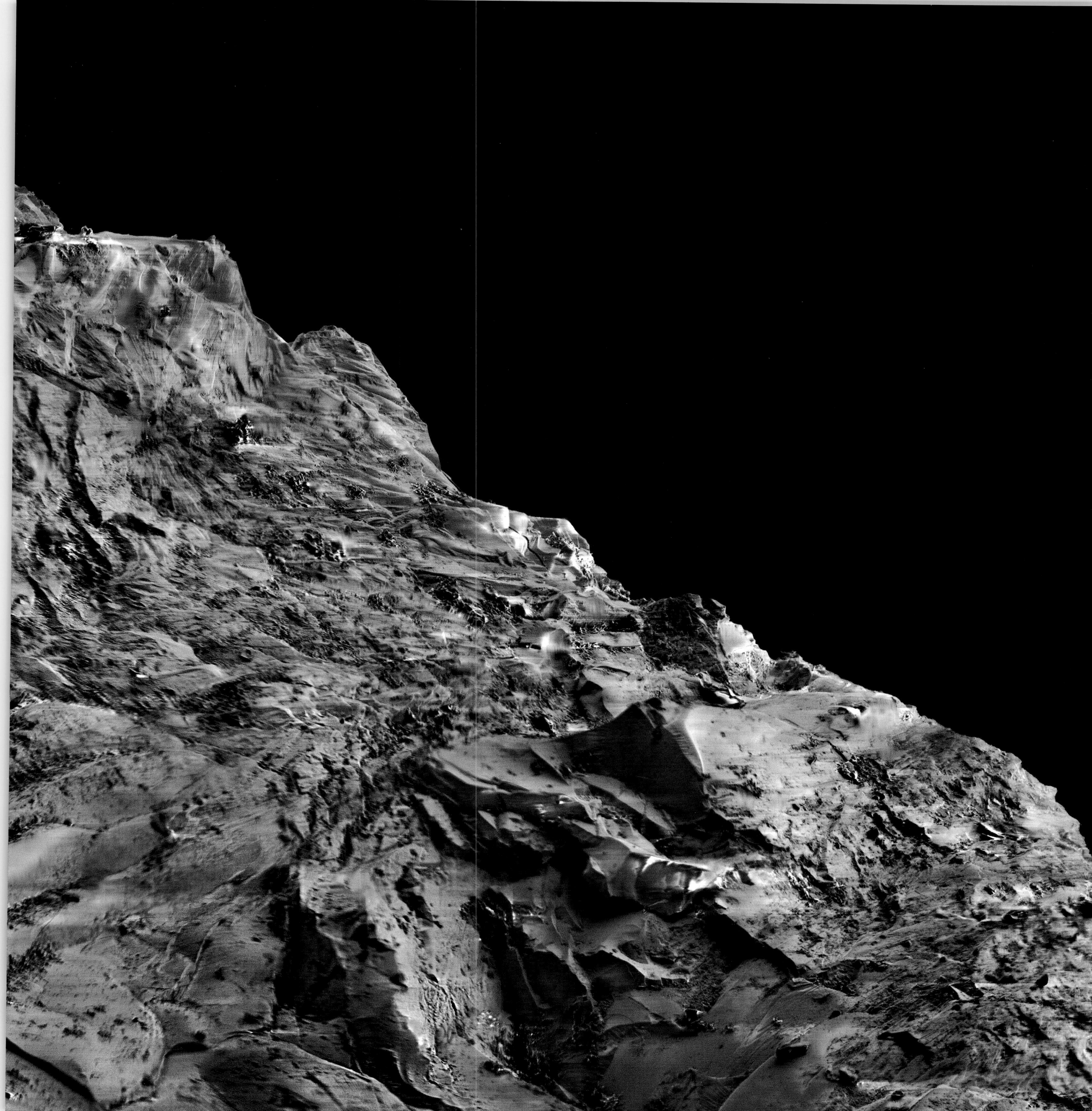

← Page 41

Lunar prominence.
250 x. About one millimeter wide.
Apollo 16 sample 60095.4

← Pages 42–43

Detail view of lunar prominence

Facing Page

Lunar impact glass with gas bubble craters and stress fractures.
307 x. About 1.5 millimeters wide.
Apollo 16 sample 60095.05

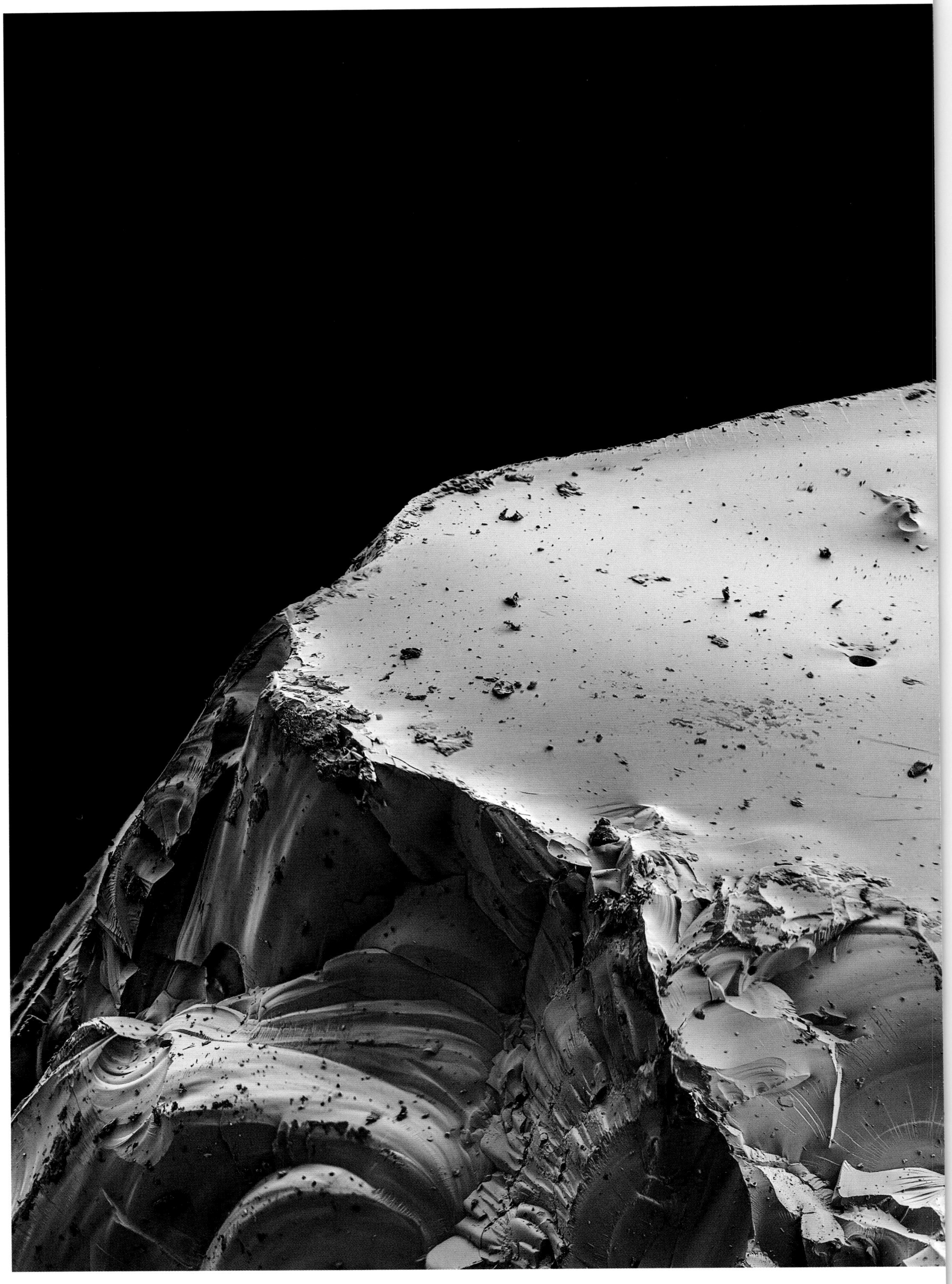

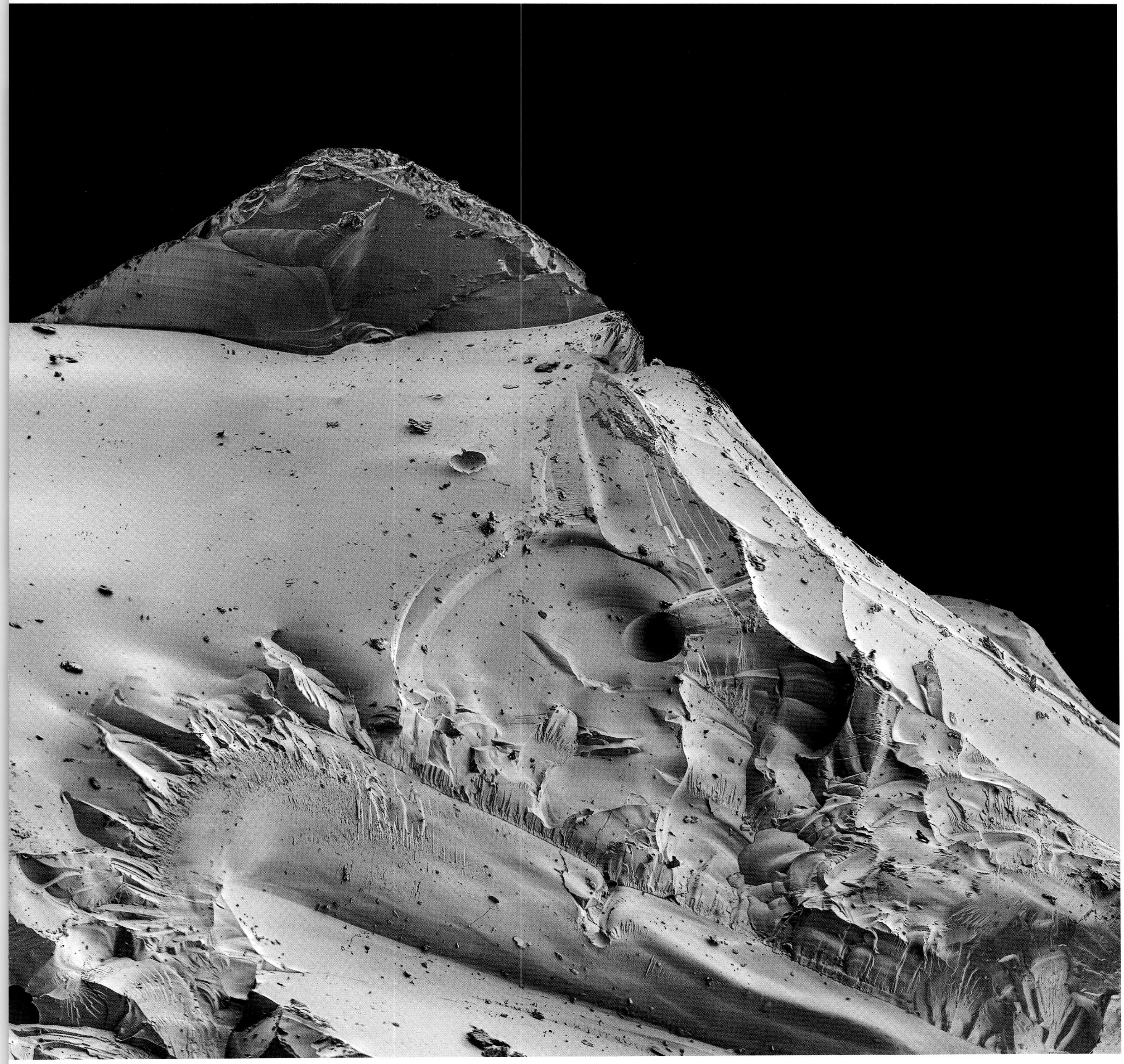

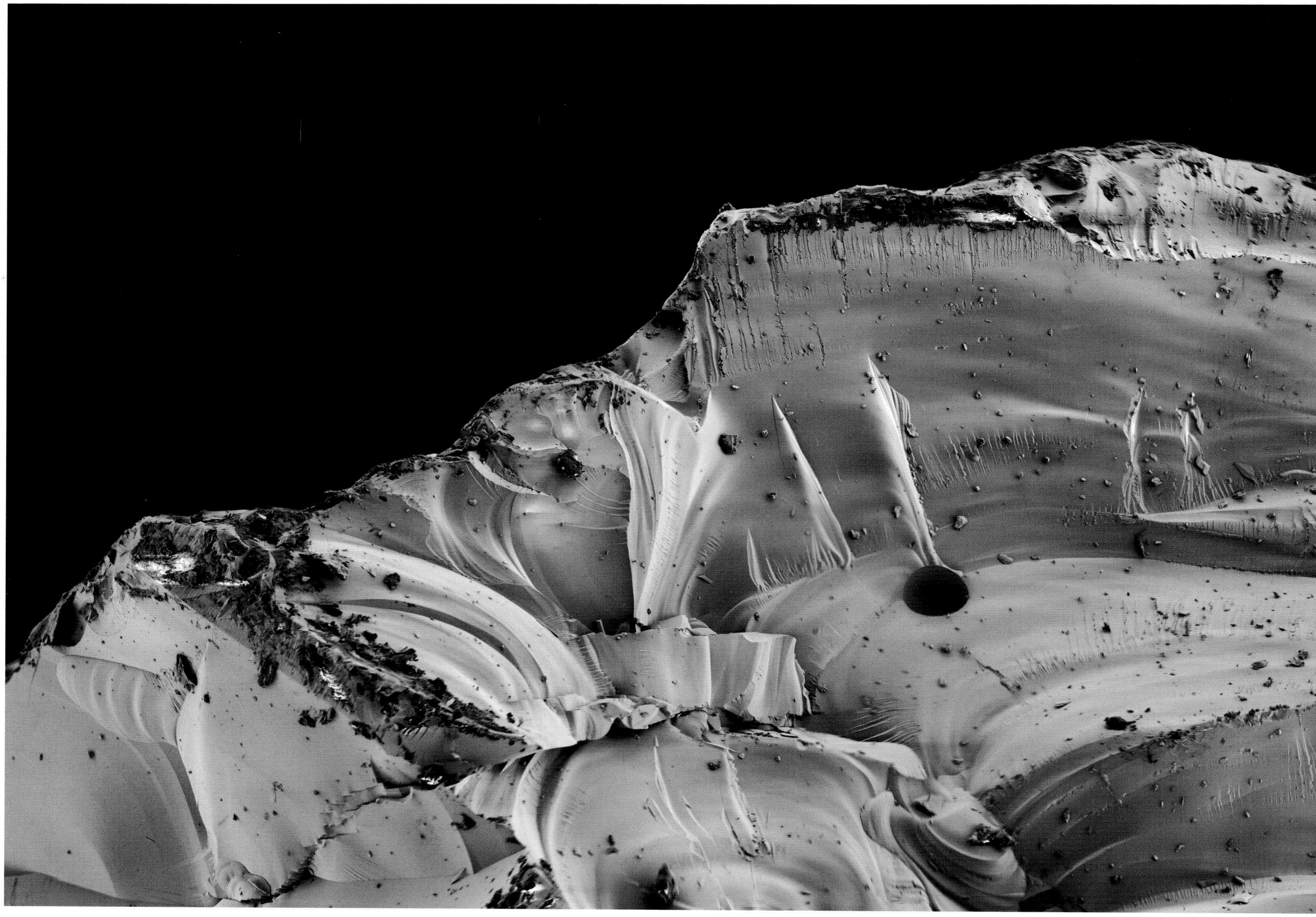

**Above**

Ridgeline with stress fractures and craters.
415 x. 1.3 millimeters wide. Apollo 16 sample 60095.05

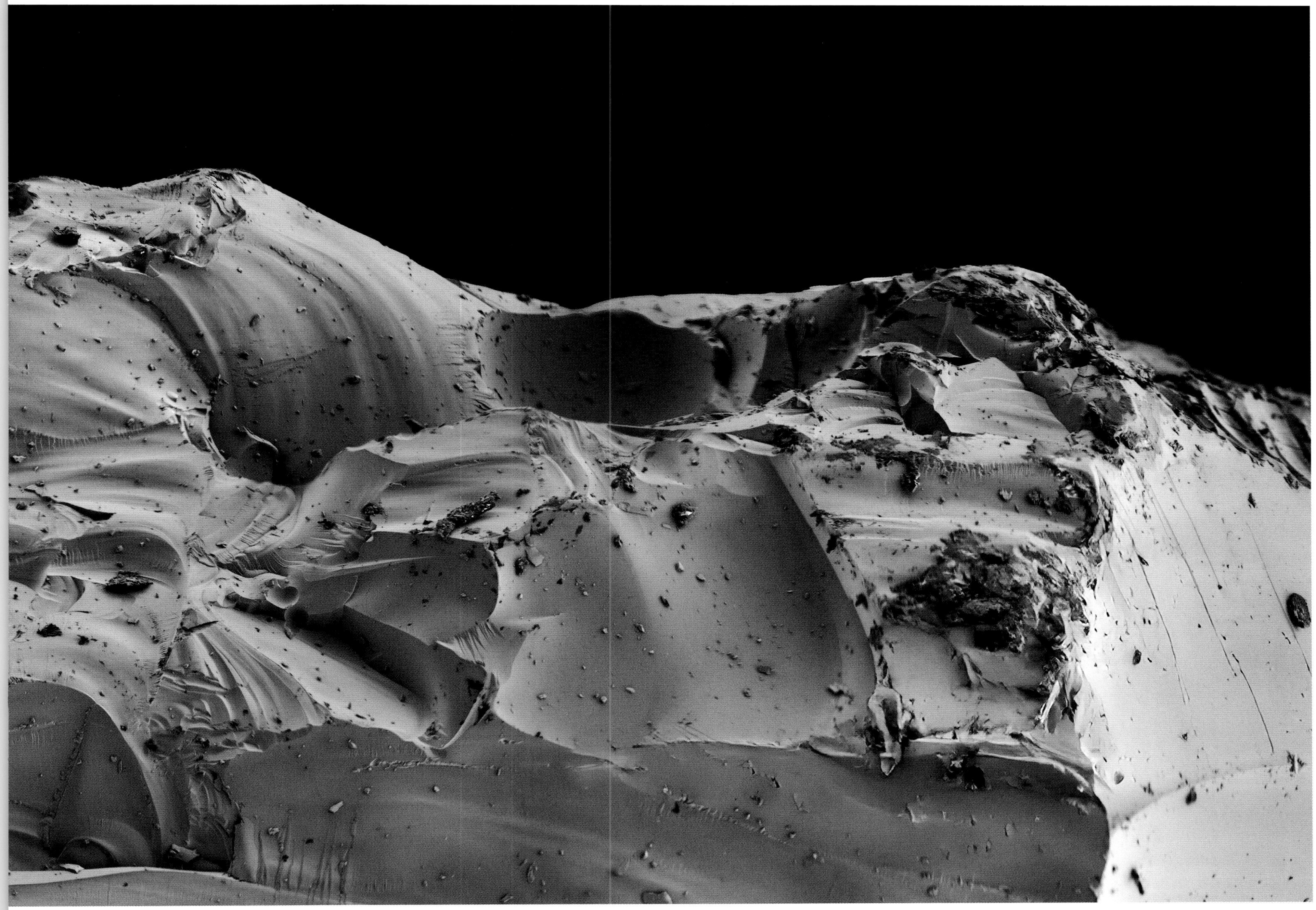

Page 48 →

Crater and stressed surface of lunar impact glass.
1100 x. About 200 microns across, or 0.2 millimeter. Apollo 16 sample 60095.05

Page 49 →

Oblique lunar landscape with crater and stress fractures.
657 x. Circa 350 microns wide, or 0.35 millimeter. Apollo 16 sample 60095.50

48

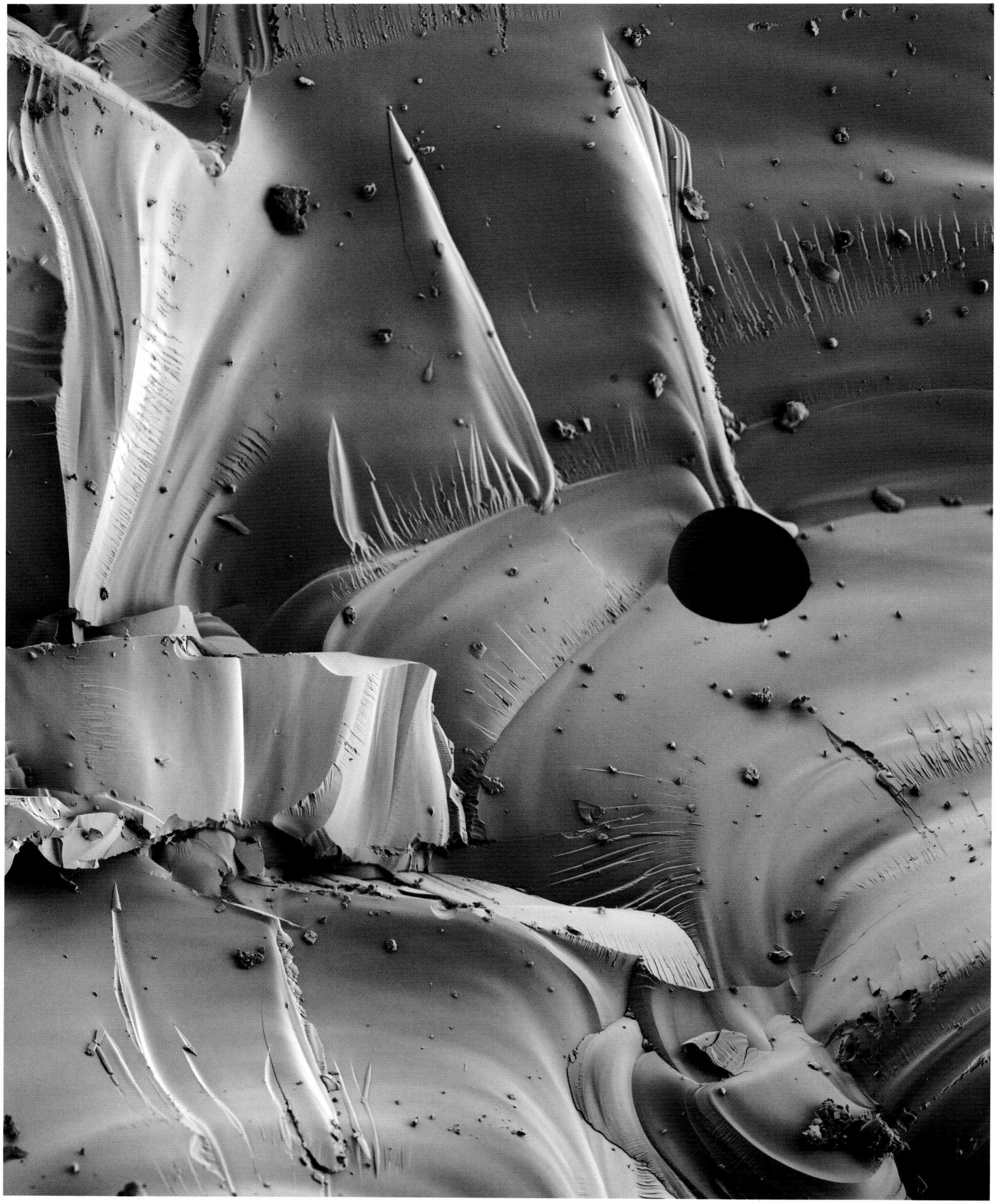

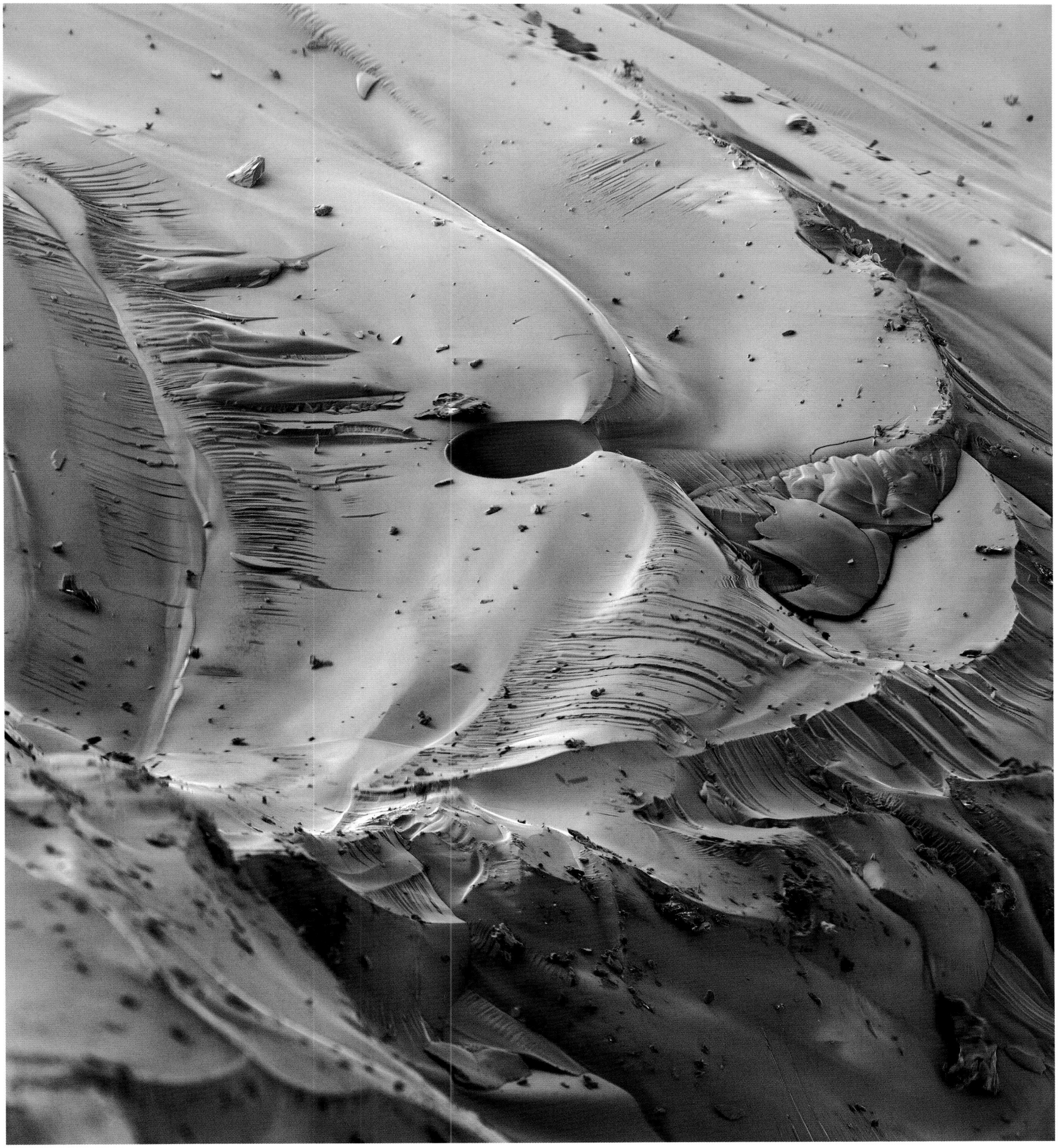

Facing Page

Shocked surface of lunar impact glass with electron beam charging
(the glowing patches on the top and left).
216 x. About 1.6 millimeter. Apollo 16 sample 60095.05

Pages 52–53 →

Lunar badlands with central crater, stress fractures, and cave-like fissure.
78 x. Slightly over a half centimeter wide. Apollo 16 sample 60095.05

Pages 54–55 →

Rugged lunar prominence with radiating spallation zones around micrometeorite craters.
57 x. Less than a centimeter wide. Apollo 16 sample 60095.6

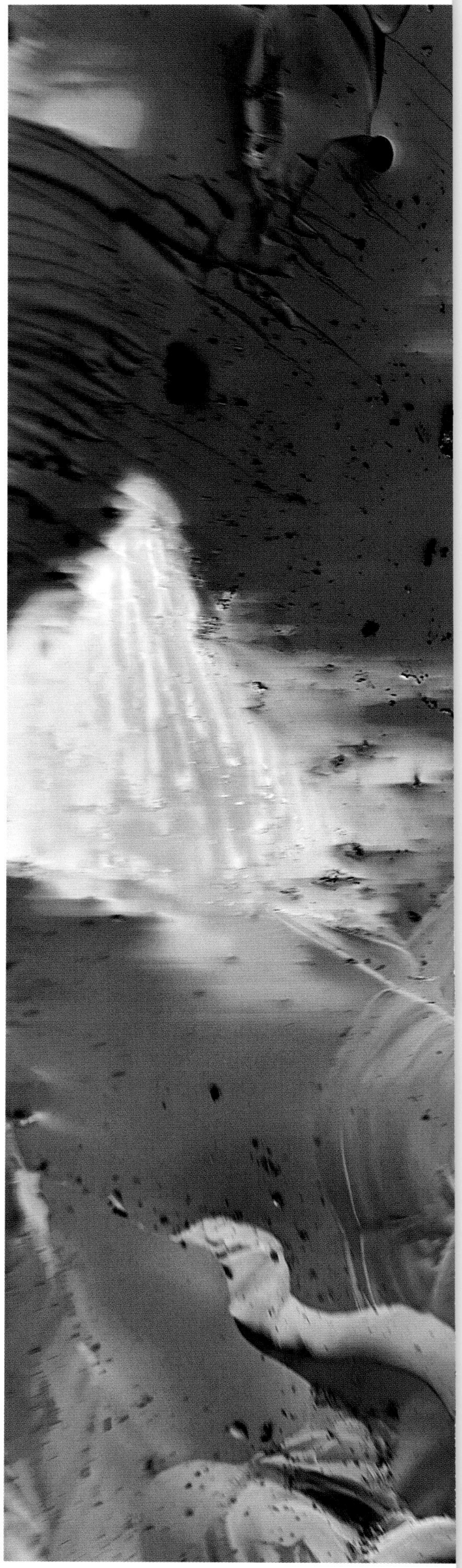

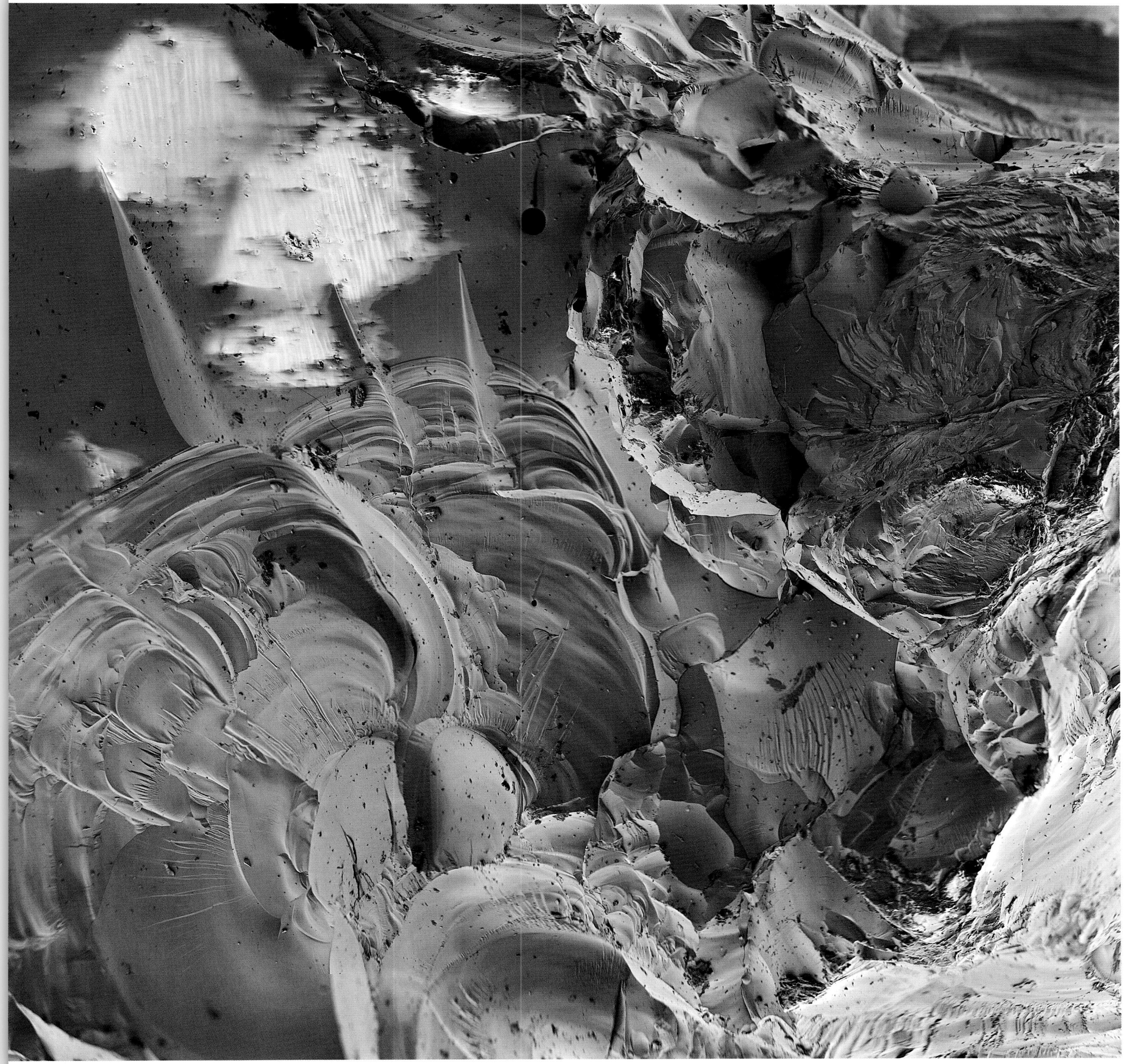

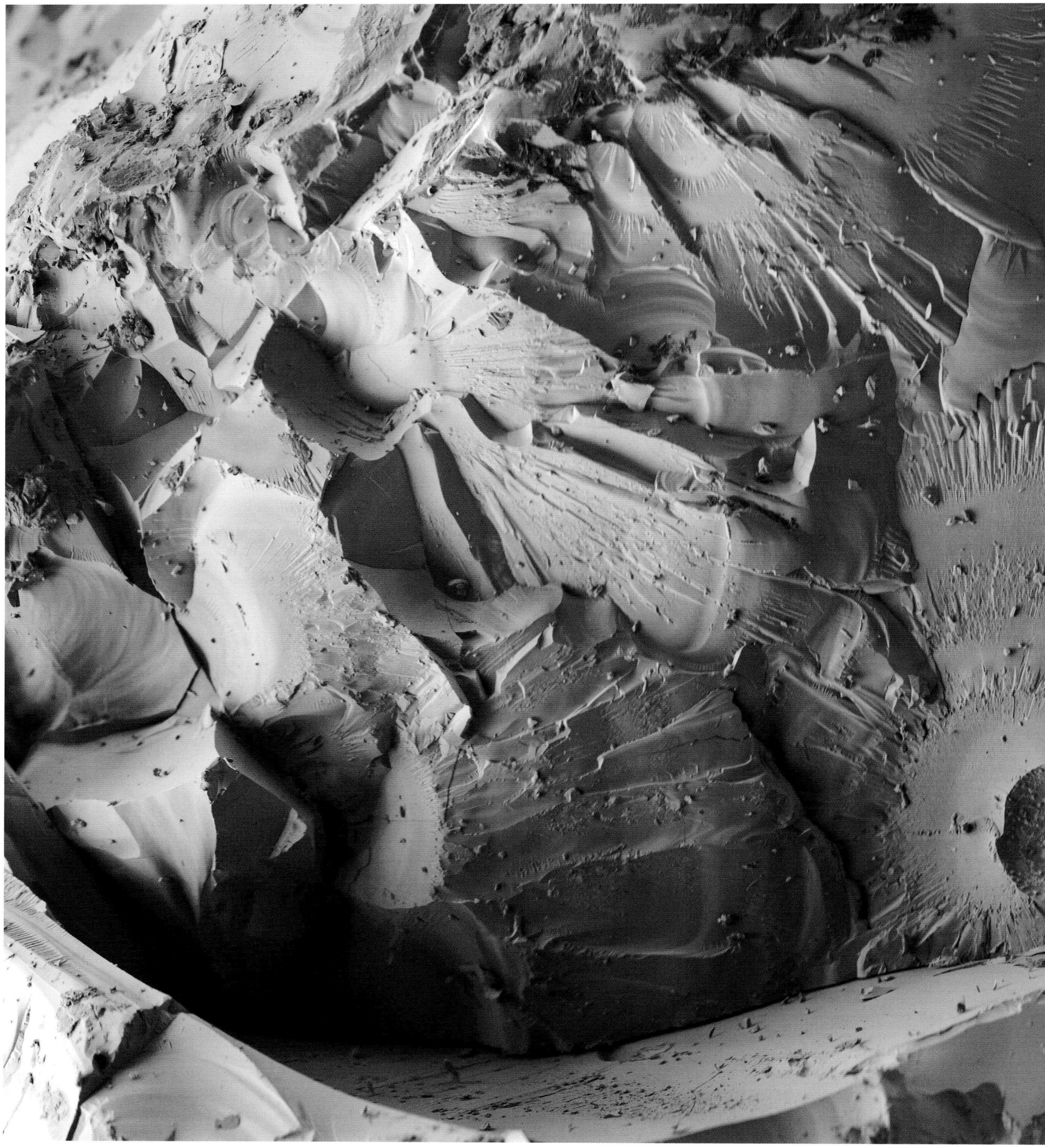

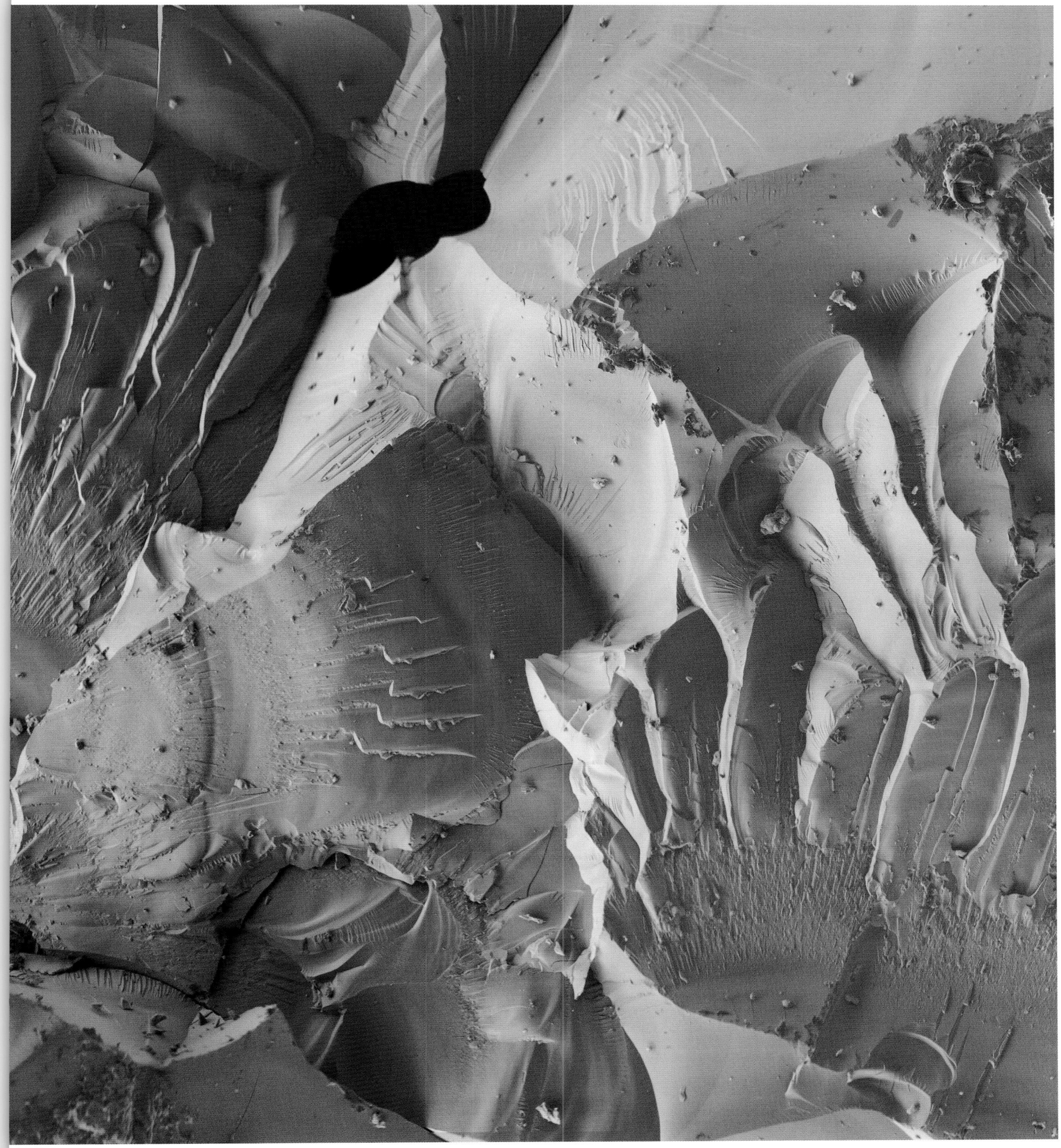

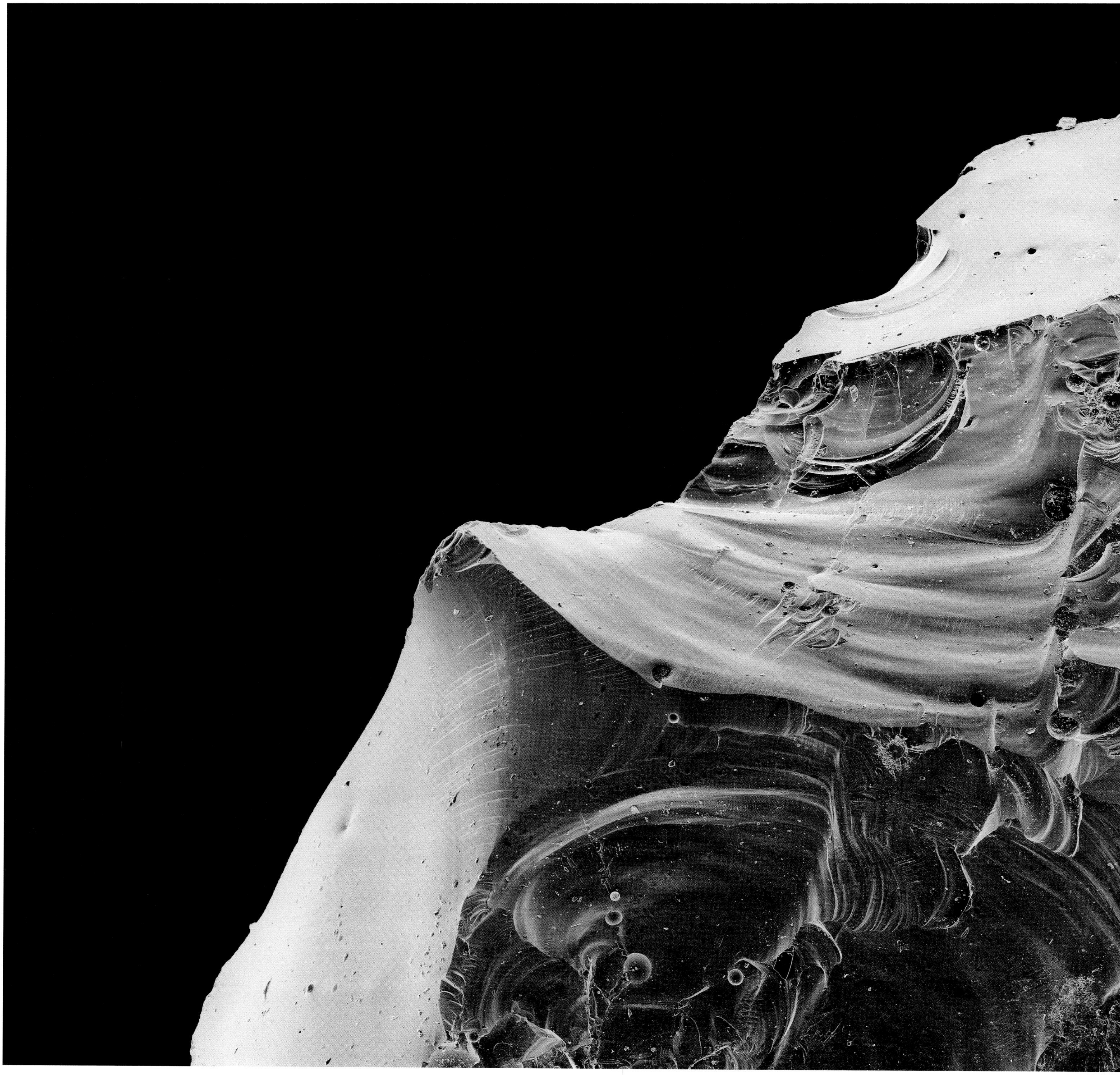

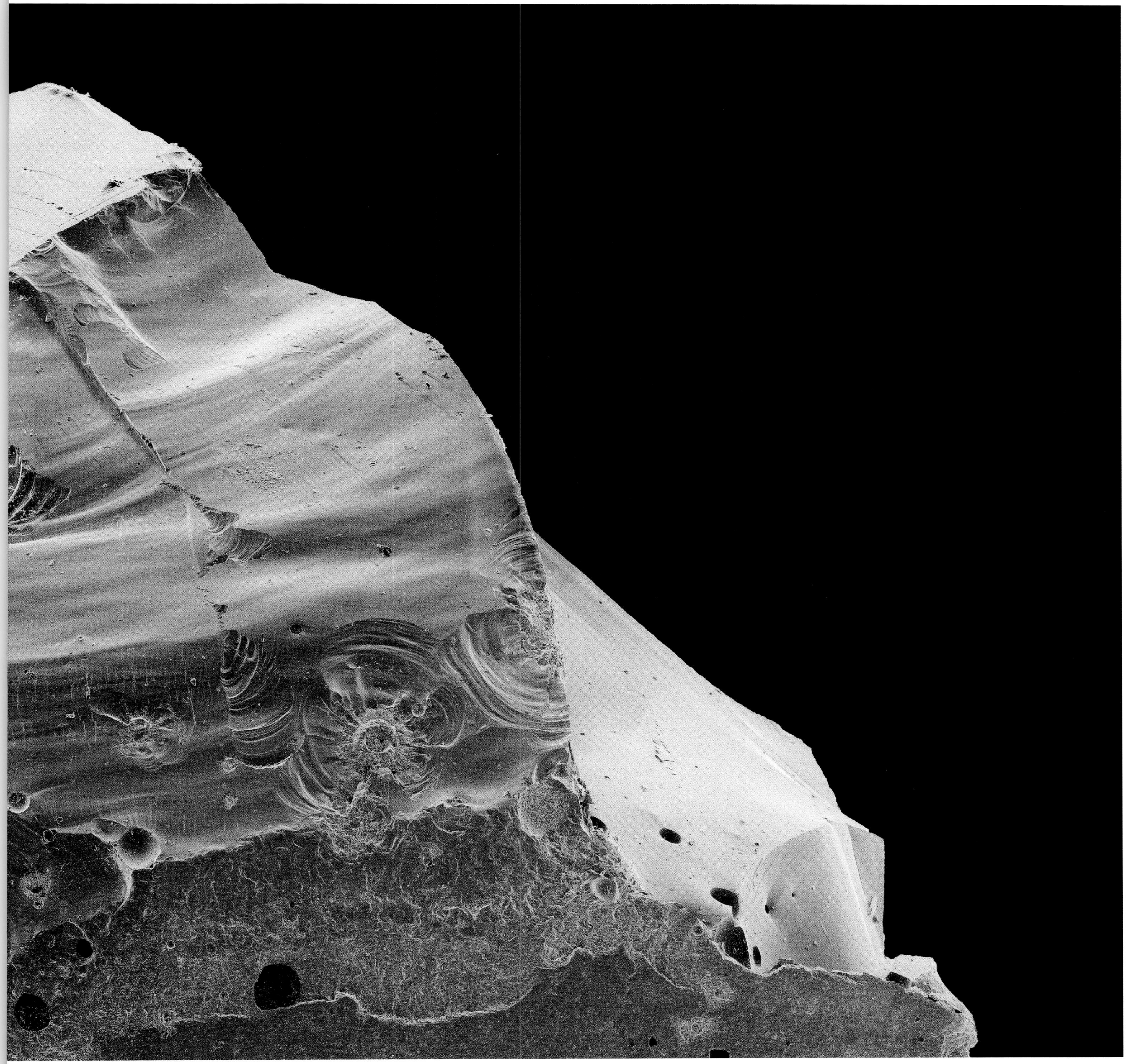

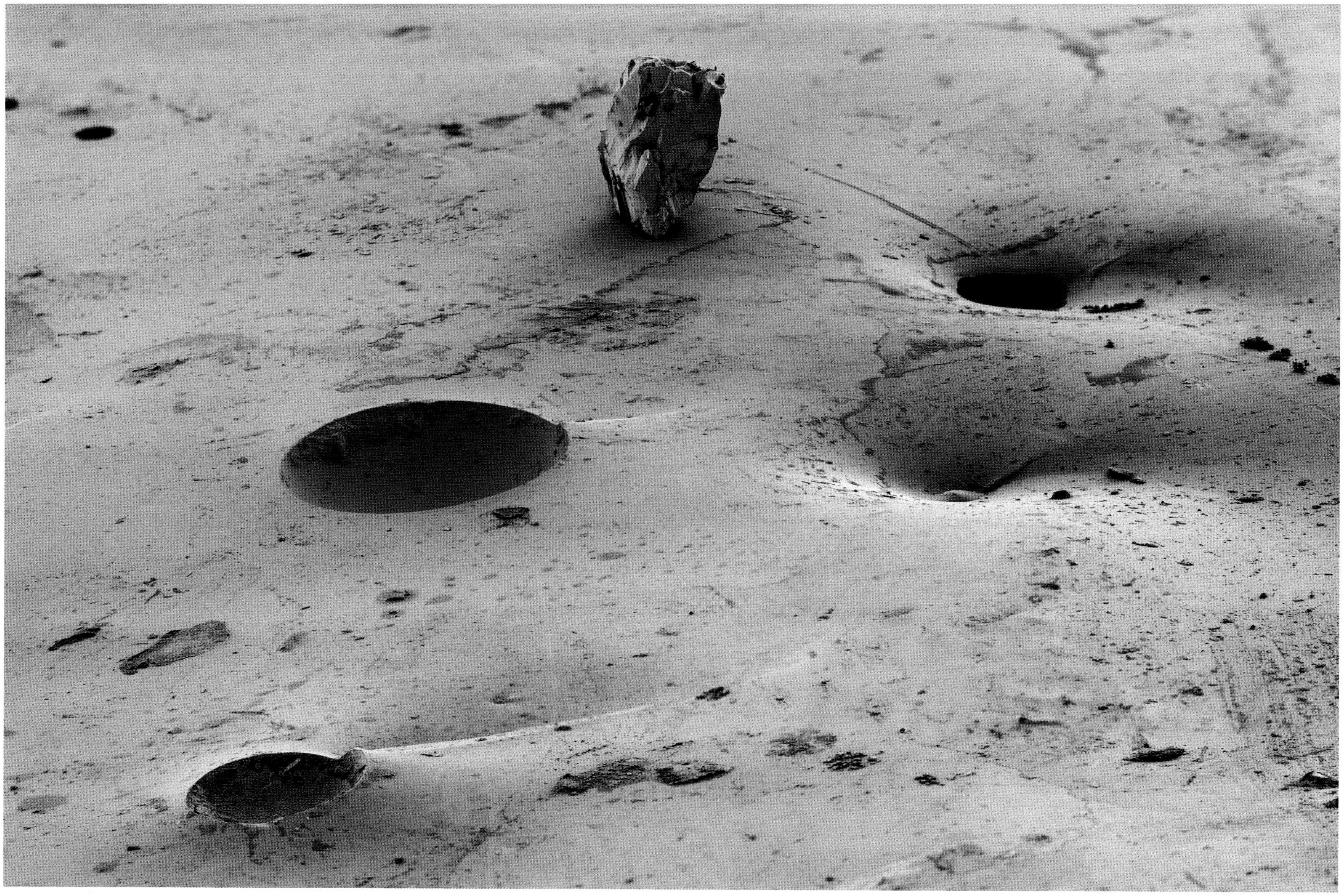

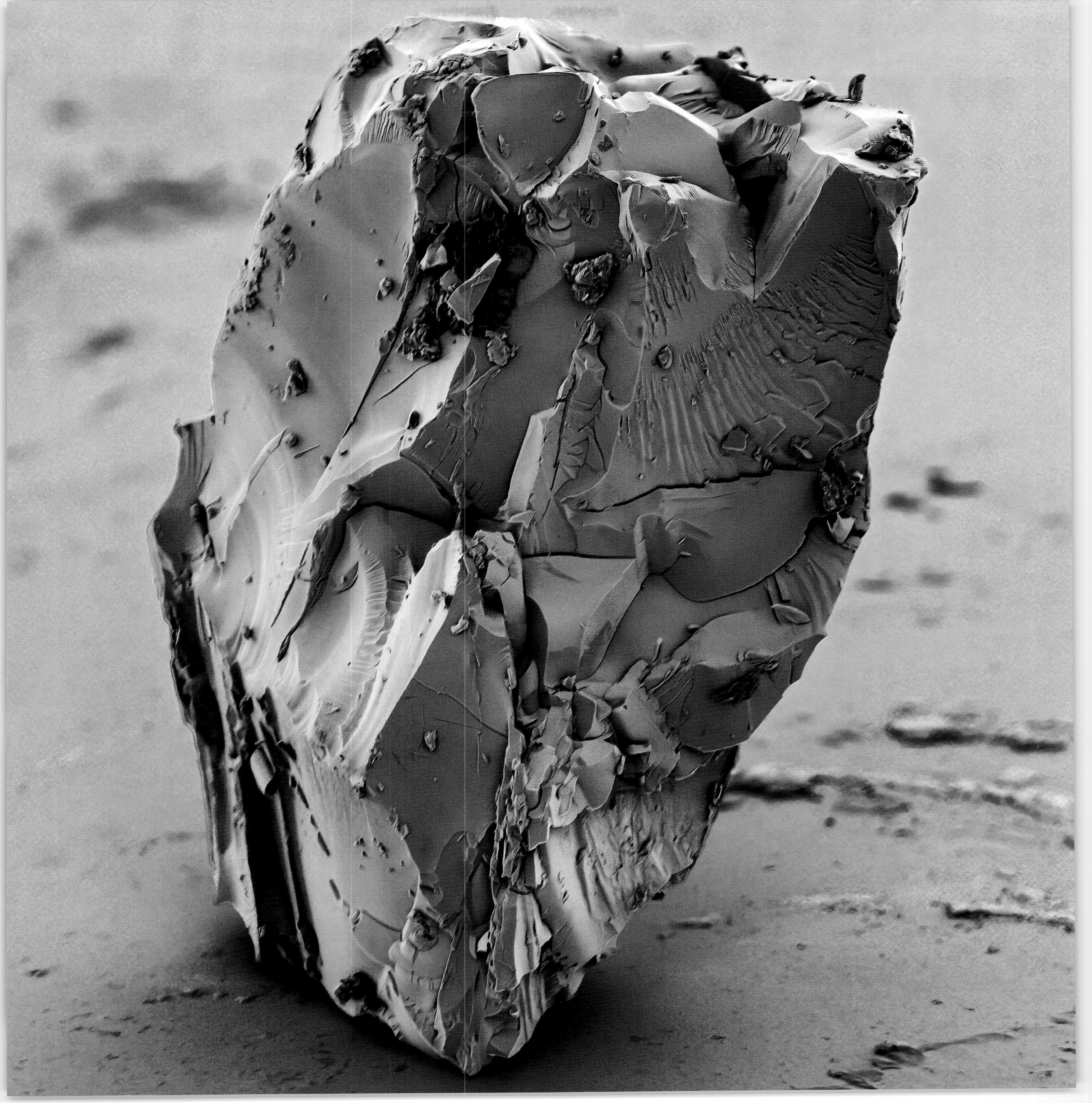

**← Pages 56, 57**

Fractured lunar rock standing on cratered surface.
The rock is about 40 microns wide, or 0.04 millimeter.
4500 x. Apollo 16 sample 60095.6

**Facing Page**

Lunar impact glass fragment with craters, fractures, and a sharp prominence.
65 x. Just under half a centimeter wide. Apollo 16 sample 60095.53

**Page 60 →**

Stress fractures and gas bubble crater in lunar impact glass.
170 x. About one millimeter wide. Apollo 16 sample 60095.53

**Page 61 →**

Lunar prominence.
145 x. About two millimeters wide at its base. Apollo 16 sample 60095.53

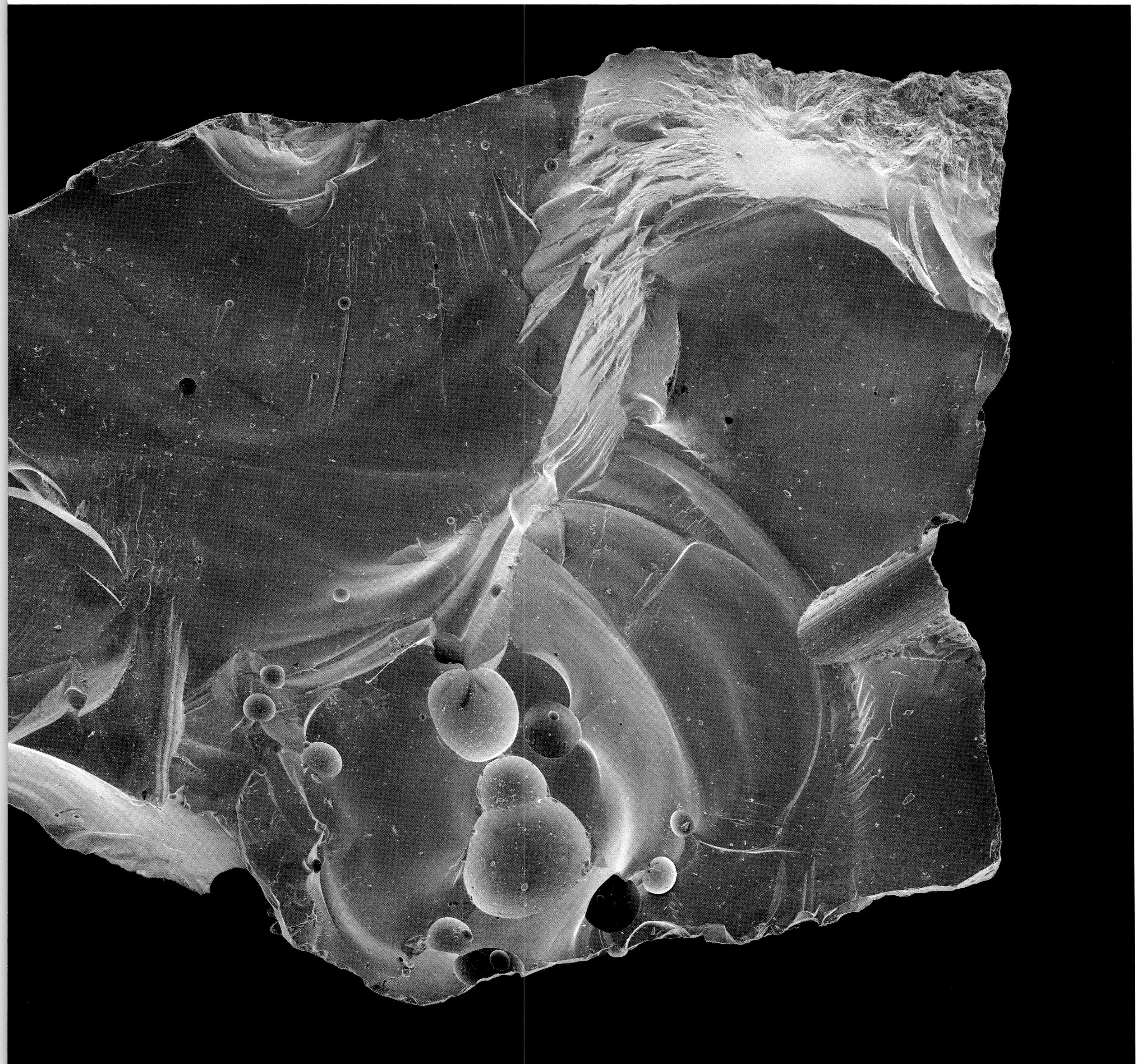

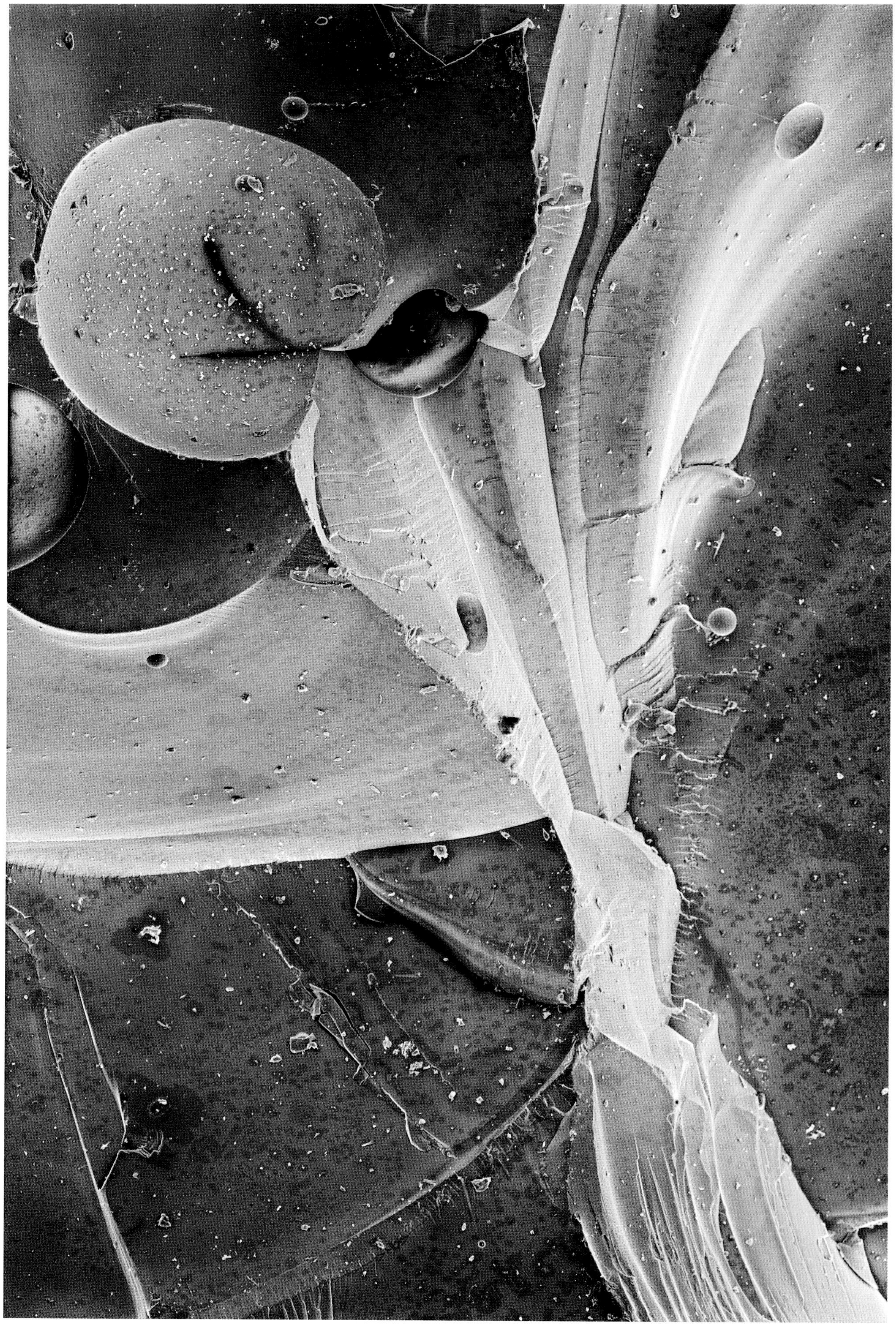

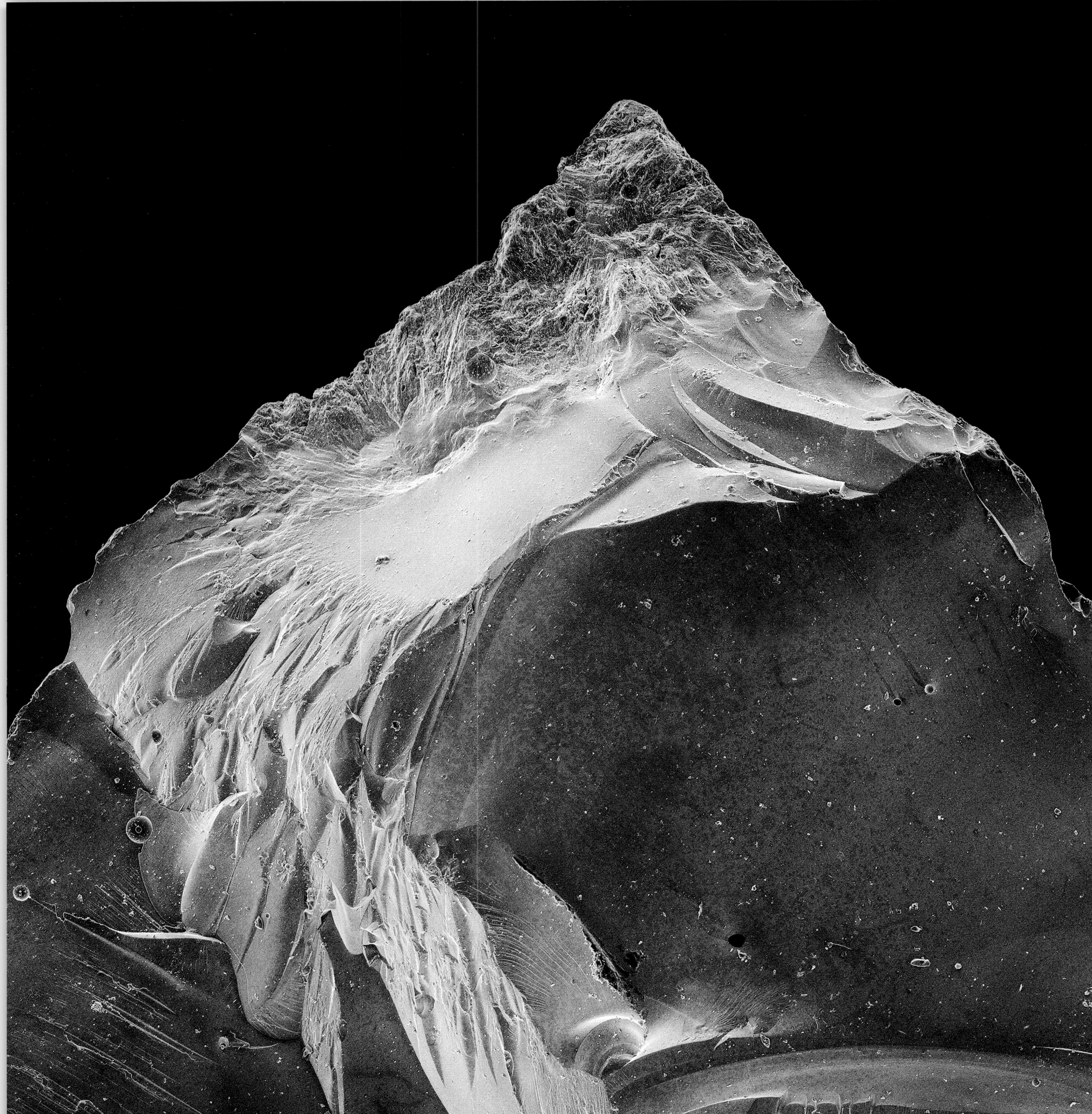

AQUA

# RADIOLARIANS

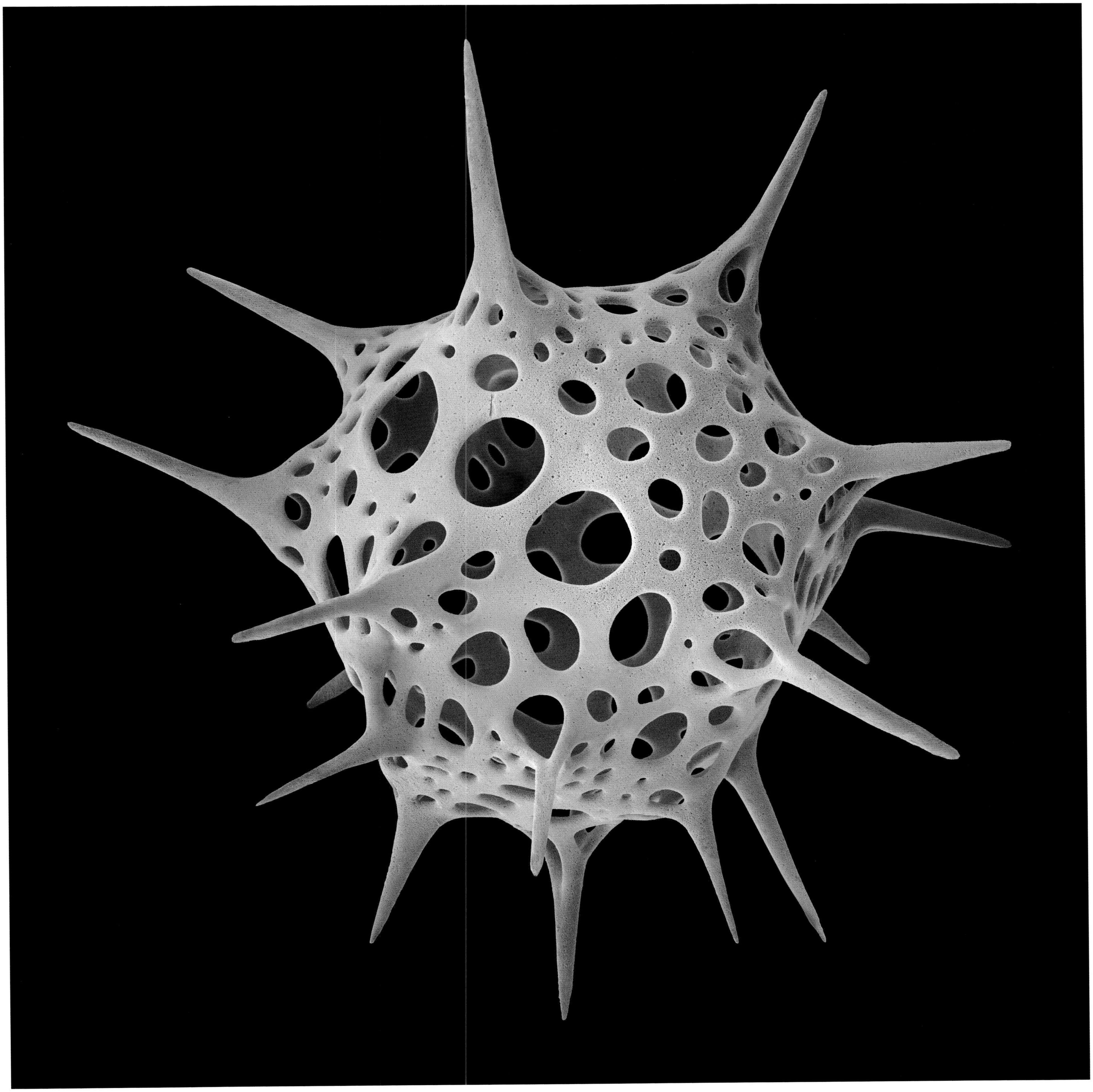

64

← **Page 63:**

*Acrosphaera spinosa fasciculopora*, equatorial Pacific.
1833 x. 120 microns wide, 0.12 of a millimeter.

**Facing Page, Pages 66–67 →**

*Clathrosphaera arachnoides*, equatorial Pacific.
700 x. 300 microns wide, 0.3 millimeter.

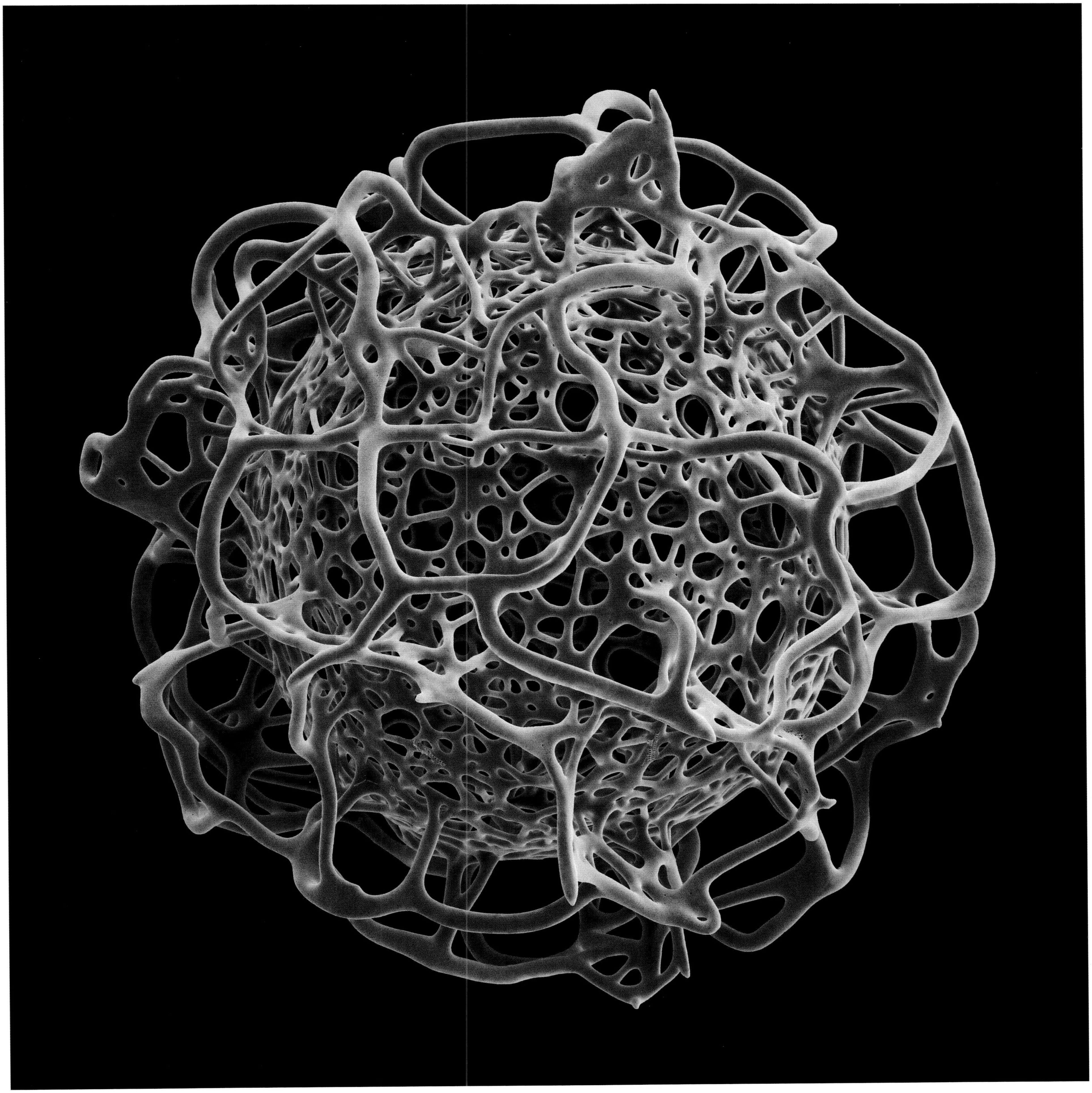

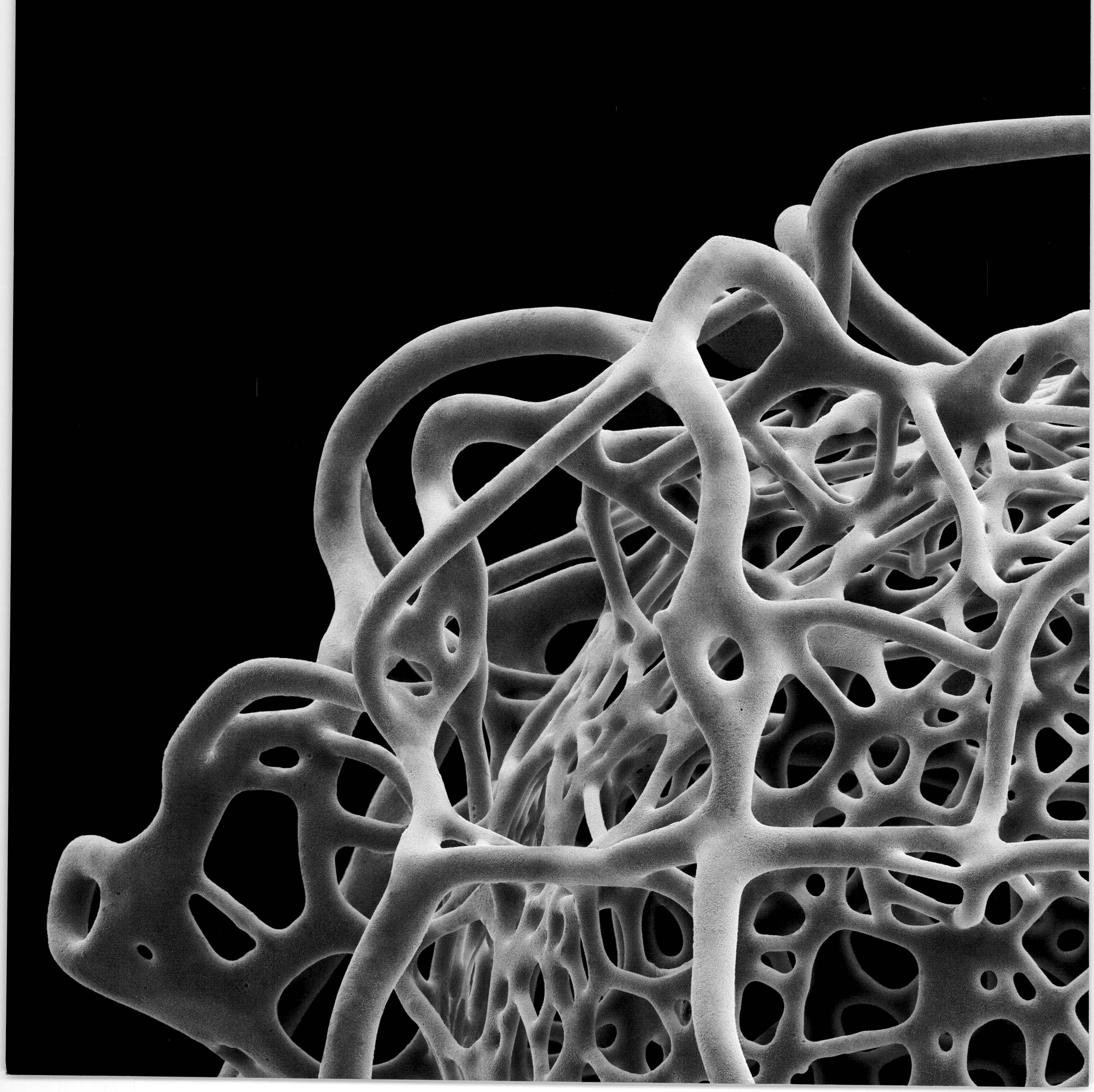

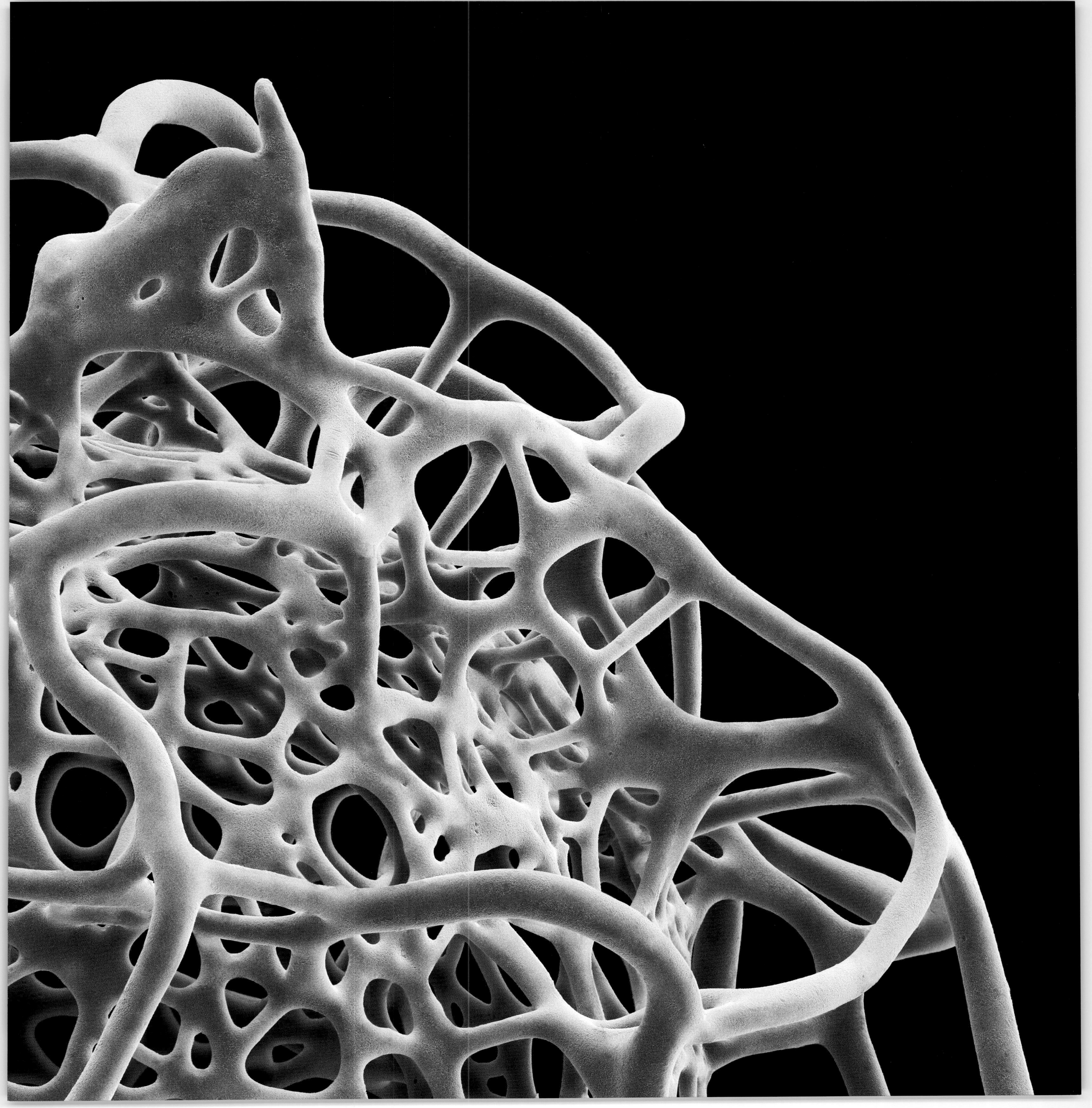

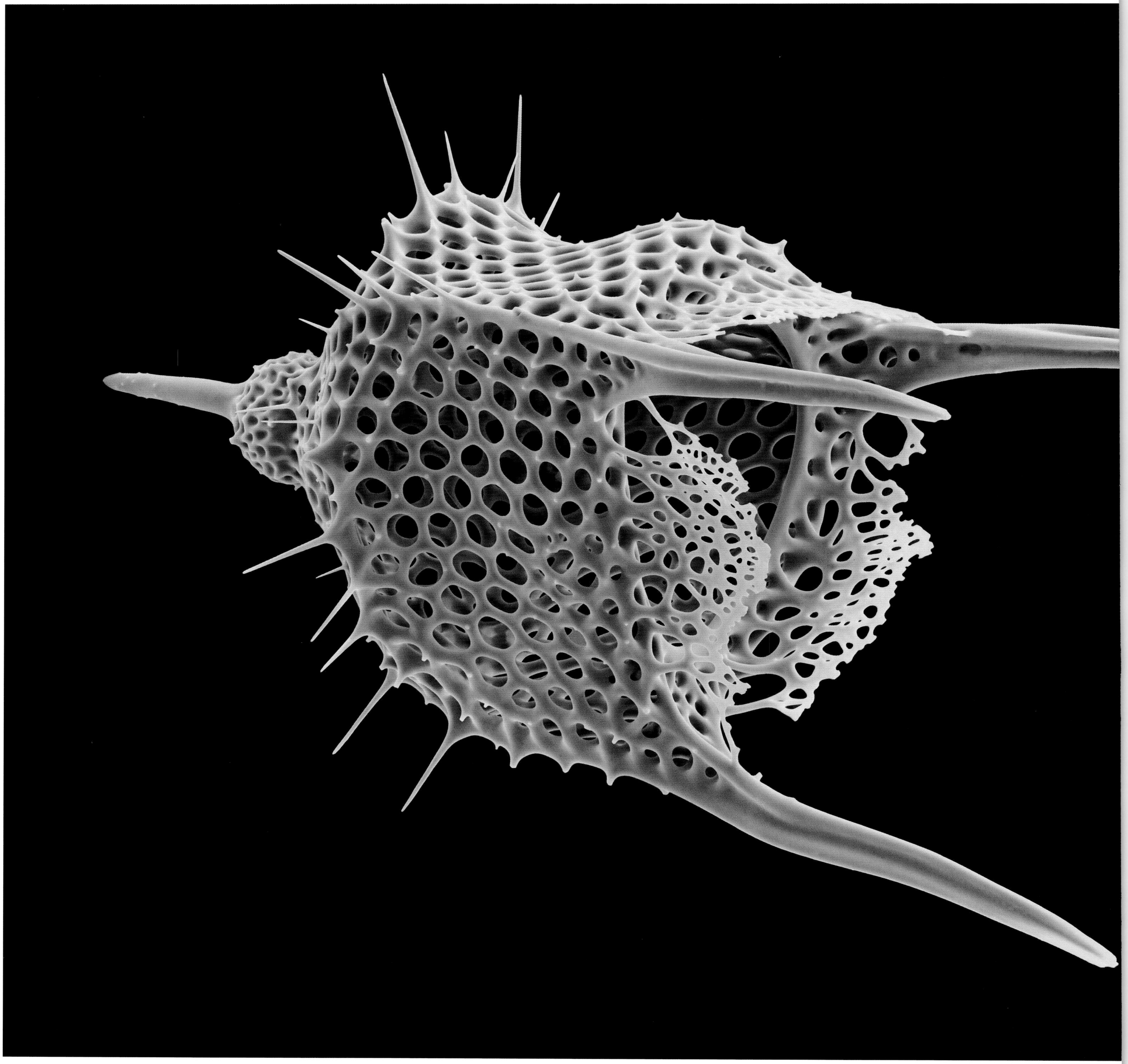

**Facing Page, Pages 70–71 →**

*Pterocanium praetextum*, equatorial Pacific.
1300 x. 200 microns wide from horn to tips, 0.2 millimeter

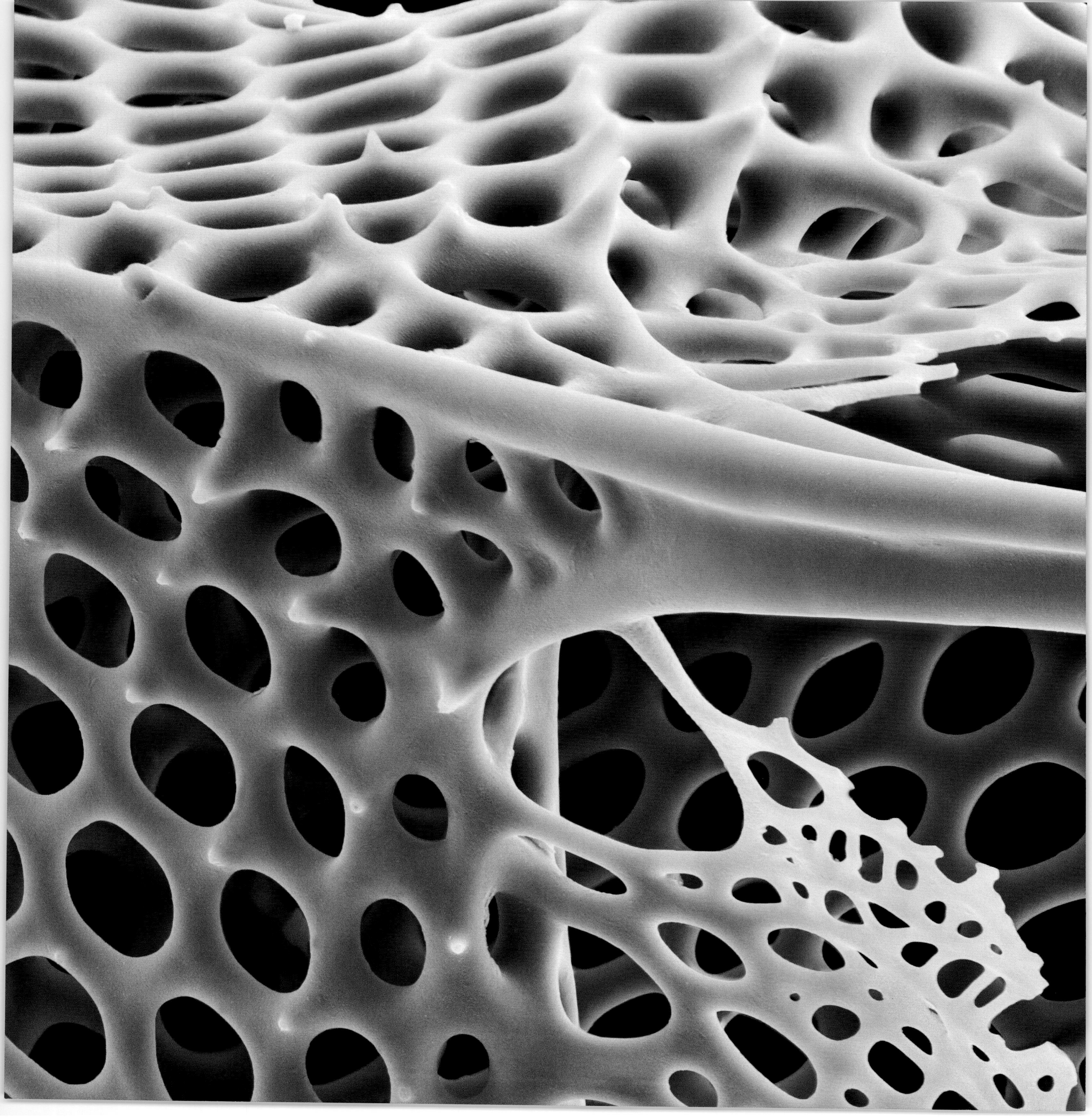

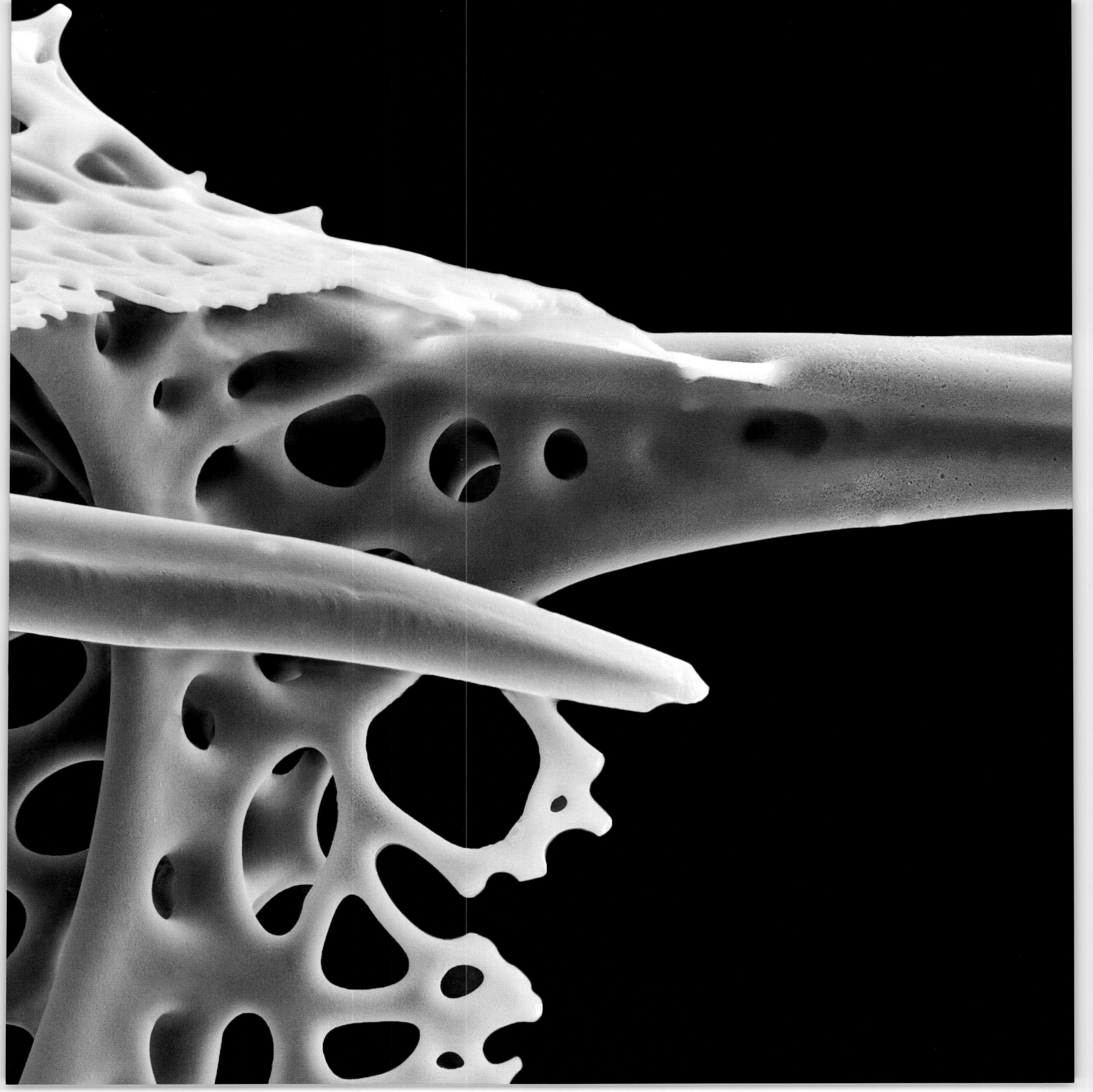

**Facing Page**

*Acrosphaera murrayana*, equatorial Pacific.
2500 x. 80 microns wide, 0.08 millimeter

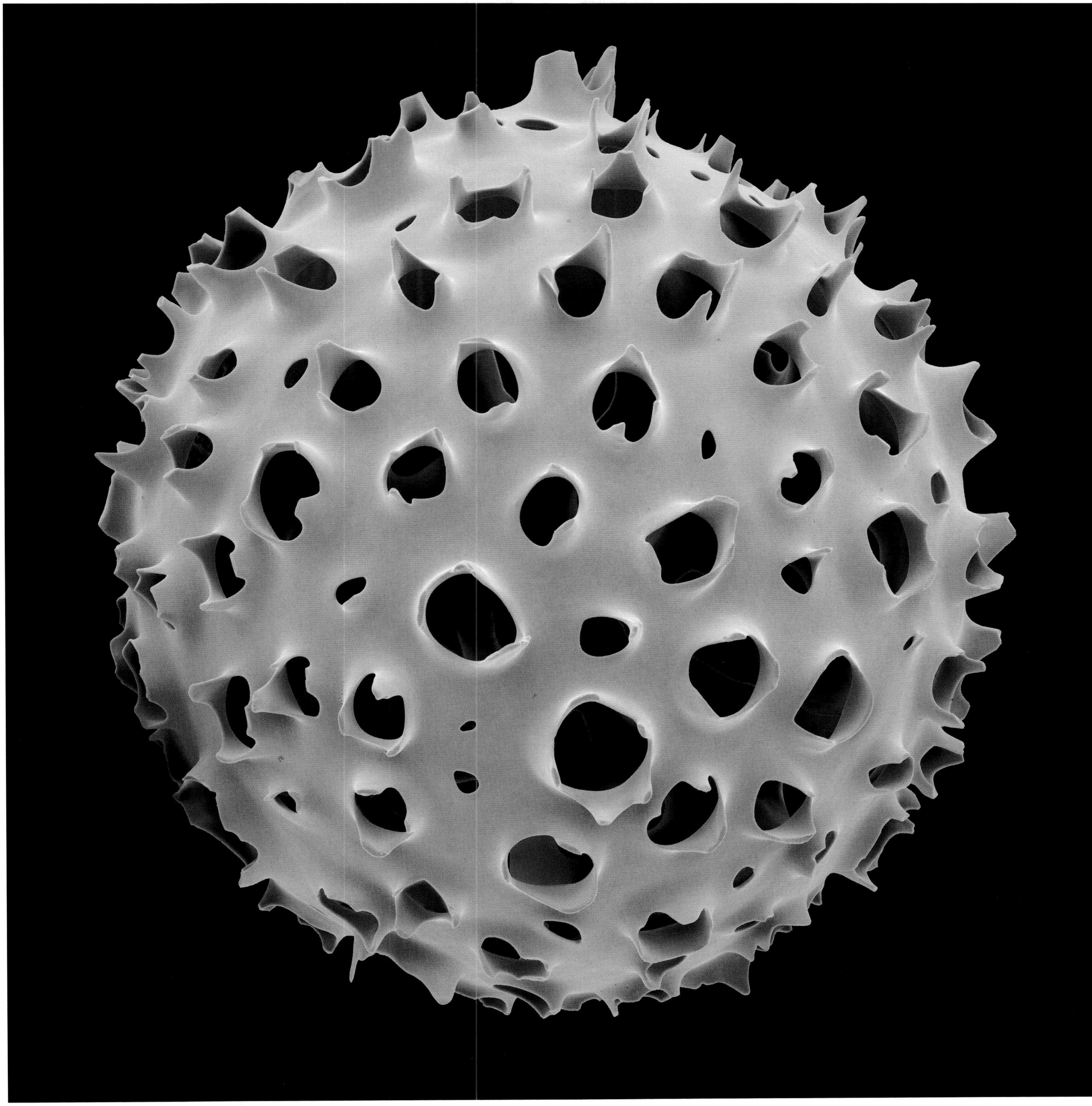

**Facing Page**

*Acrosphaera spinosa fasciculopora*, equatorial Pacific.
1433 x. About 150 microns from point to point, 0.15 millimeter

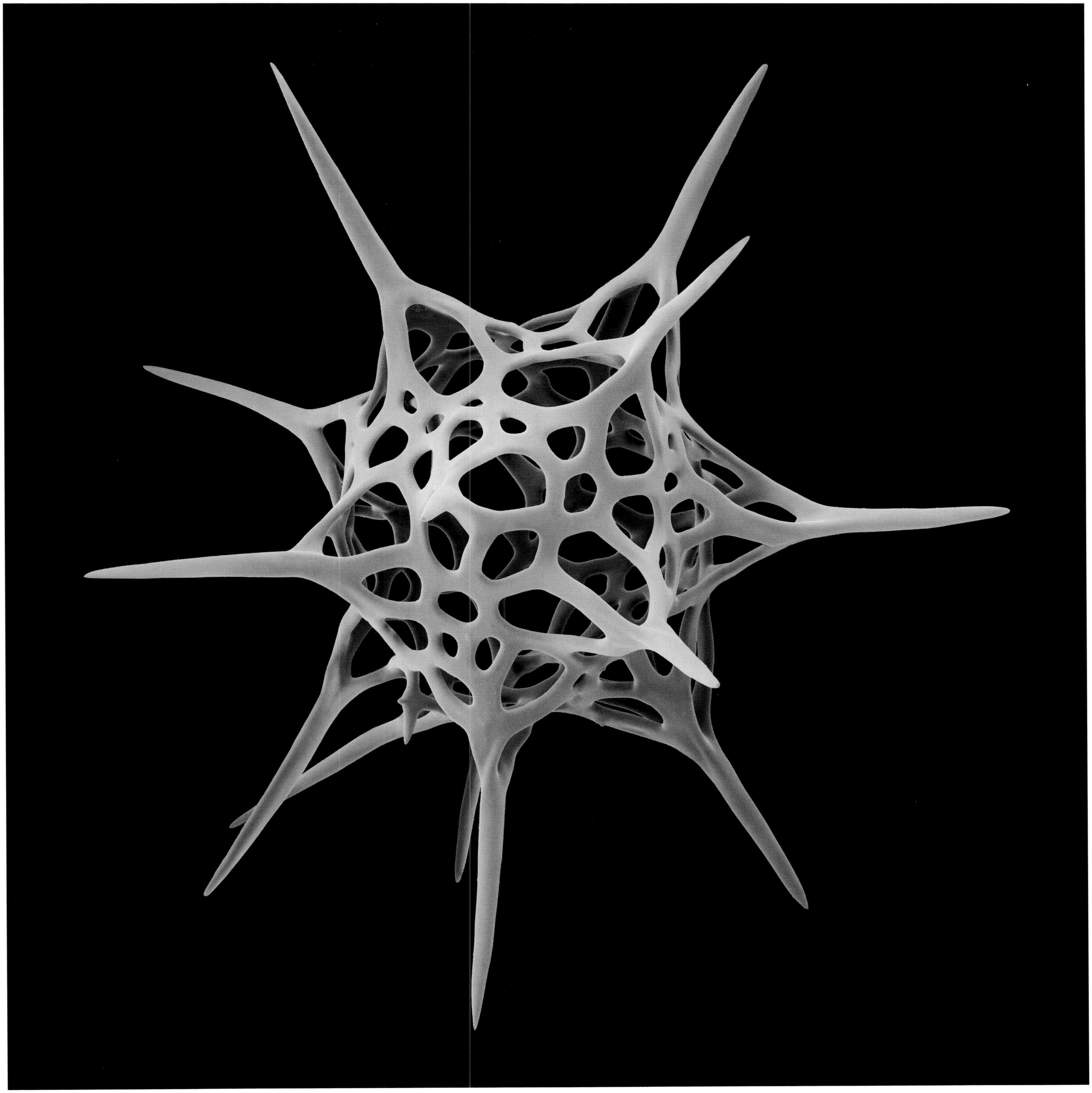

76

**Facing Page**

*Acrosphaera collina*, equatorial Pacific.
1400 x. 150 microns across, 0.15 millimeter

**Page 78, 79 →**

Two *Acrosphaera* genus radiolarians, equatorial Pacific.
733 x. Each is about 300 microns from top to bottom, 0.3 millimeter

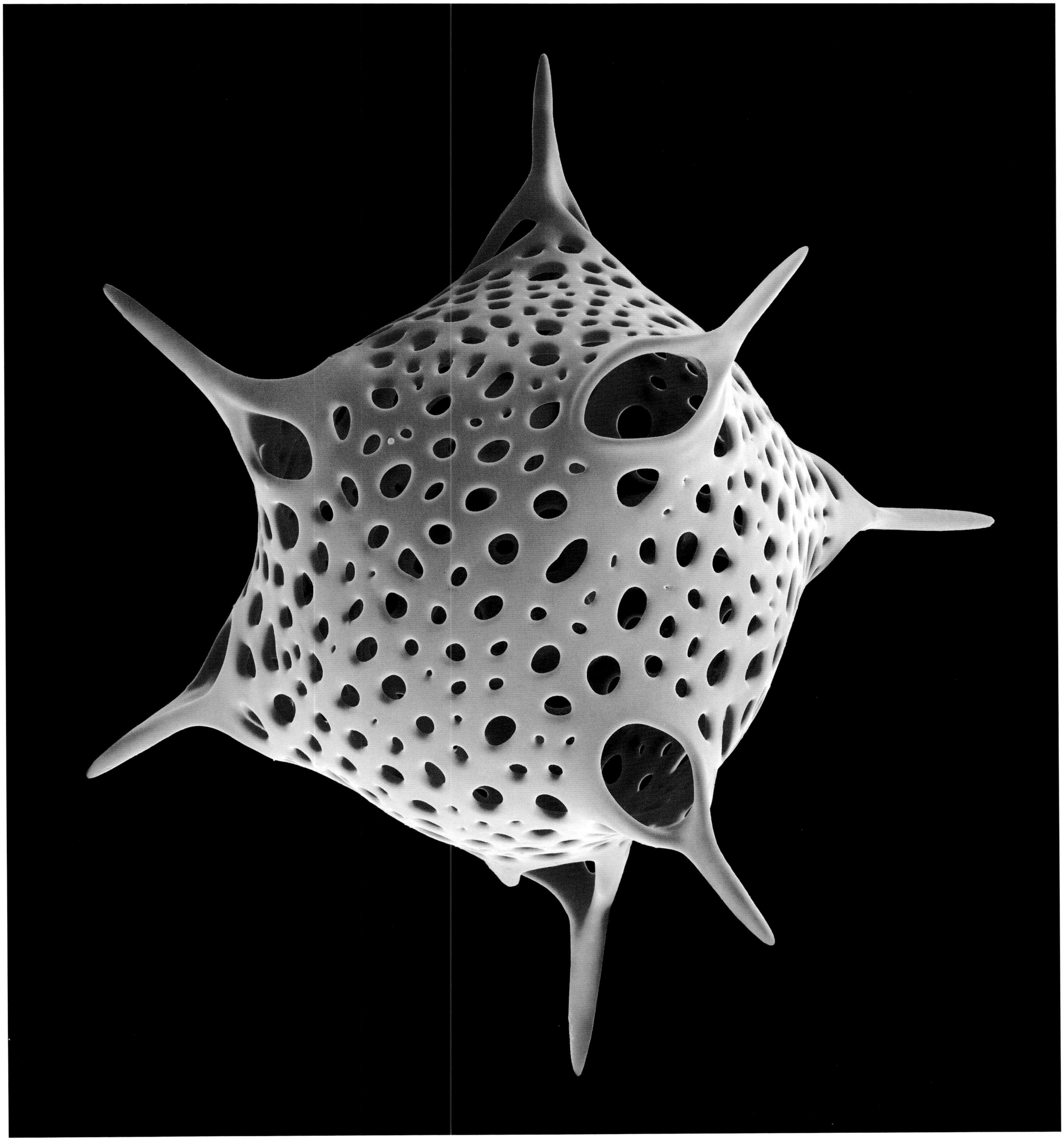

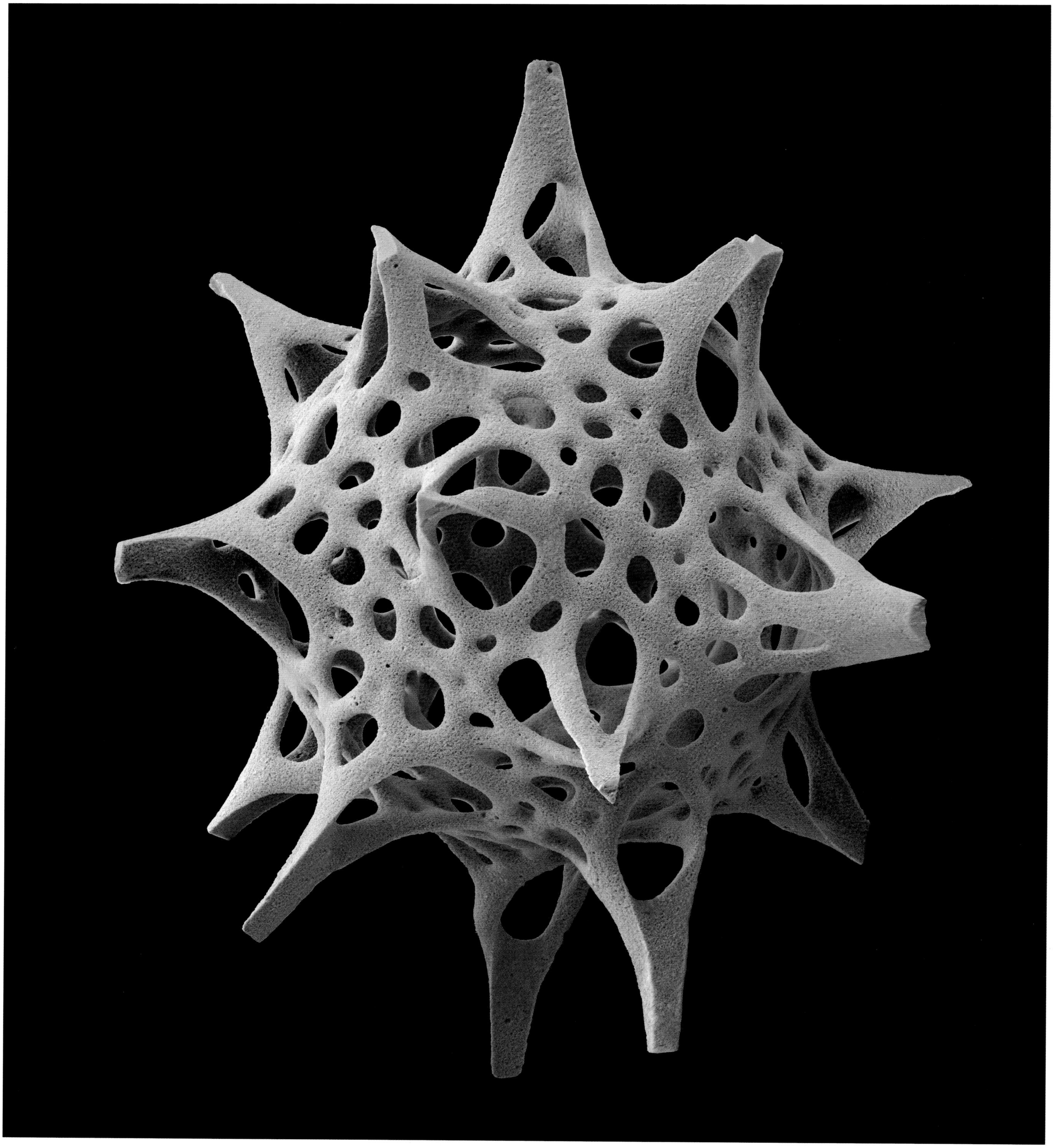

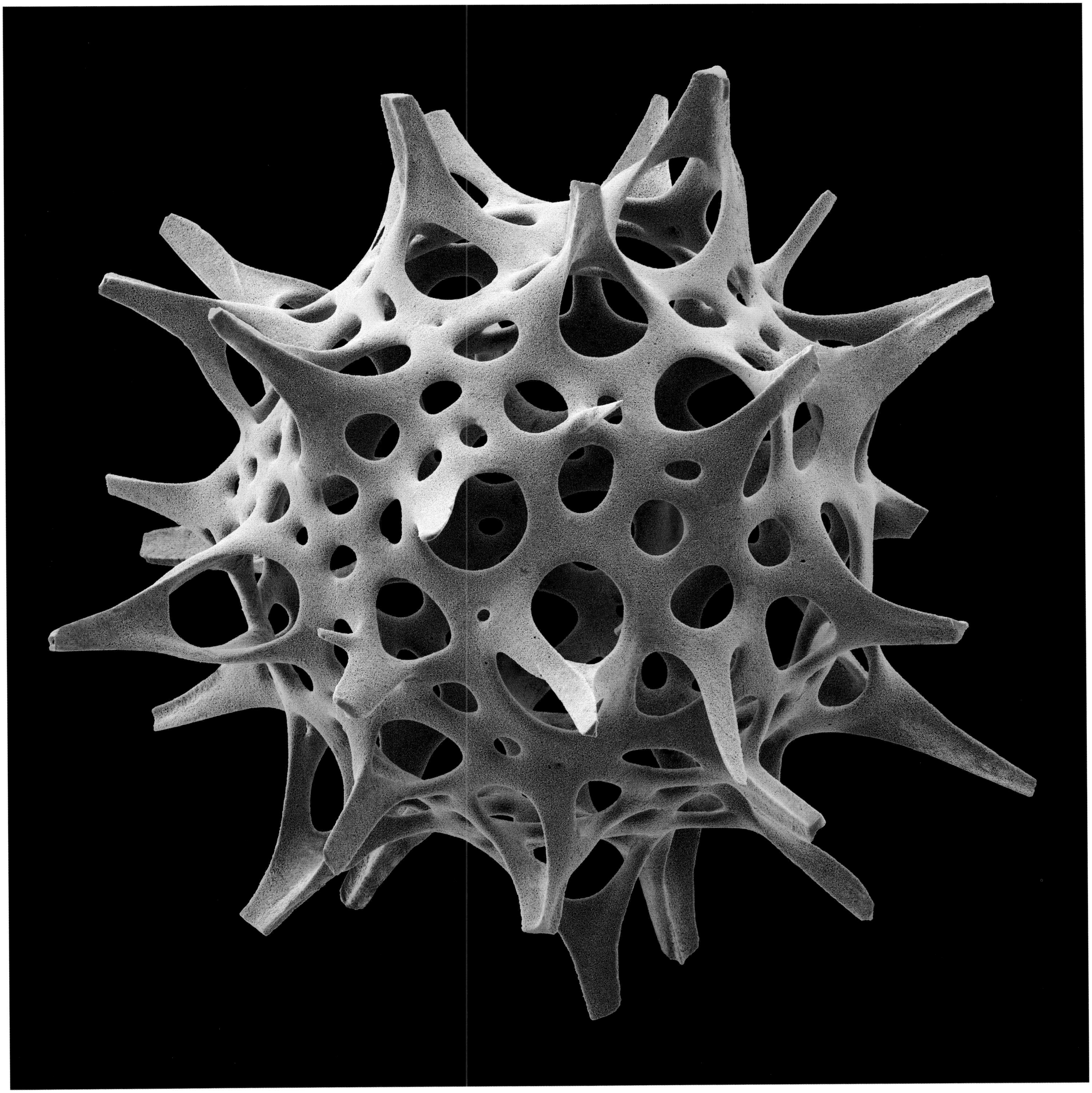

**Facing Page**

*Cladococcus abietinus*, equatorial Pacific.
420 x. 500 microns across, 0.5 millimeter

**Page 82 →**

*Thecosphaera* genus radiolarian, probably *Thecosphaera concentrica*, equatorial Pacific.
2100 x. 100 microns wide, 0.1 millimeter

**Page 83 →**

*Lamprocyclas maritalis*, equatorial Pacific.
1050 x. 200 microns from tip to end, 0.2 millimeter

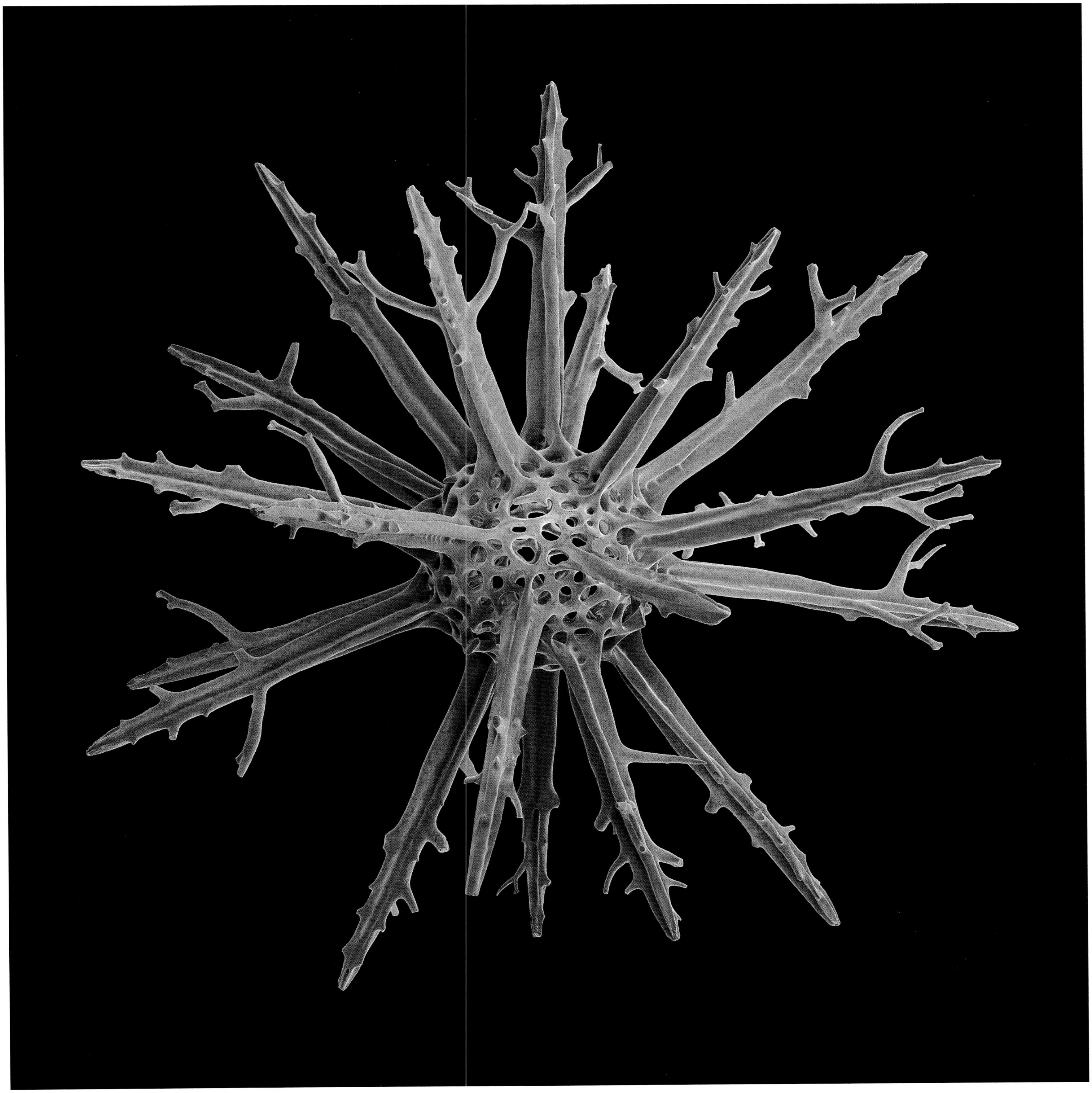

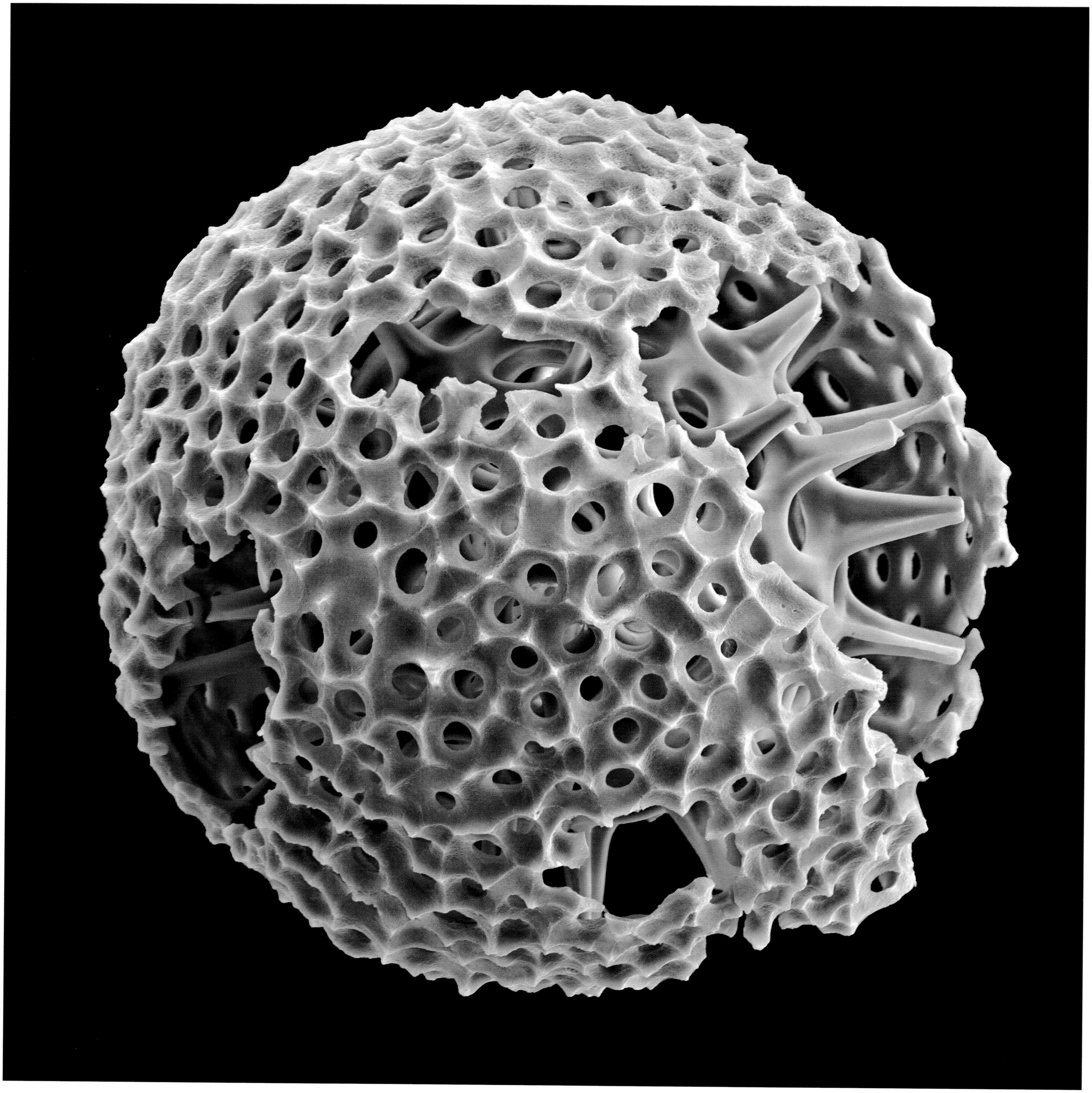

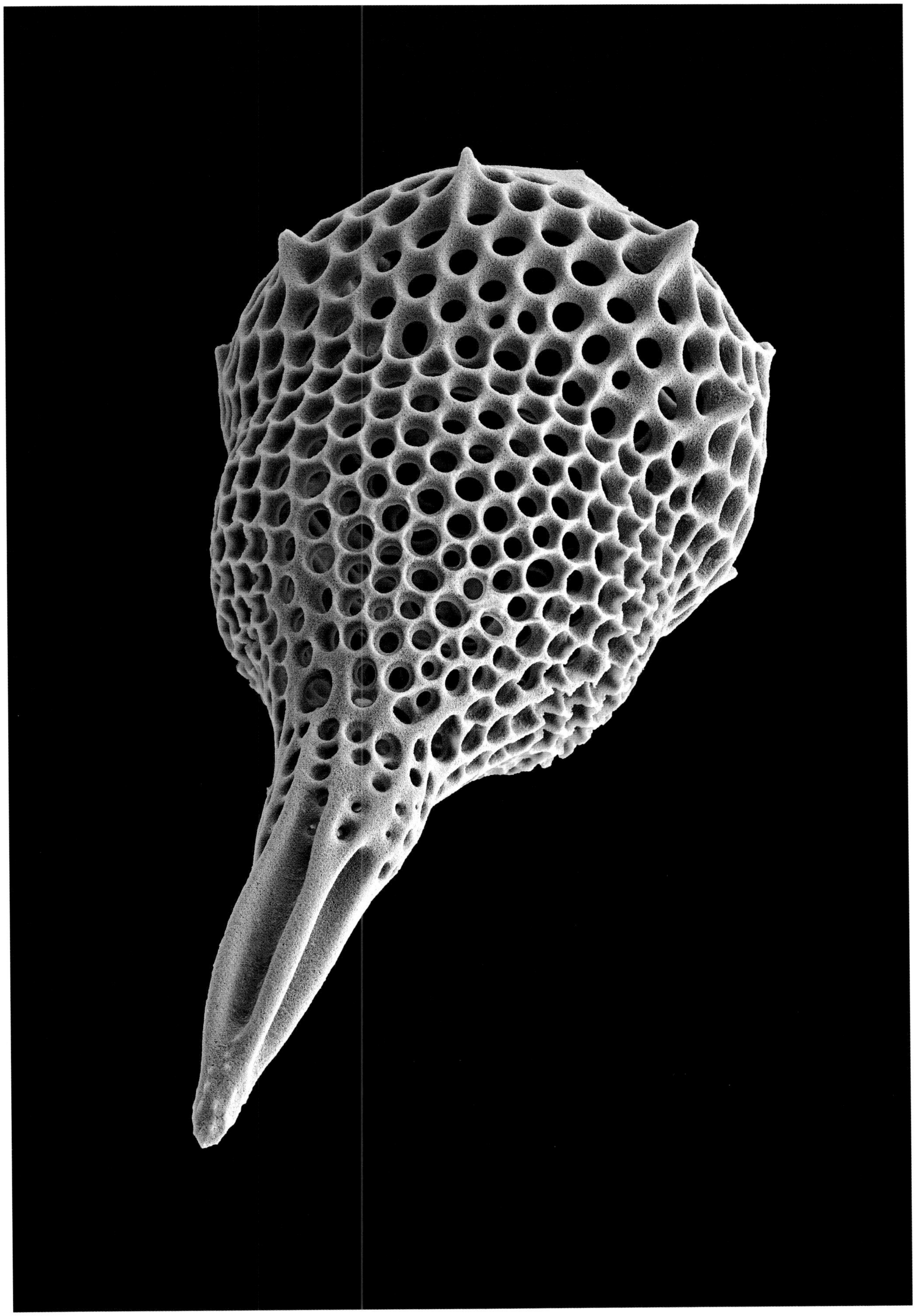

Facing Page, Pages 86–87 →

*Haliomma circumtextum*, equatorial Pacific.
433 x. Central oval is about 300 microns wide, 0.3 millimeter

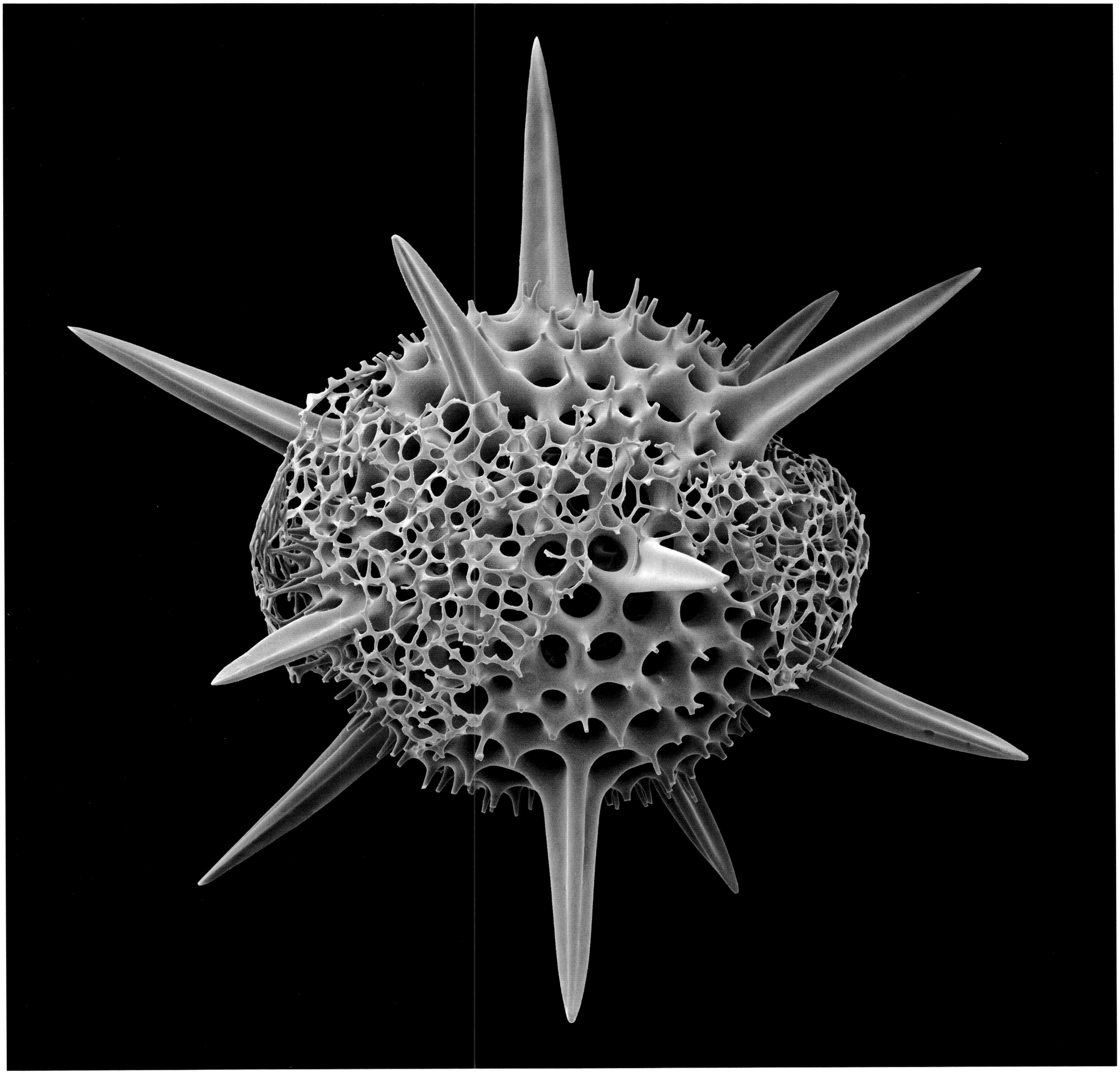

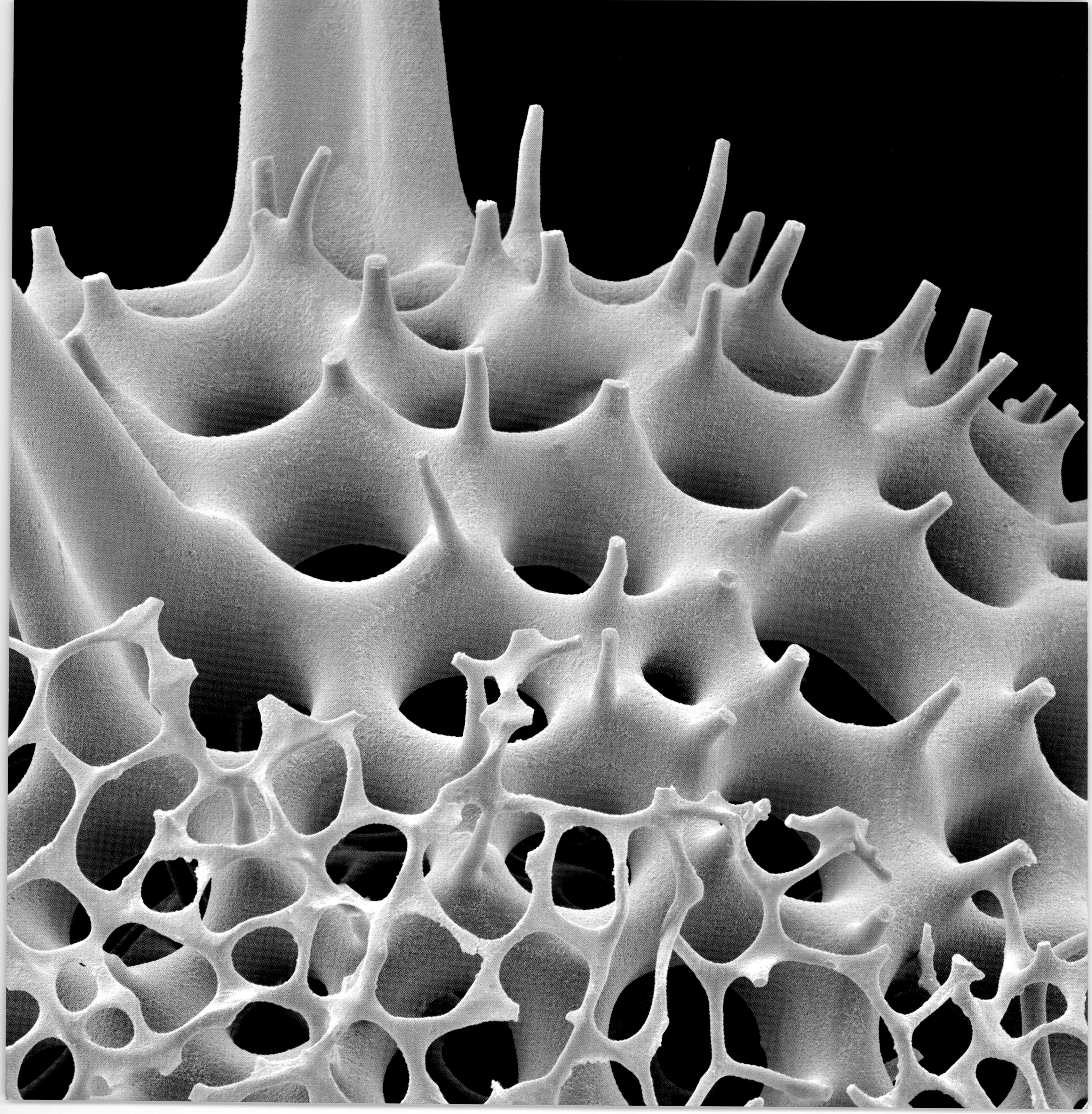

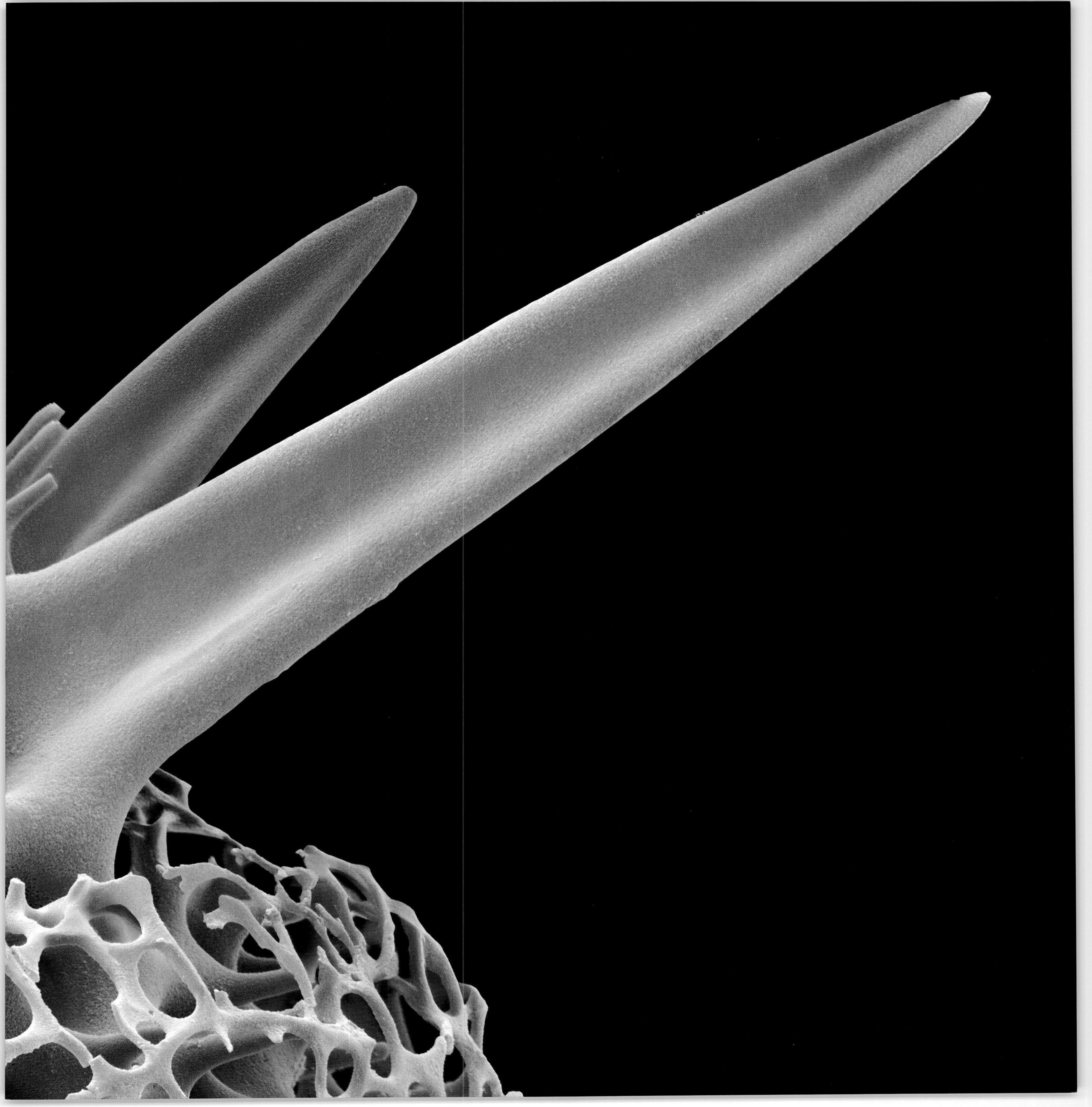

**Facing Page, Pages 90–91 →**

*Gondwanaria campanulaeformis*, top, and a sea sponge spicule, equatorial Indian Ocean. 880 x. The pairing is about 250 microns from top to bottom, 0.25 millimeter

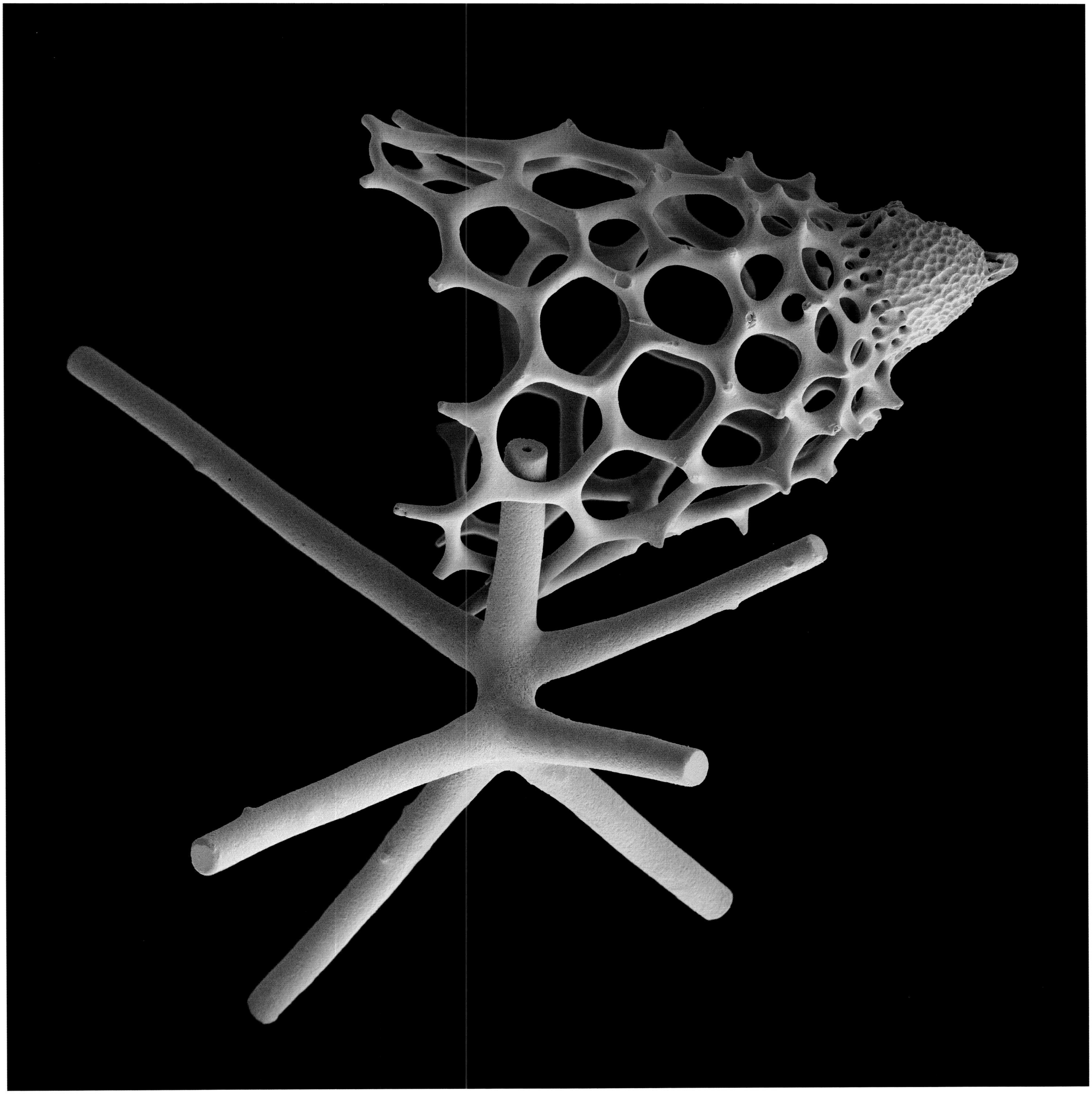

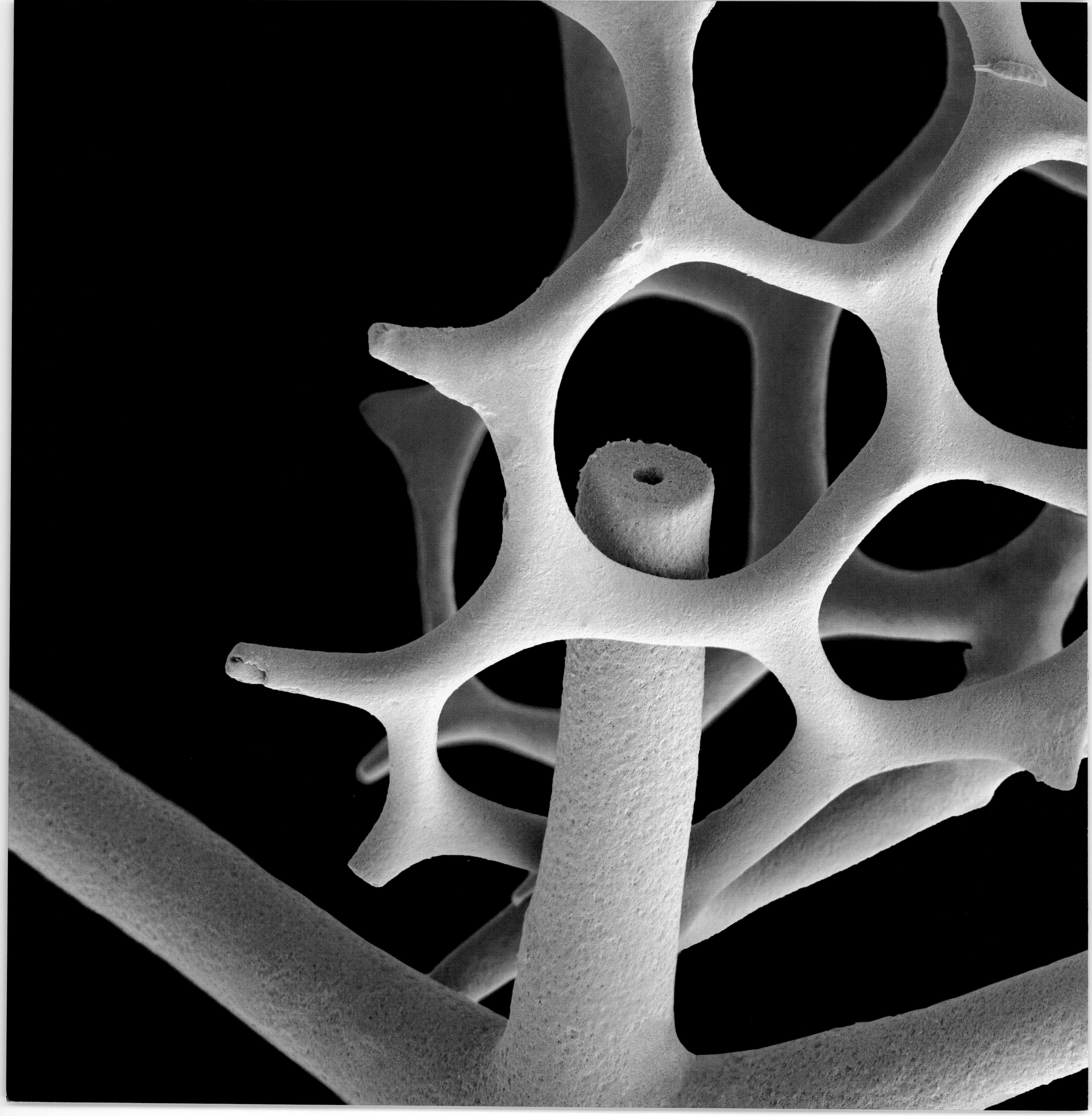

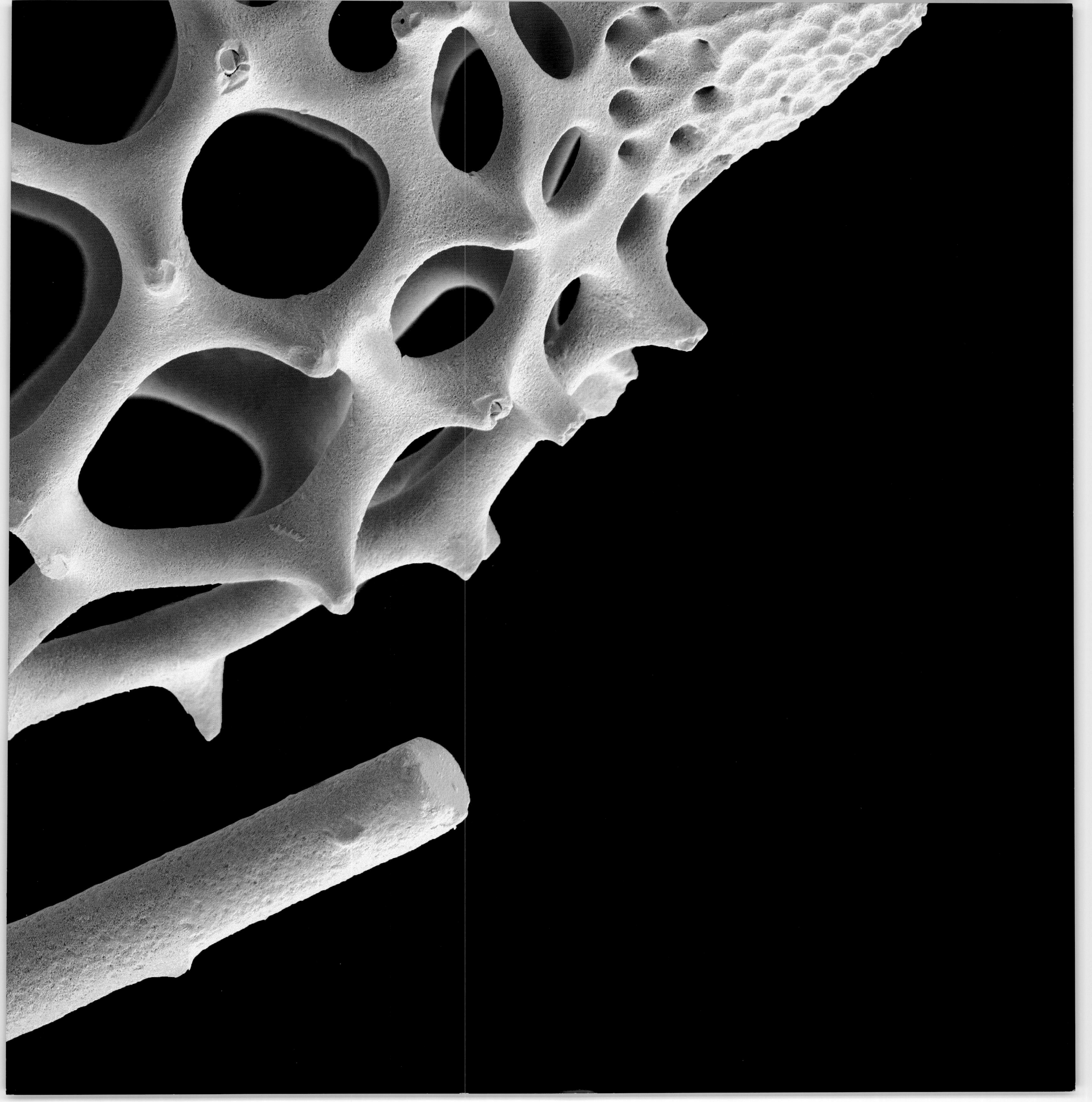

**Facing Page, Pages 94–95 →**

*Hexalonche philosophica*, equatorial Pacific.
889 x. 225 microns from tip to tip, 0.2 millimeter

**Page 96 →**

*Hexastylus* genus radiolarian, equatorial Pacific.
1075 x. About 200 microns from tip to tip, 0.2 millimeter

**Page 97, Pages 98–99 →**

Radiolarian of the *Hexastylus* genus, equatorial Pacific.
850 x. About 200 microns wide, 0.2 millimeter

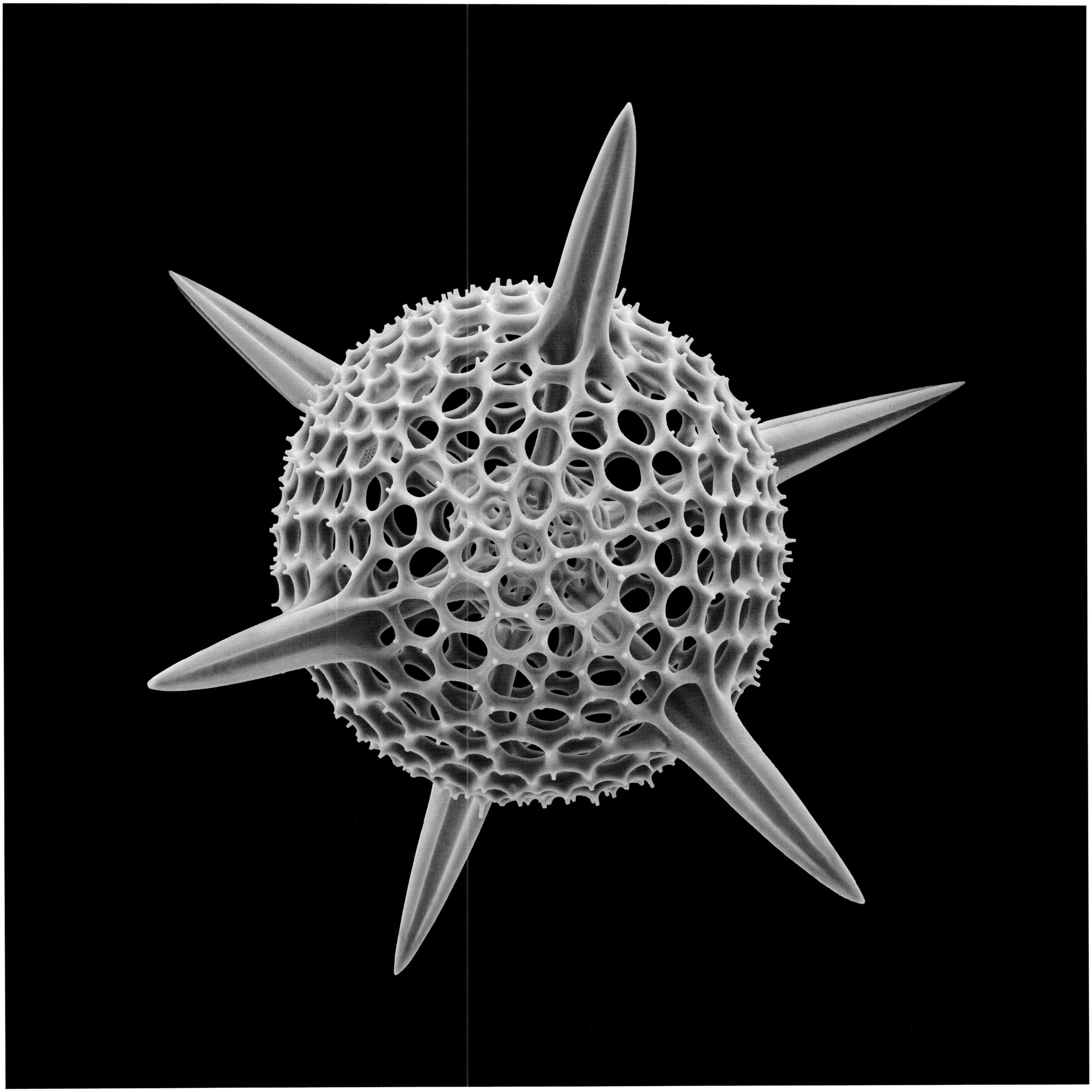

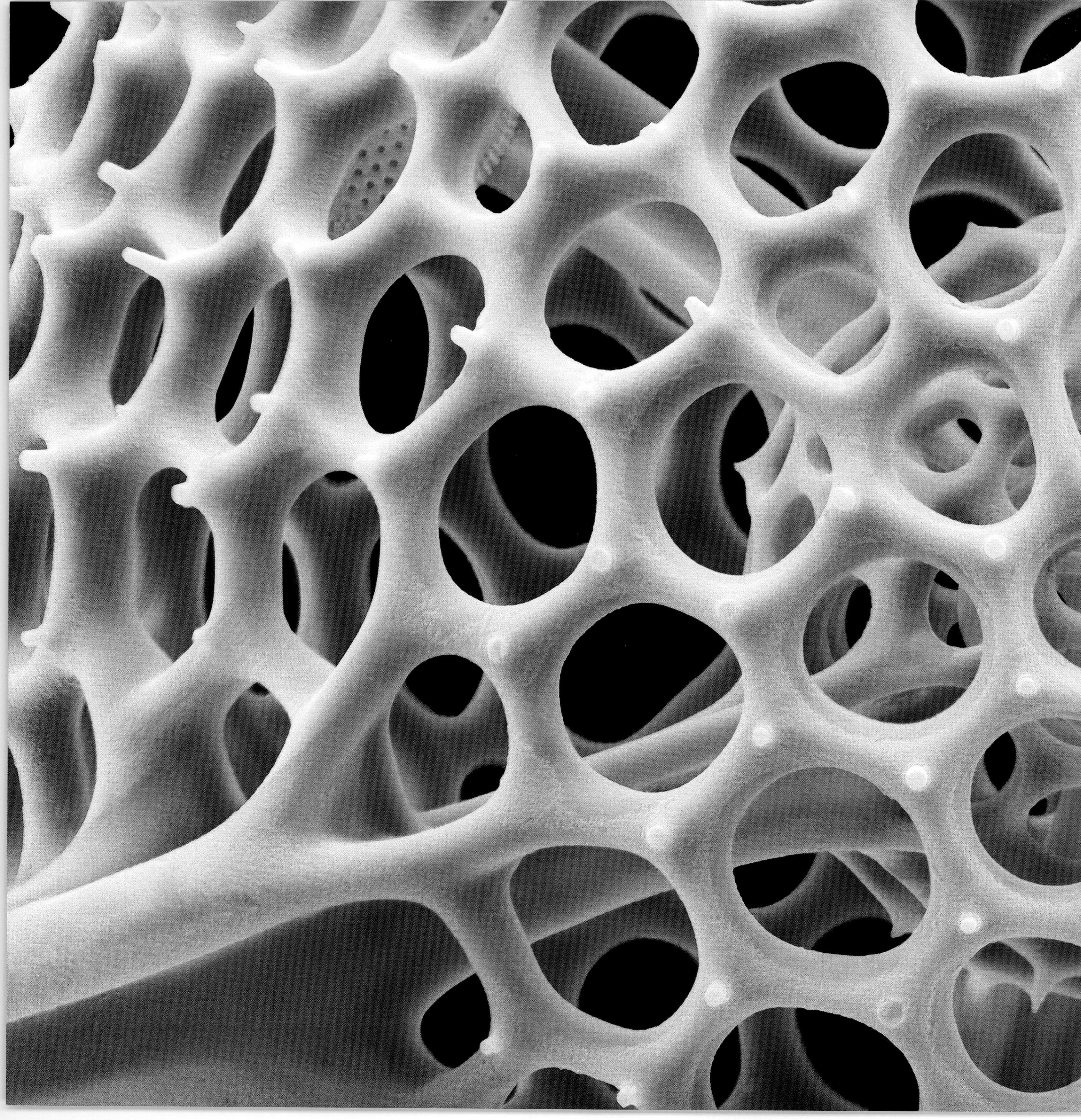

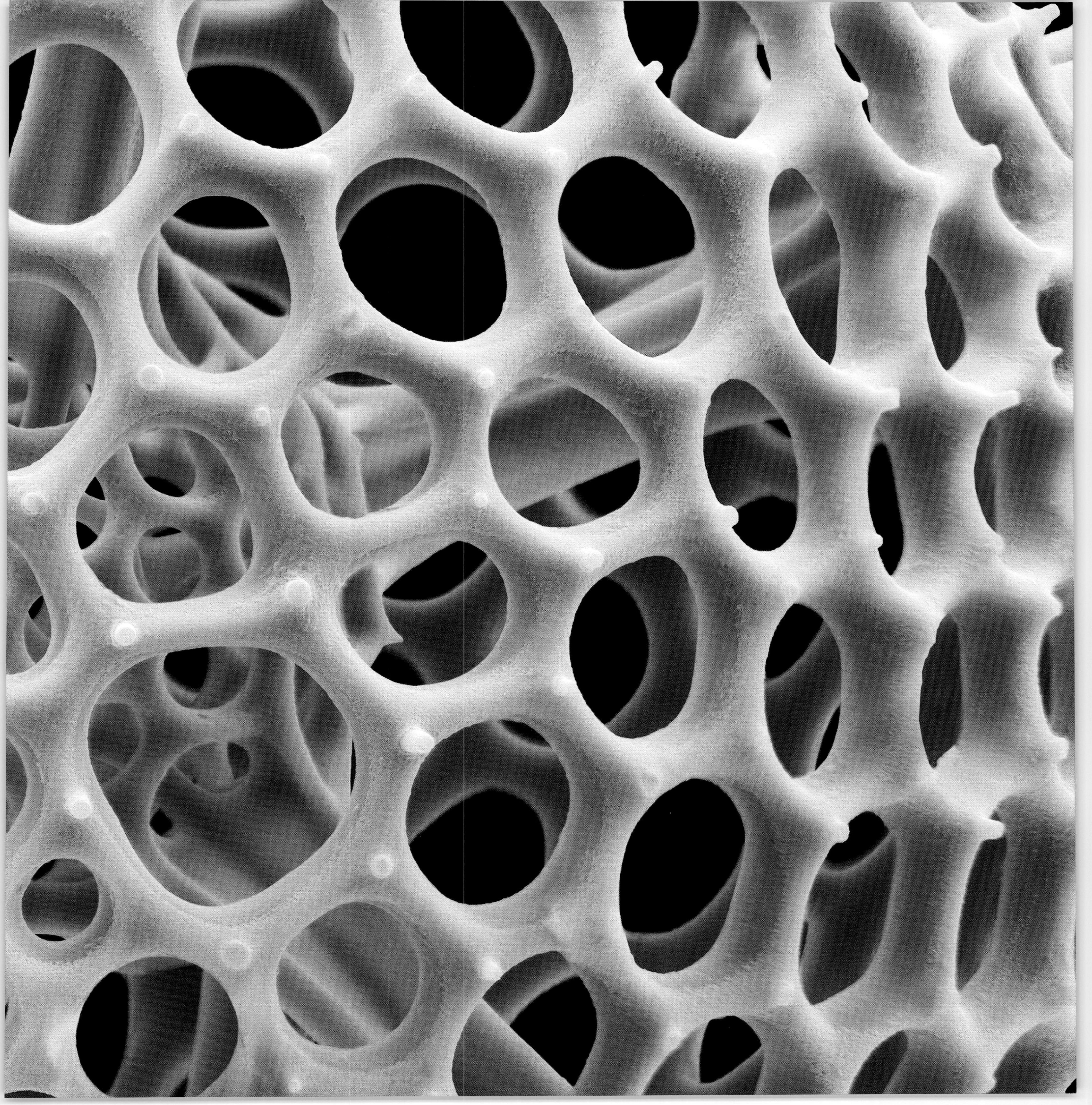

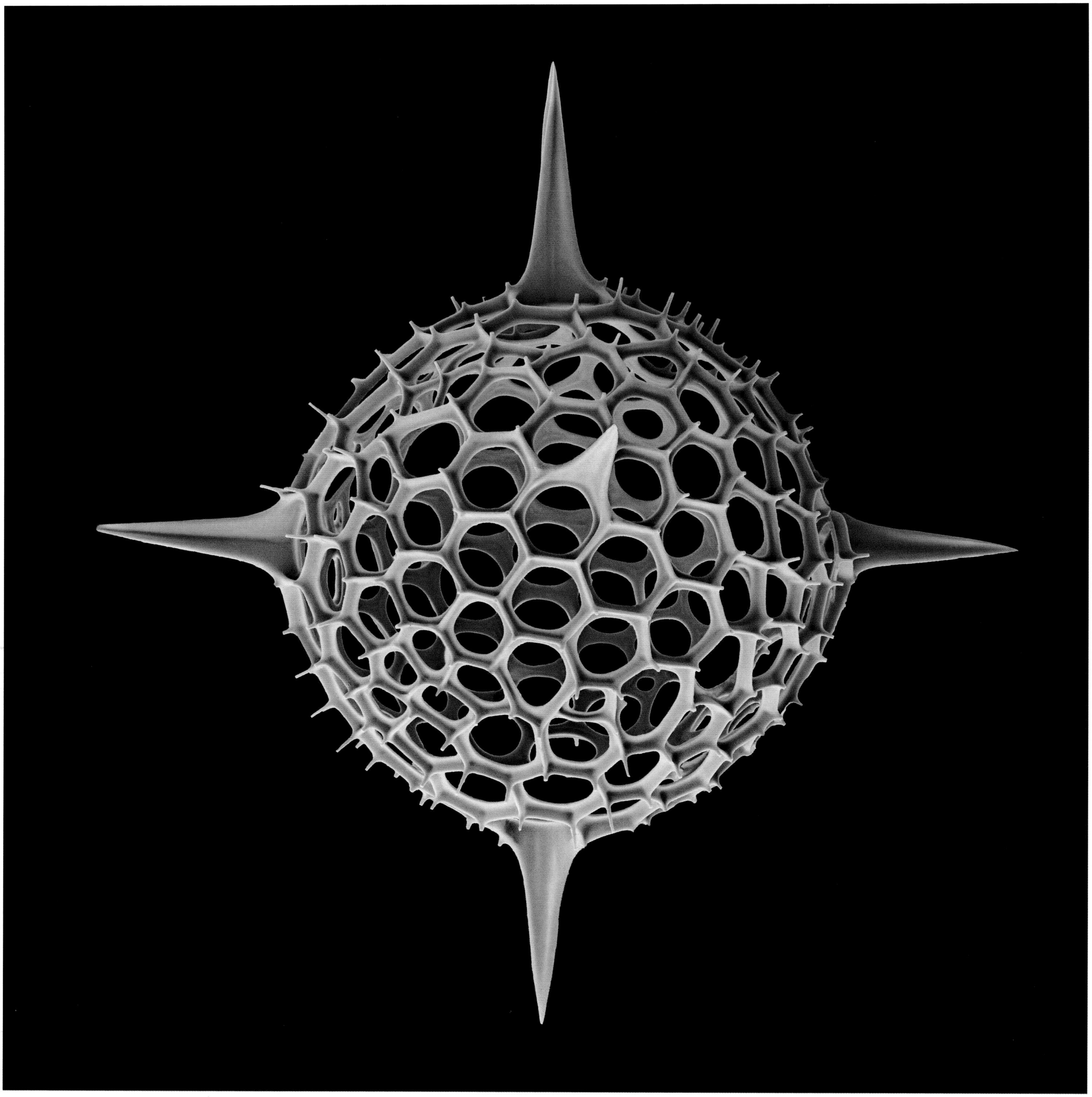

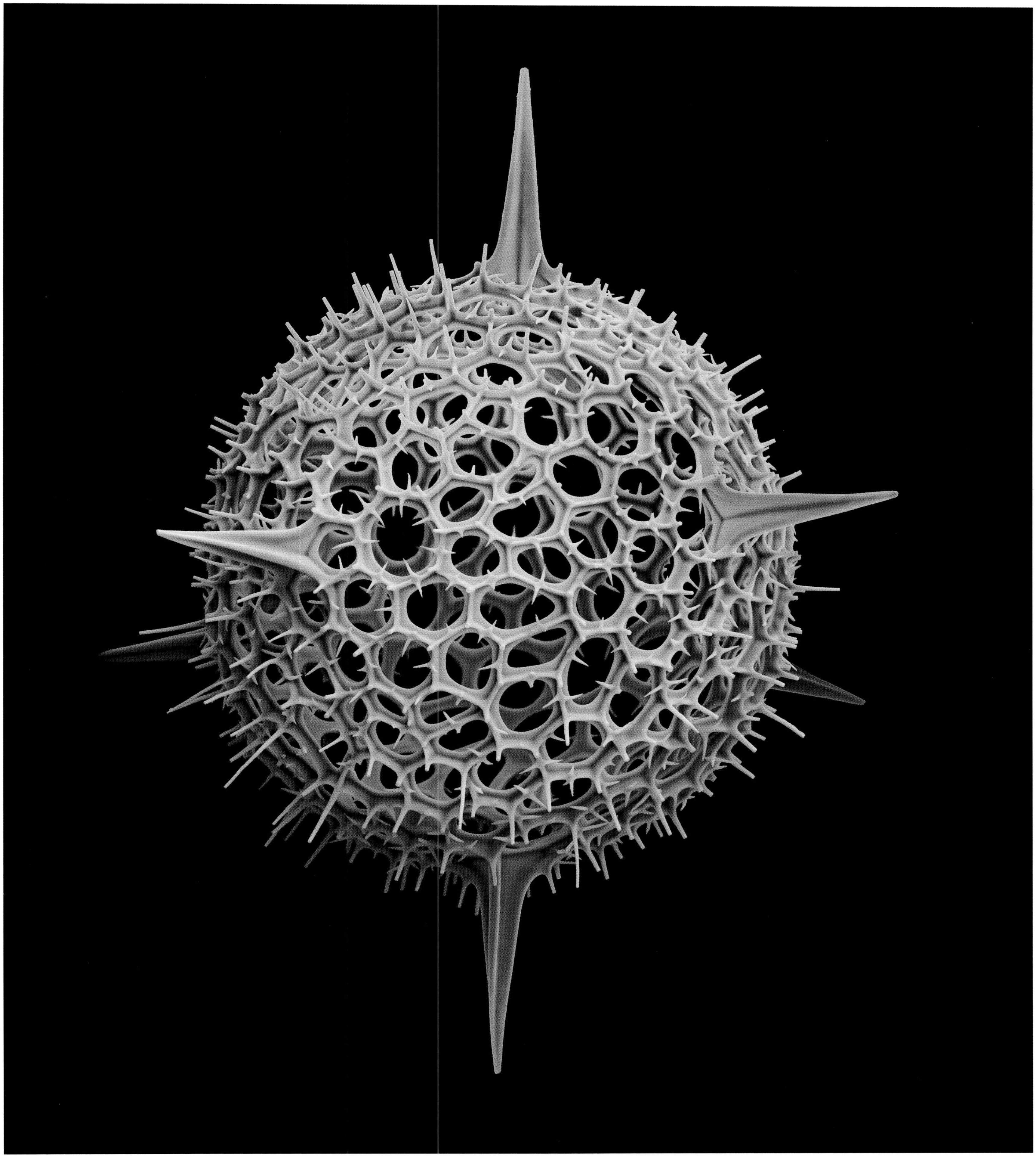

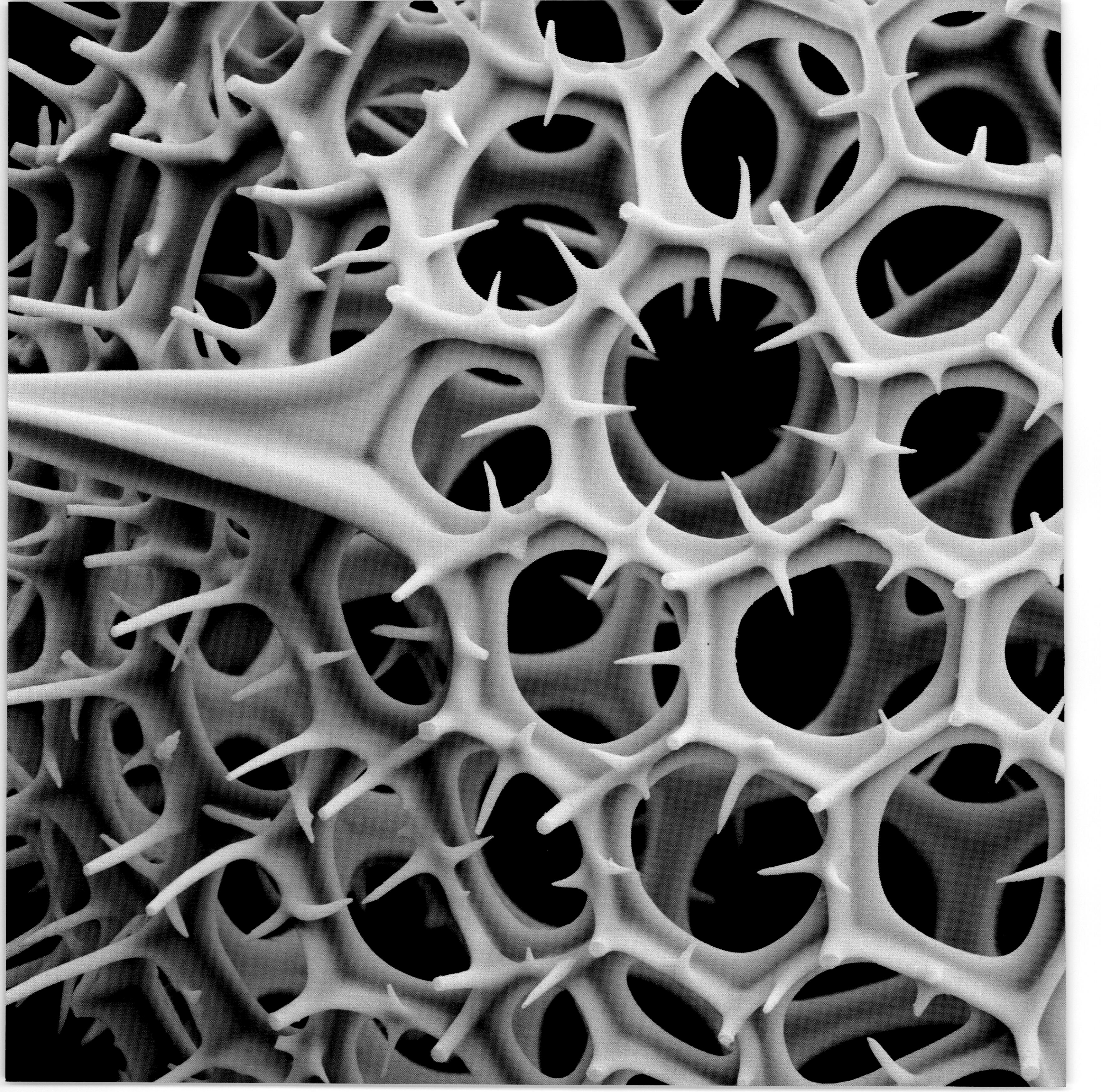

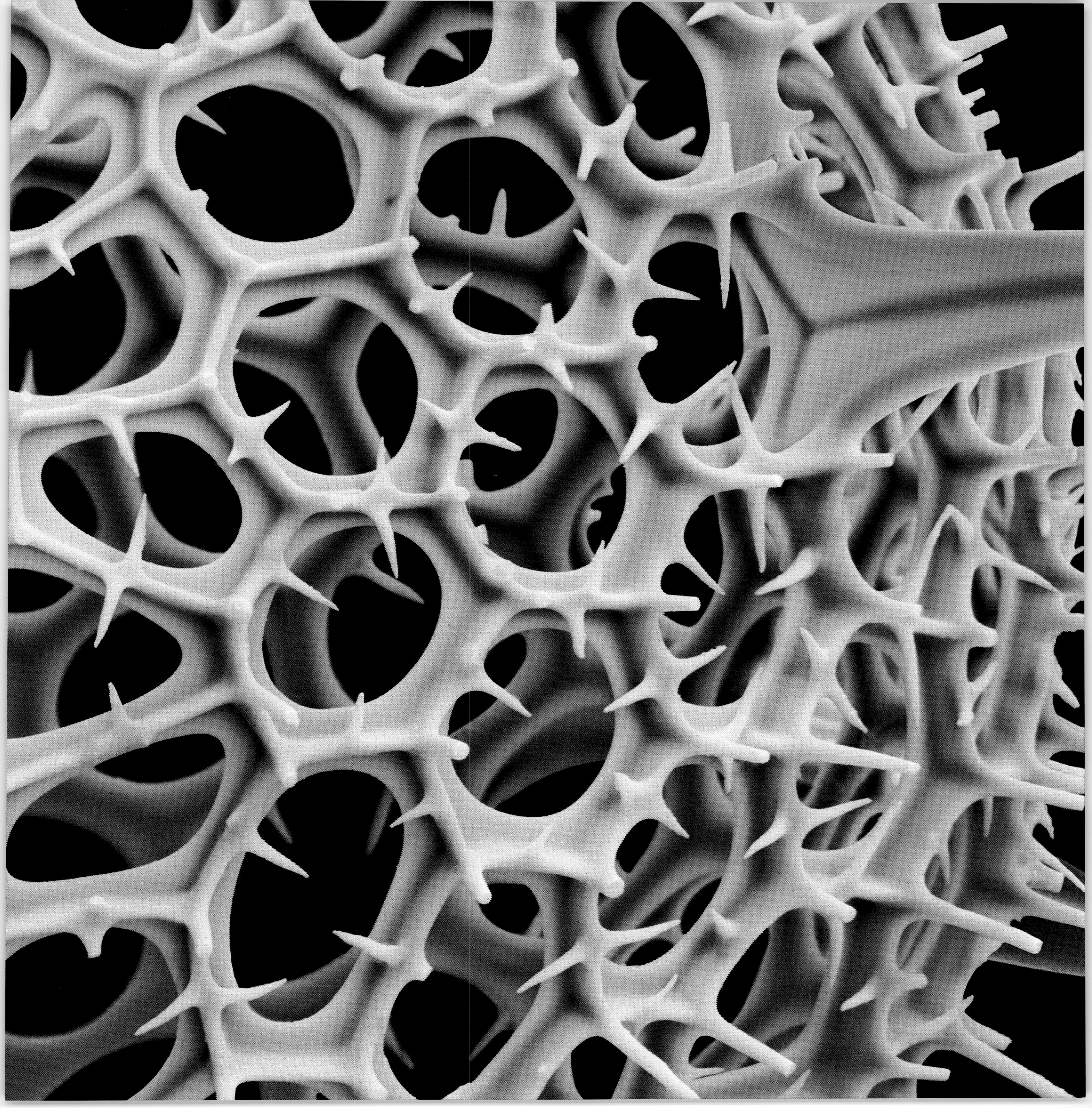

Facing Page, Pages 102–103 →

*Dictyophimus crisiae*, equatorial Pacific.
733 x. About 300 microns in length, 0.3 millimeter

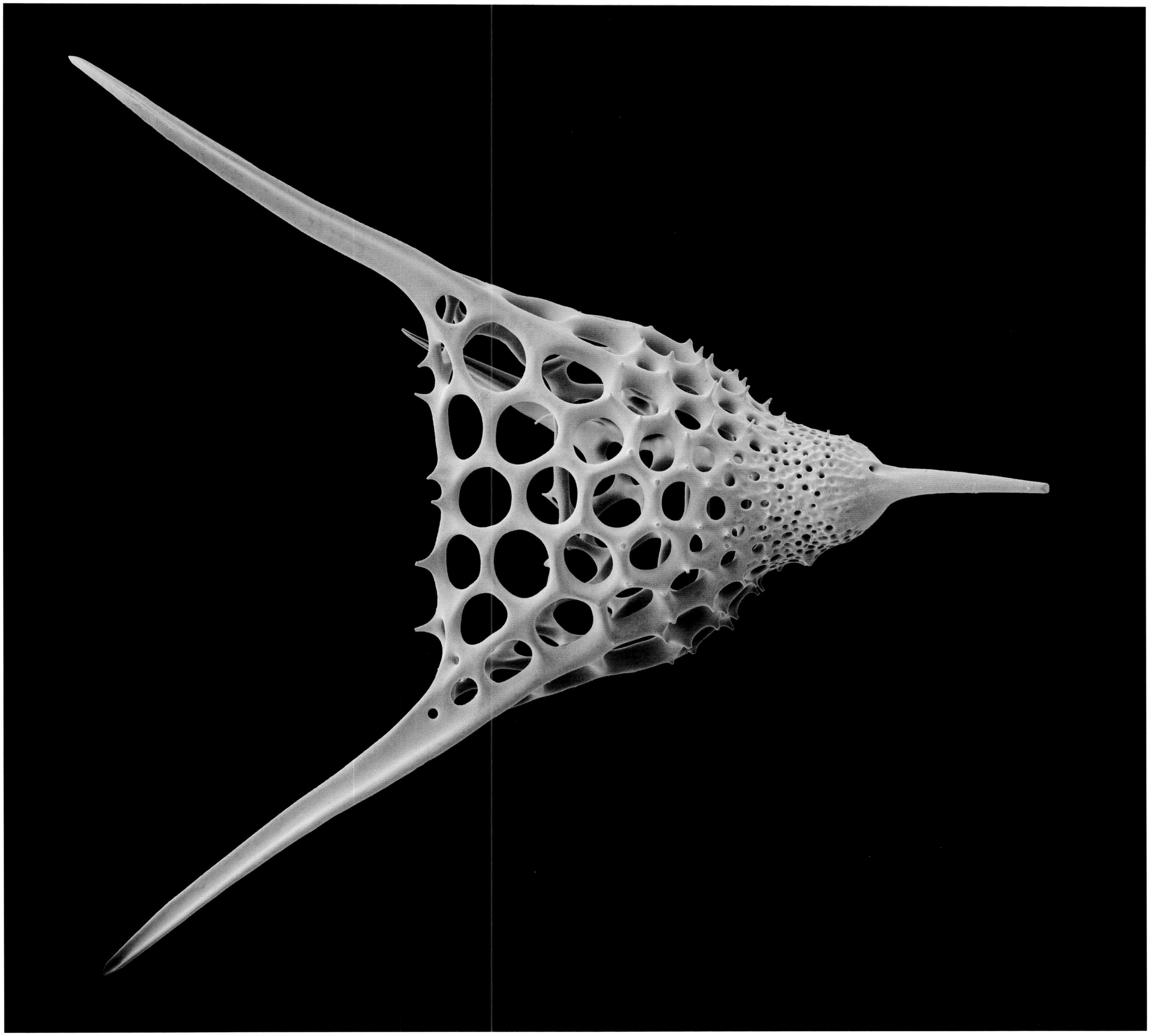

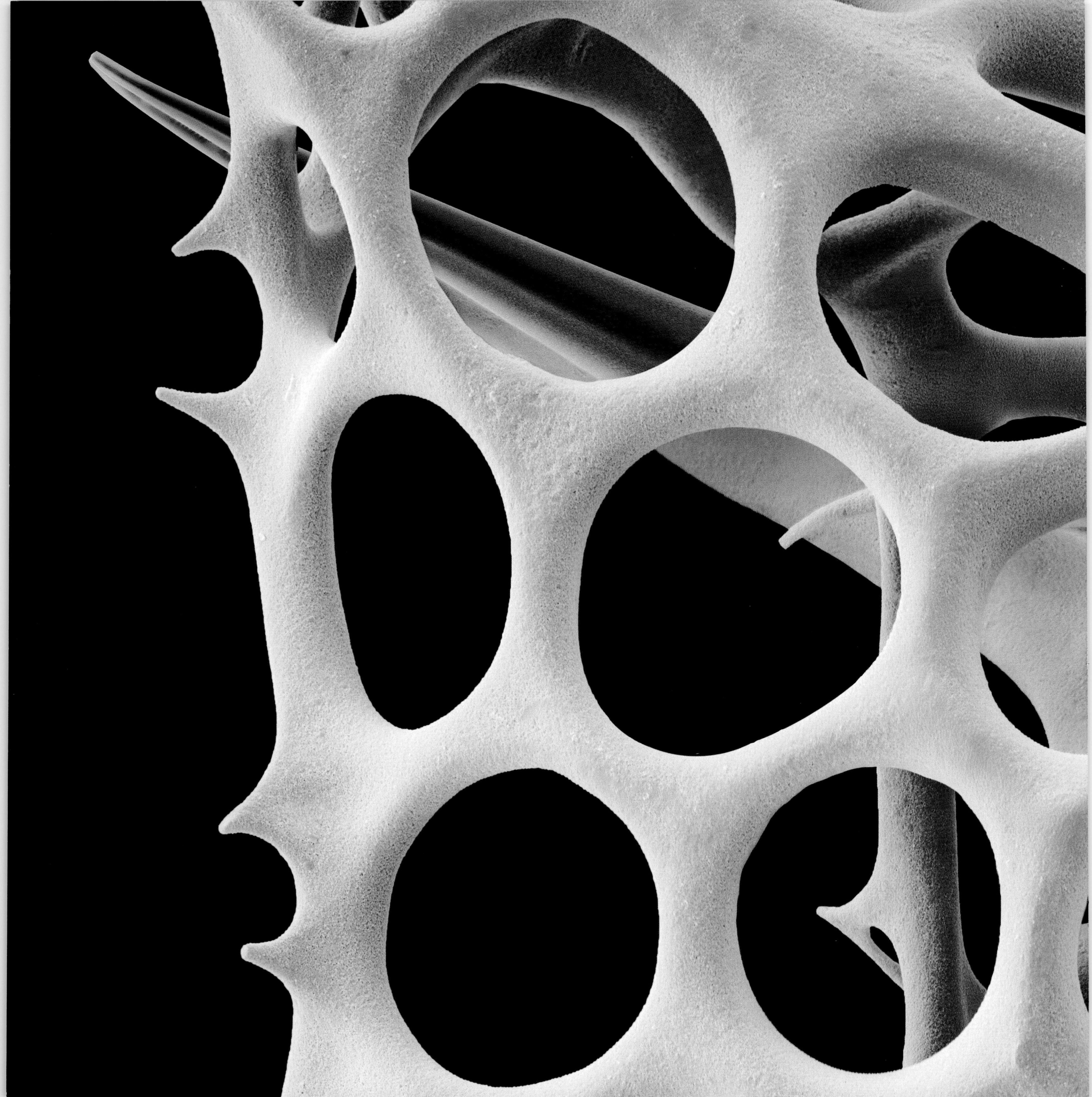

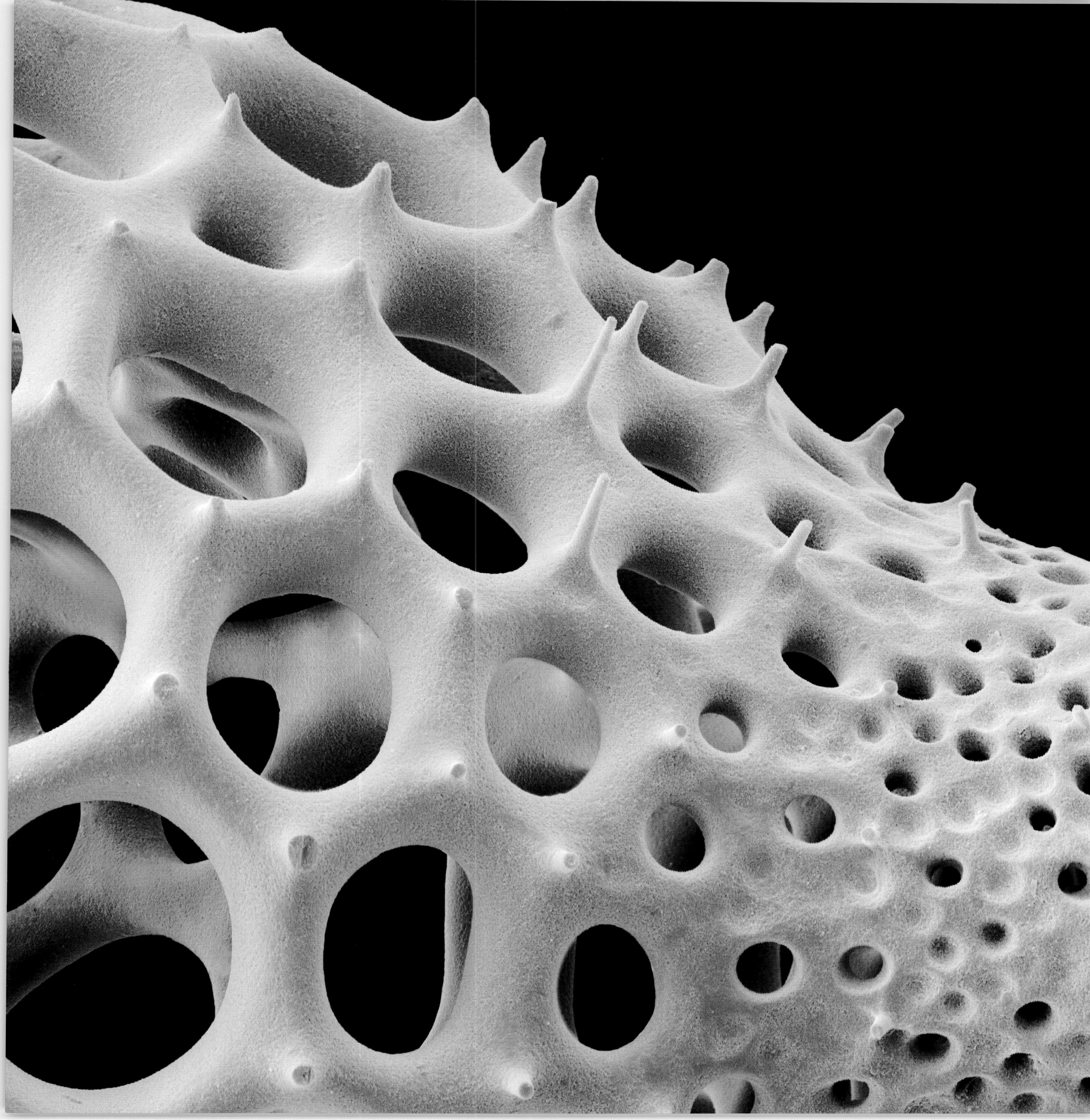

104

**Facing Page**

*Clathrosphaera arachnoides*, equatorial Pacific.
2000 x. 100 microns wide, 0.1 millimeter

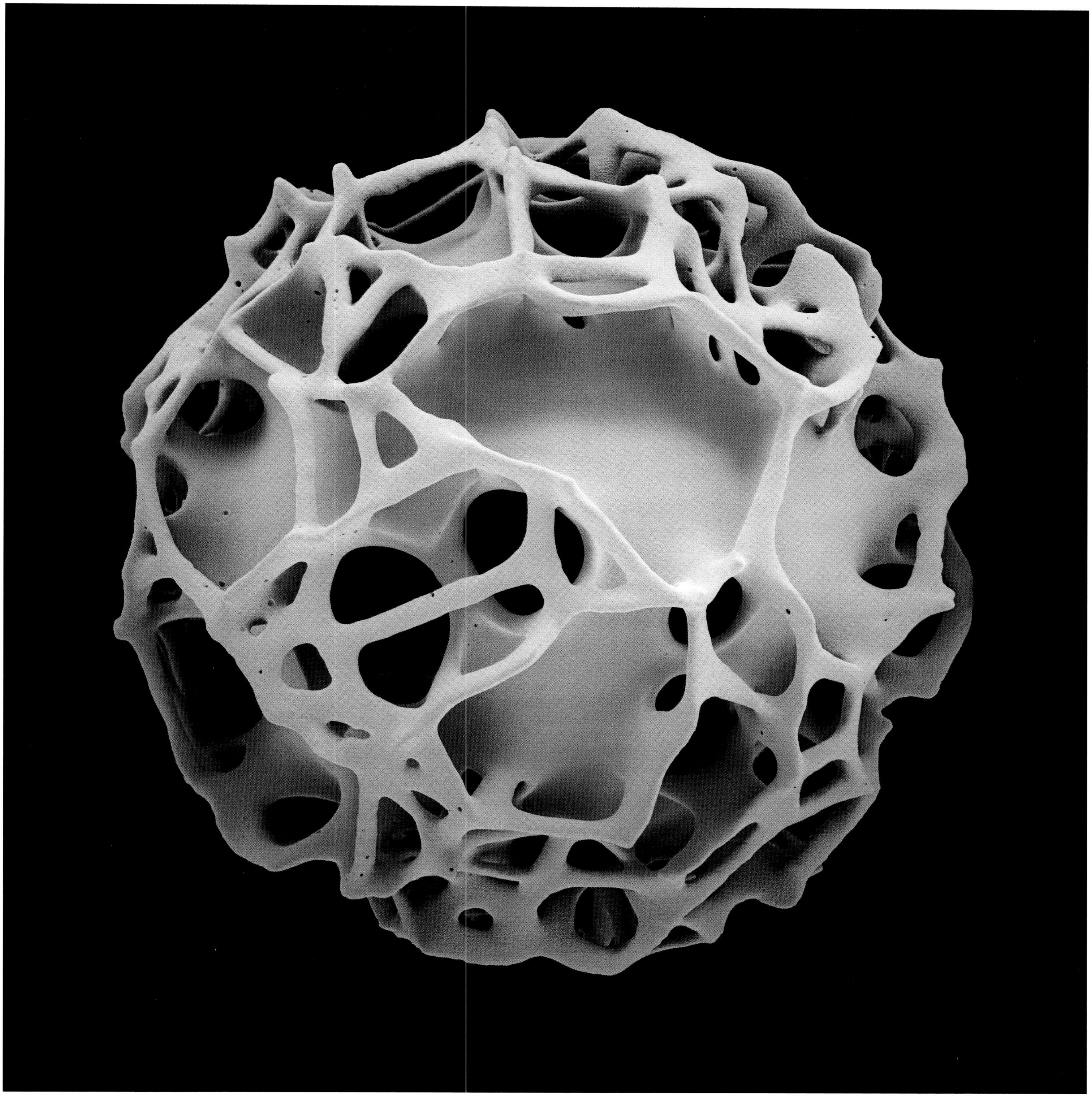

Facing Page, Pages 108–111 →

*Hexicontum pachydermum*, equatorial Pacific.
733 x. 250 microns wide, 0.25 millimeter

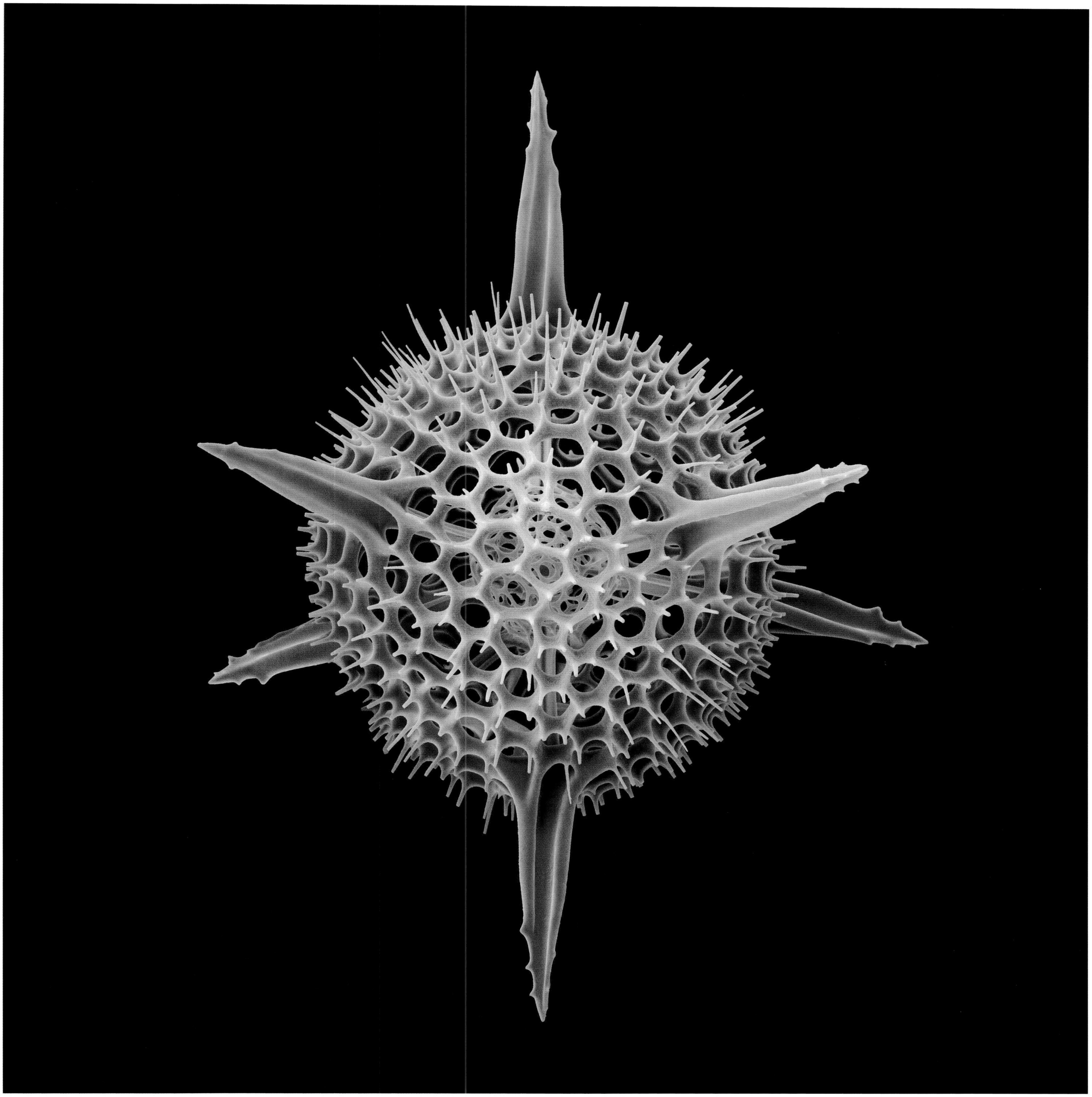

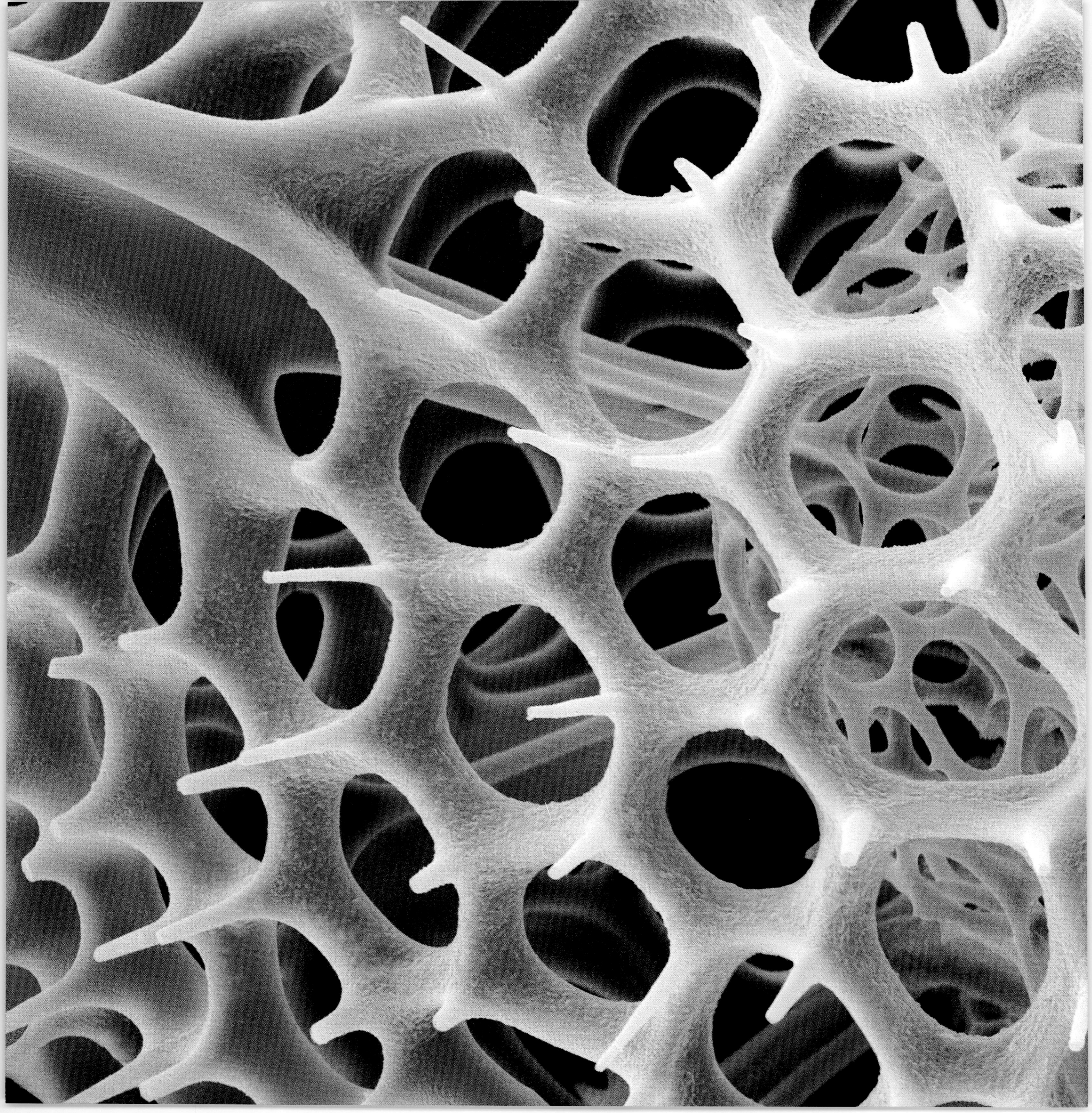

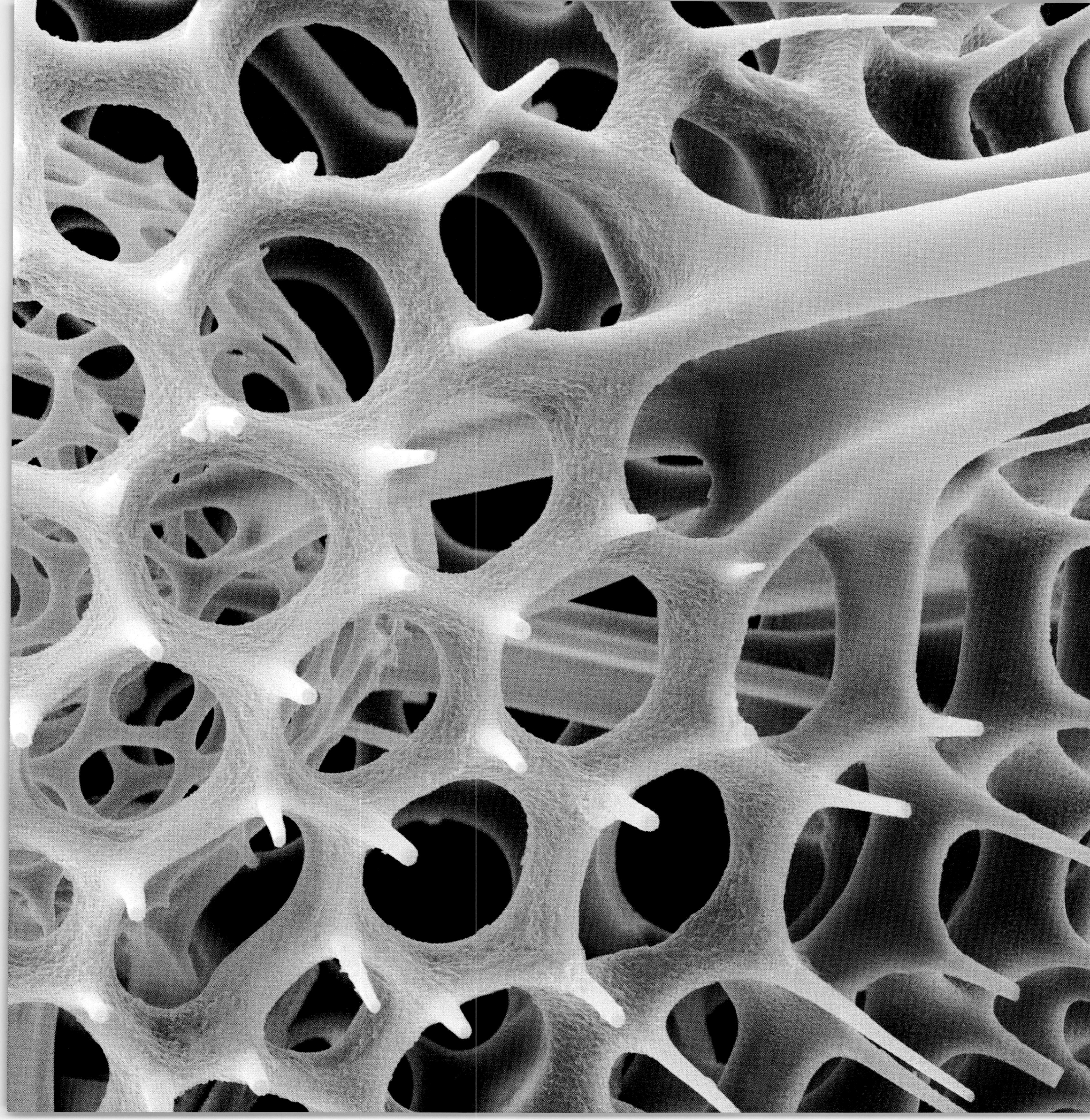

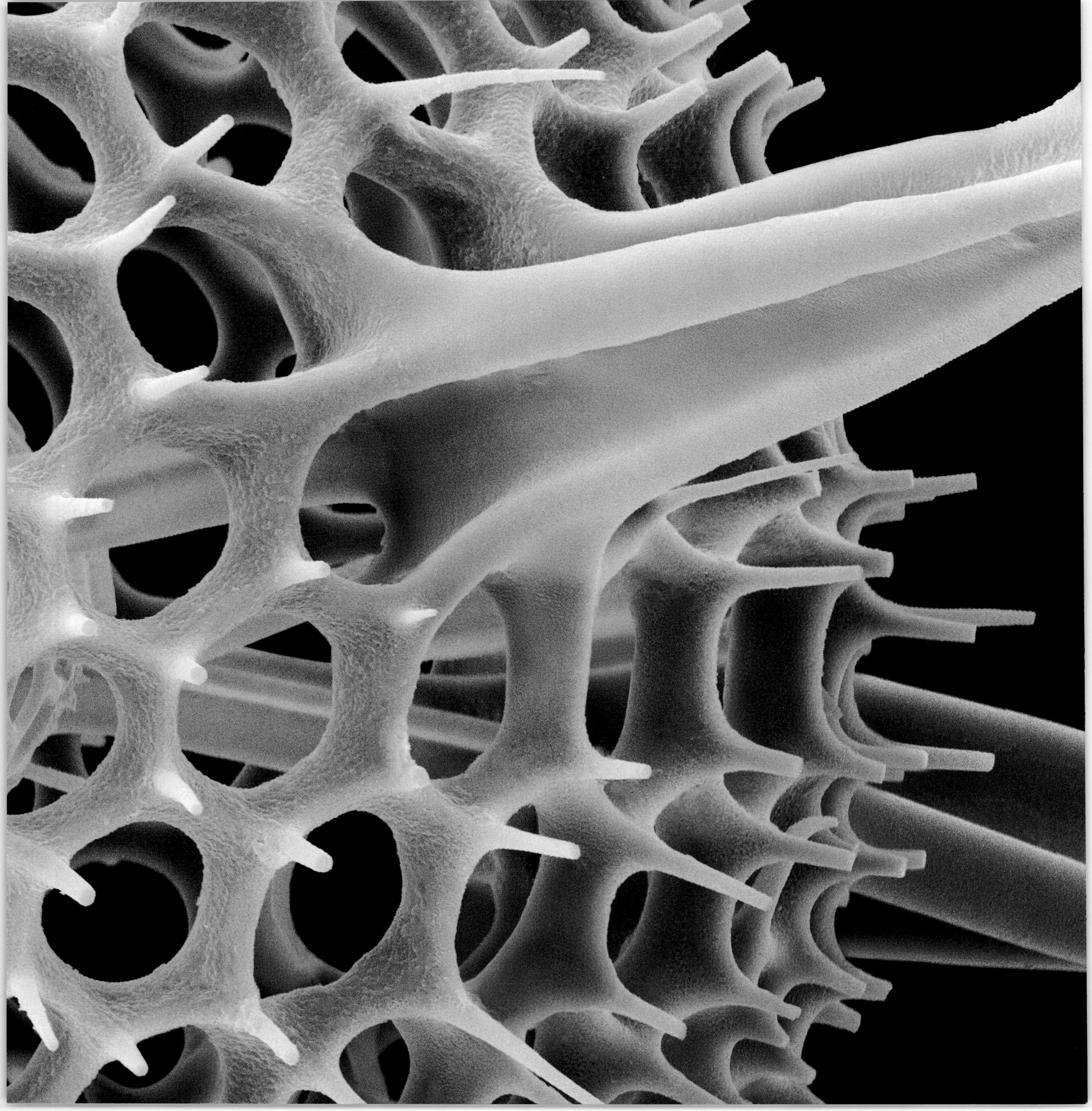

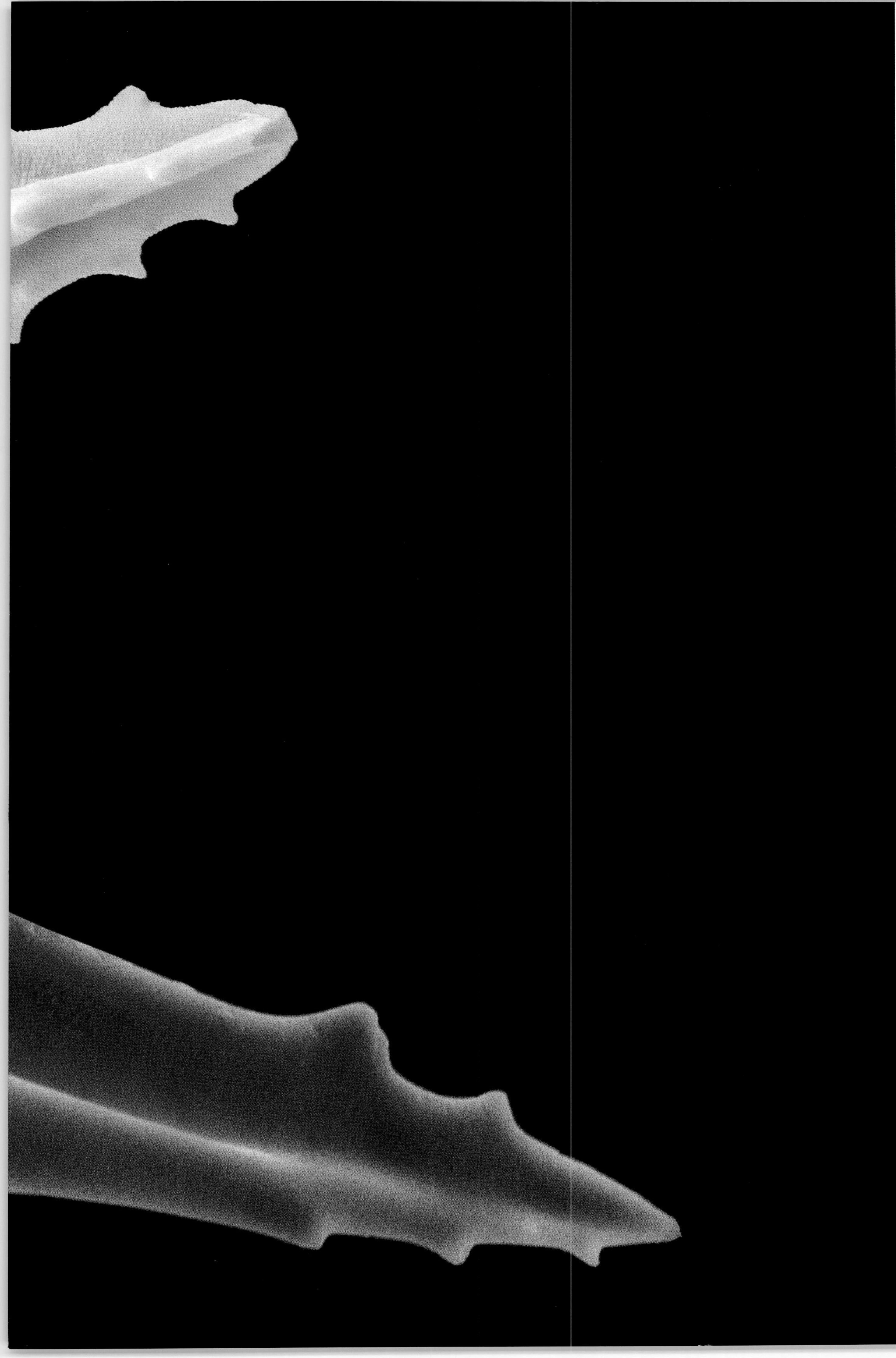

Facing Page, Pages 114–115 →

*Acrosphaera spinosa*, equatorial Pacific.
933 x. 225 microns across, 0.22 millimeter

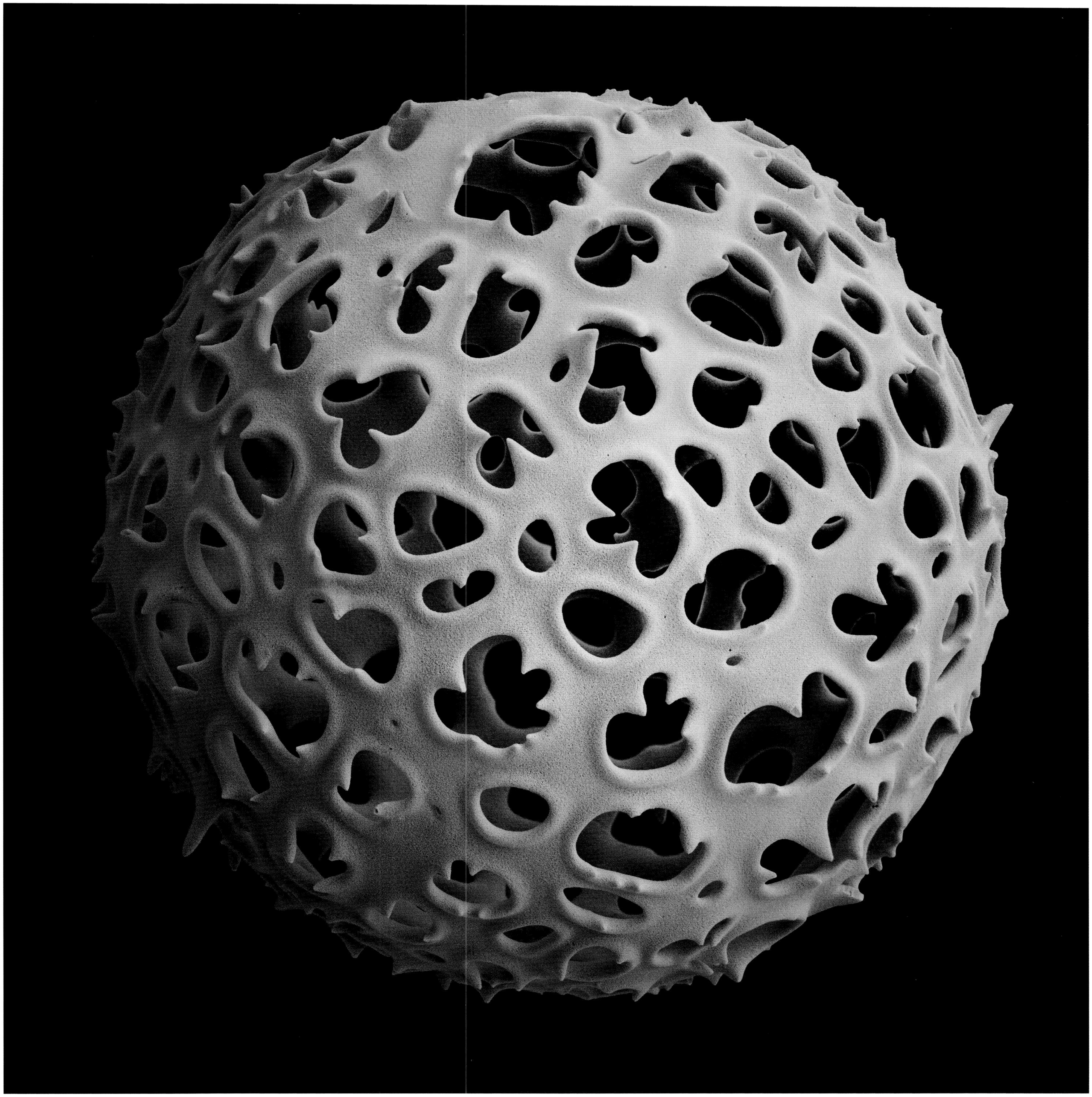

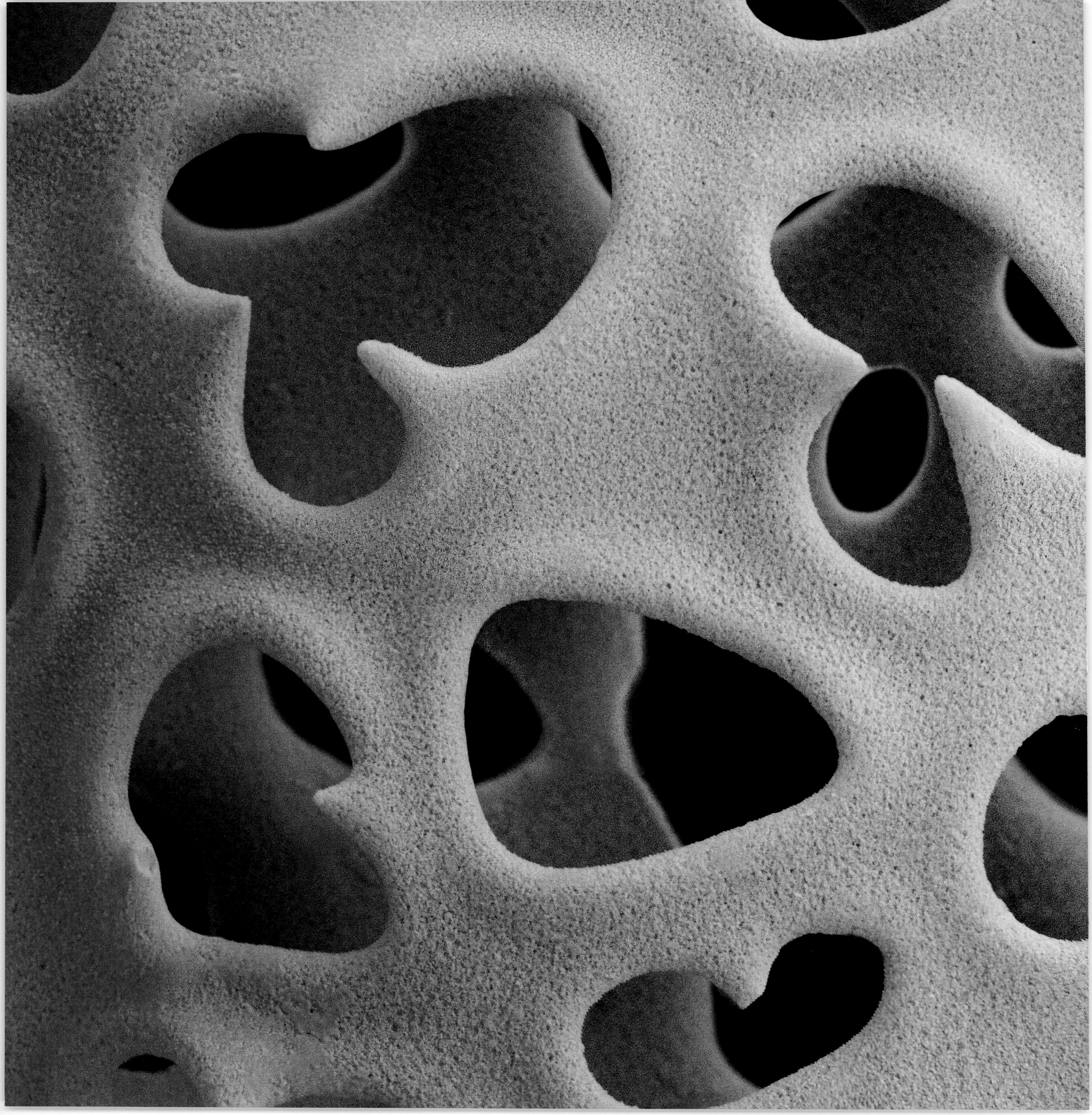

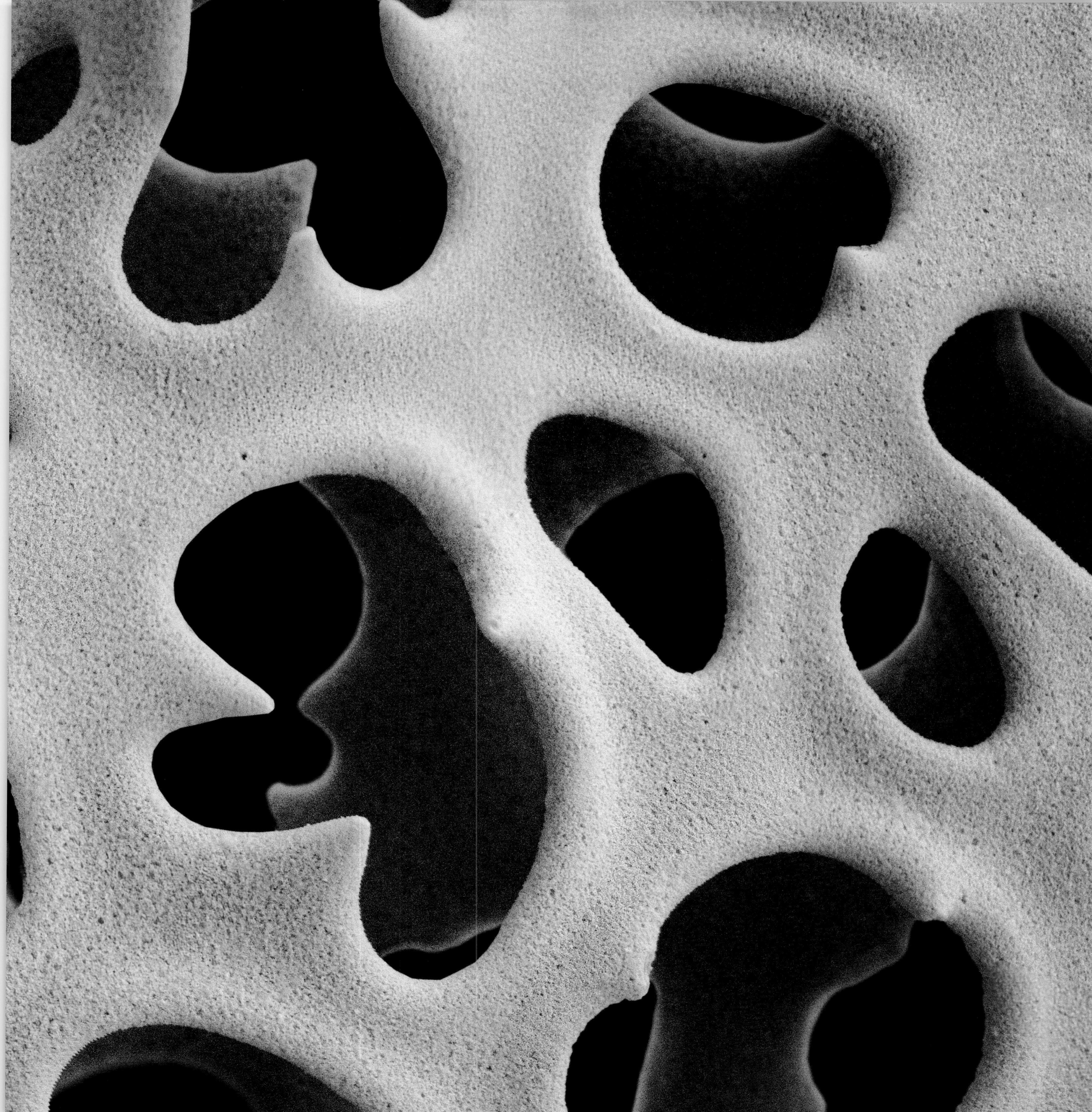

AQUA

# DIATOMS

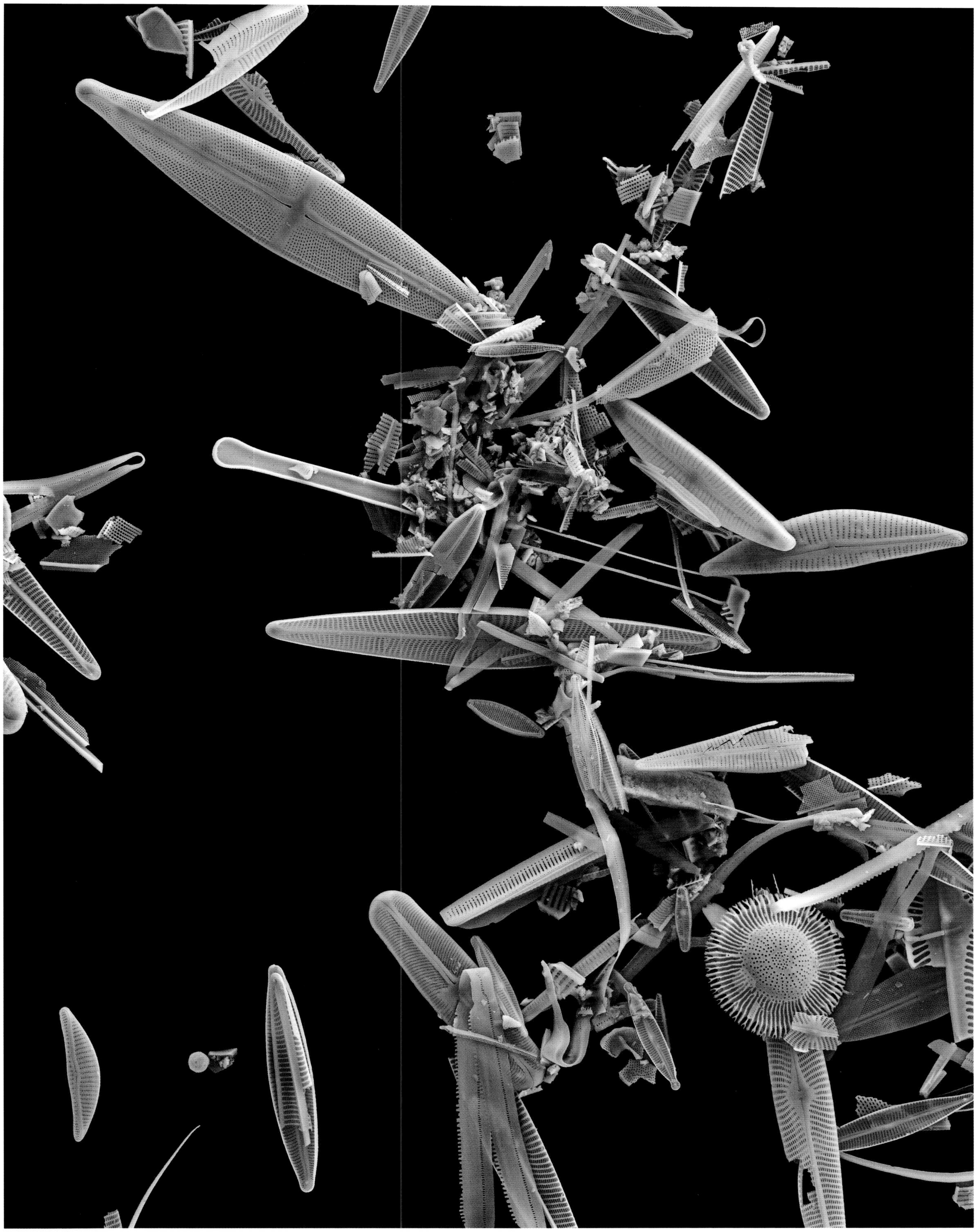

← Page 117

Intact and fractured freshwater diatoms
from Wapusk National Park, Hudson Bay.
1250 x. About 160 microns across, 0.16 millimeter

**Facing Page**

*Navicula radiosa* and other freshwater diatoms with two
spherical chrysophyte cysts, Wapusk National Park, Hudson Bay.
5000 x. About 80 microns across, 0.08 millimeter

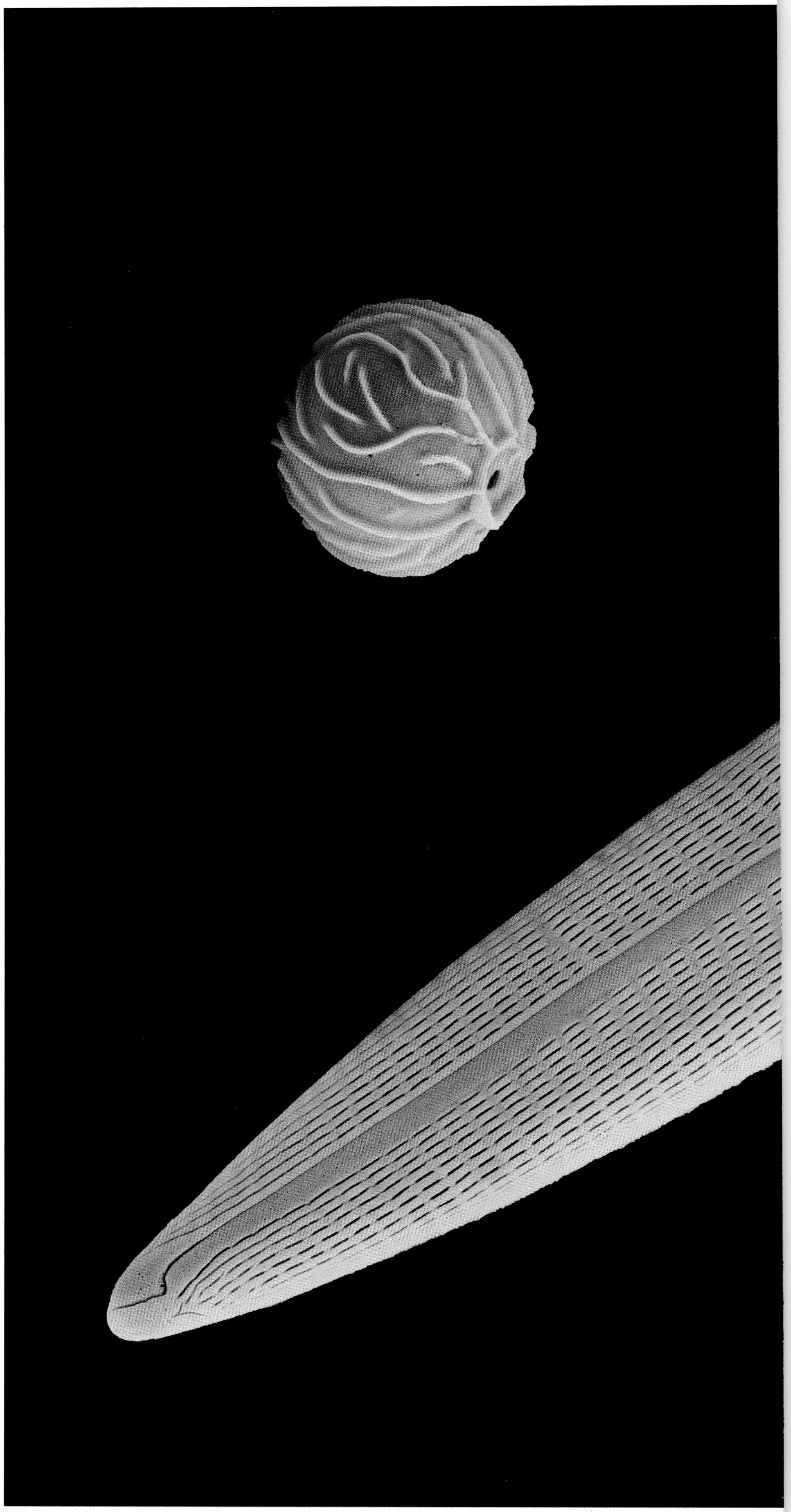

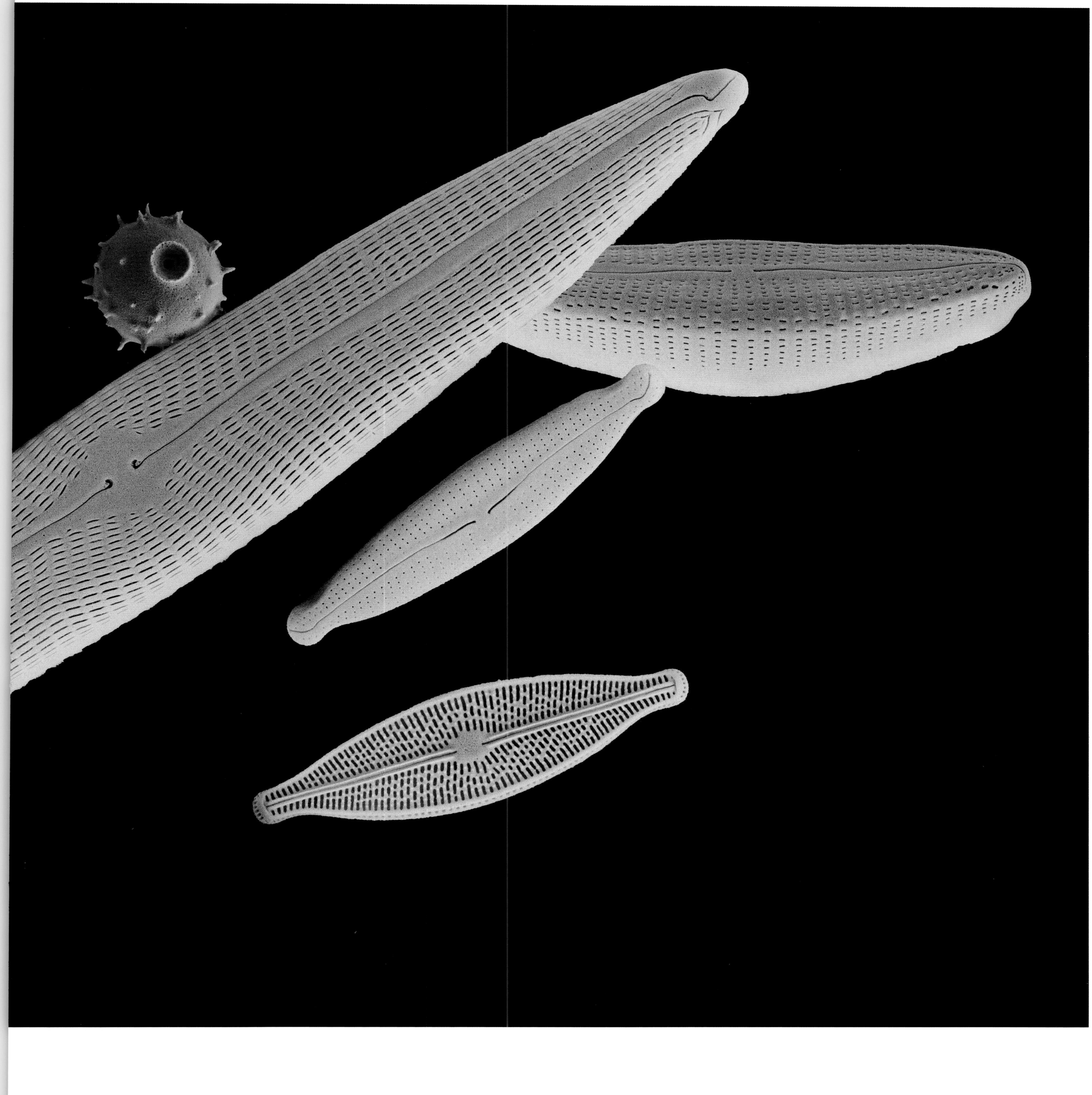

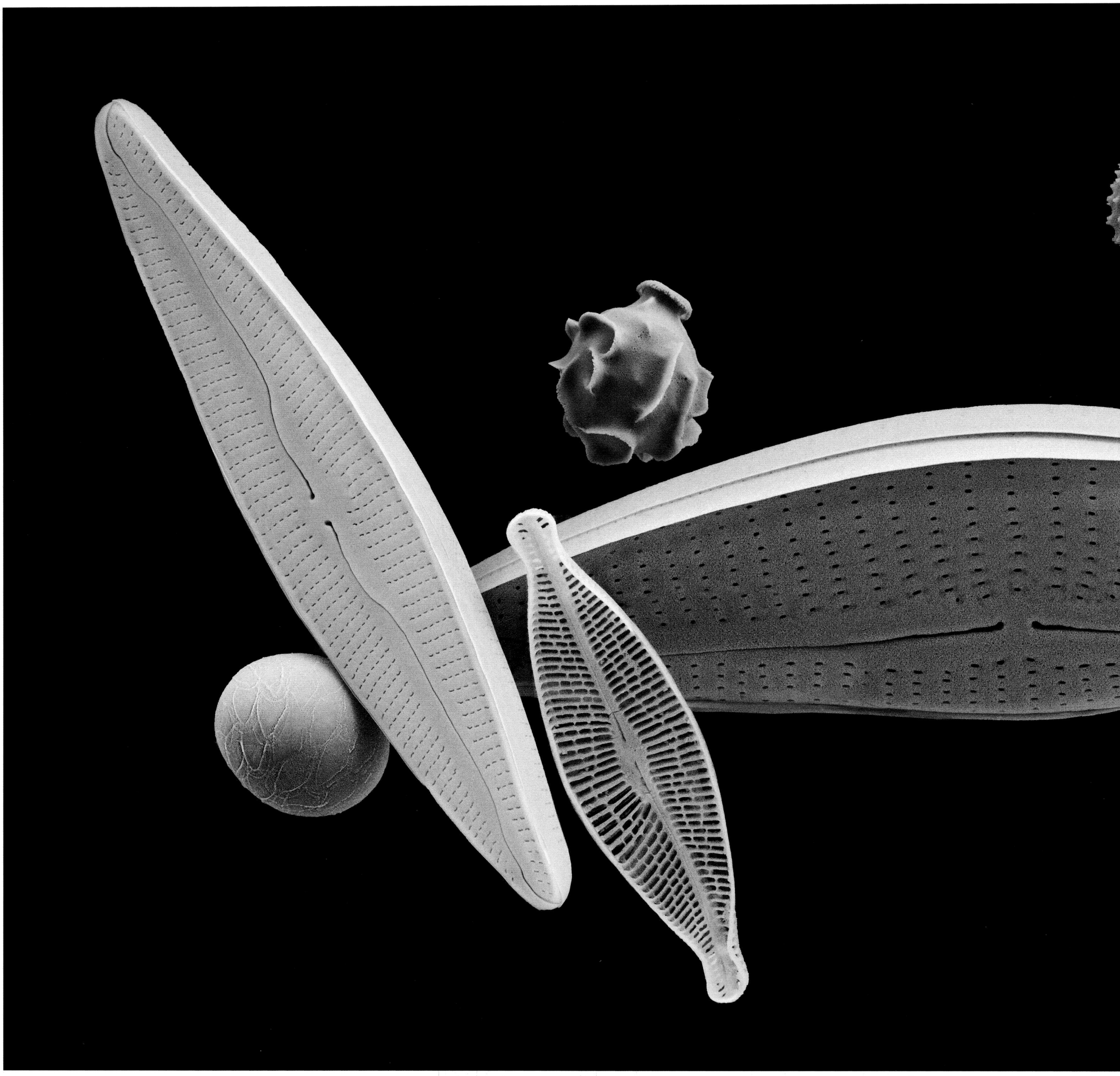

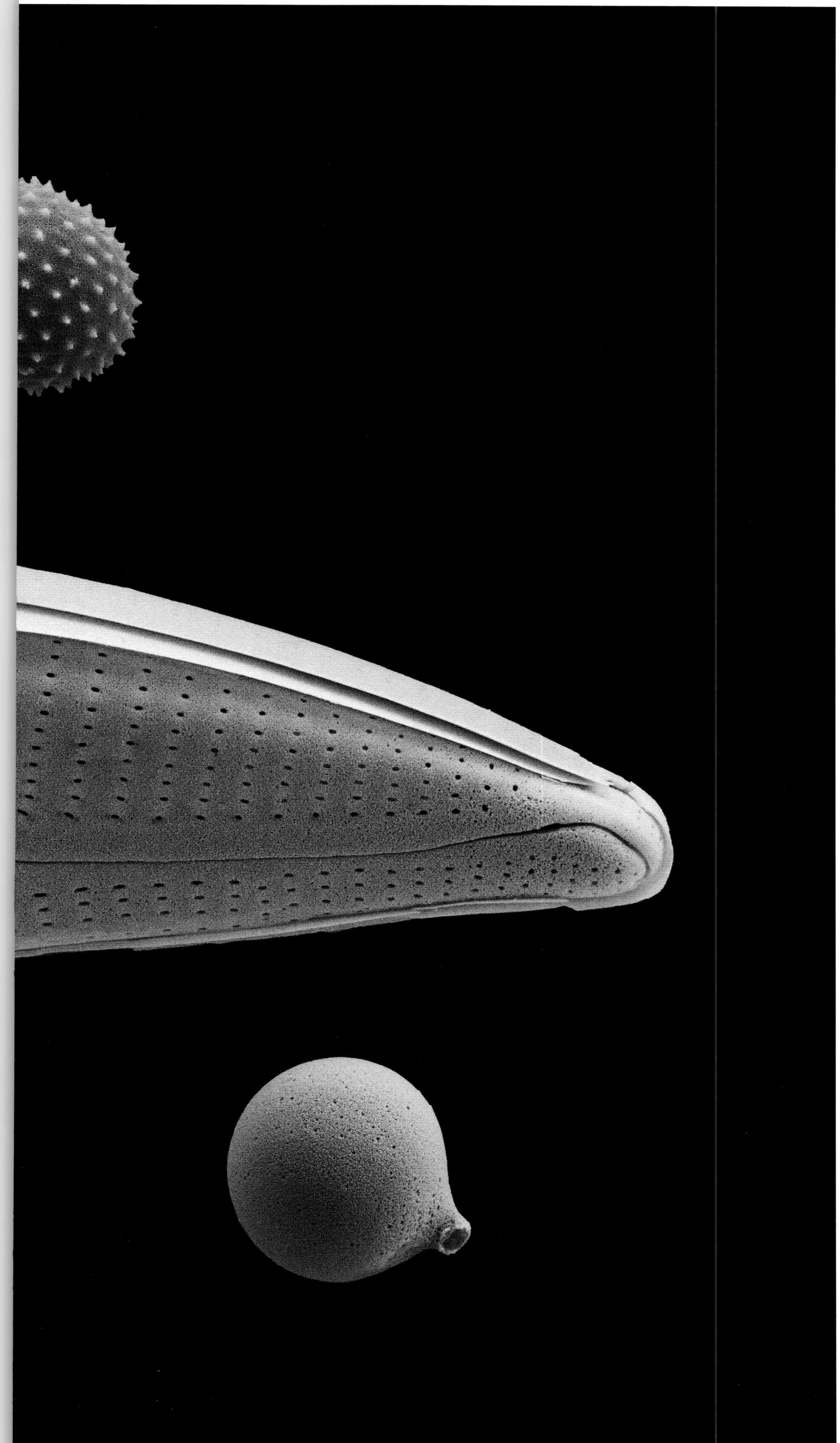



**Facing Page**

*Encyonena silesiacum* and other freshwater diatoms
and chrysophyte cysts, Wapusk National Park, Hudson Bay.
7000 x. 60 microns across, 0.06 millimeter

**Facing Page**

Diatom of the *Rhopalodia* genus, Wapusk National Park, Hudson Bay.
8700 x. The visible part is 40 microns wide, 0.04 millimeter

**Pages 124–125 →**

Two views of *Rhopalodia gibba*, Wapusk National Park, Hudson Bay.
8000 x. About 60 microns from end to end, 0.06 millimeter

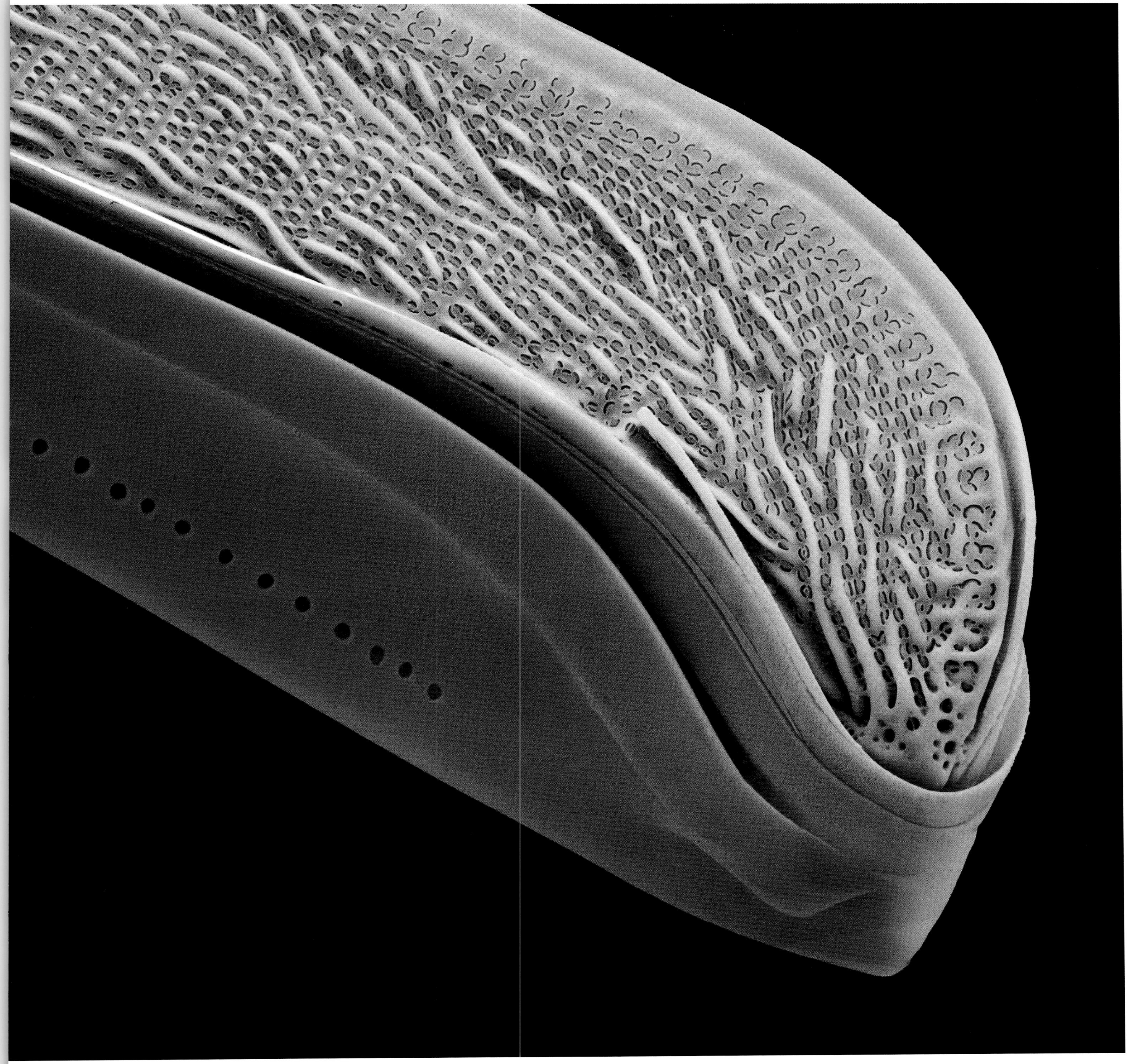

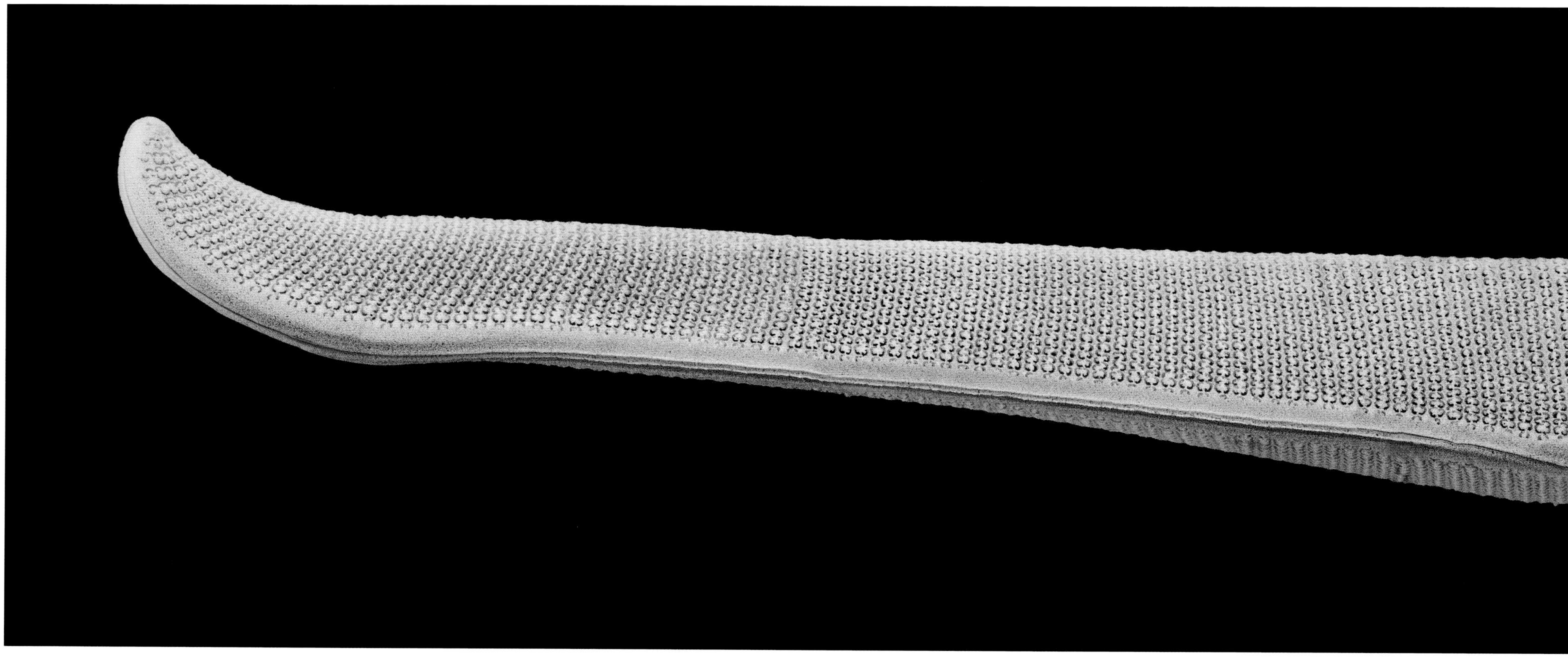

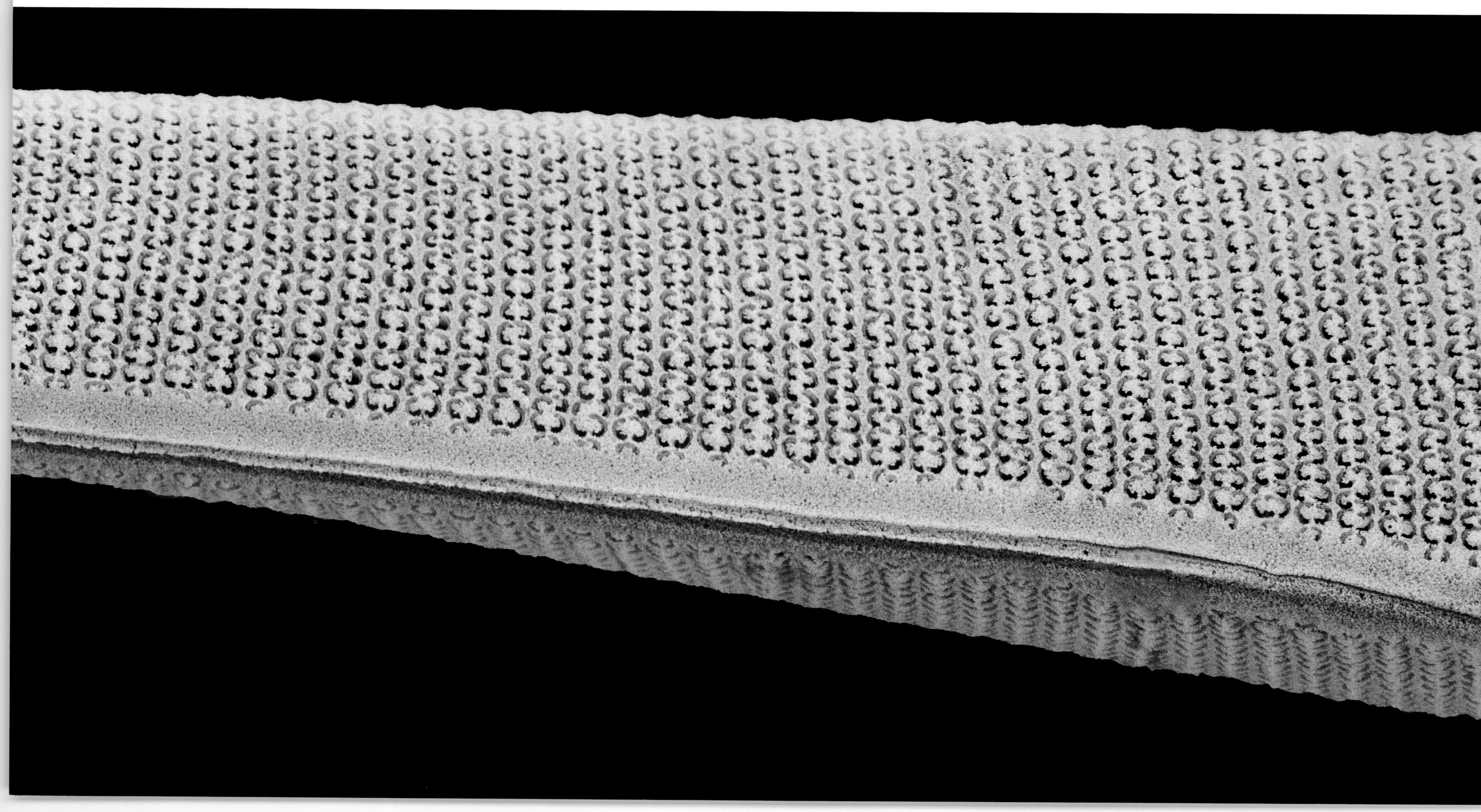

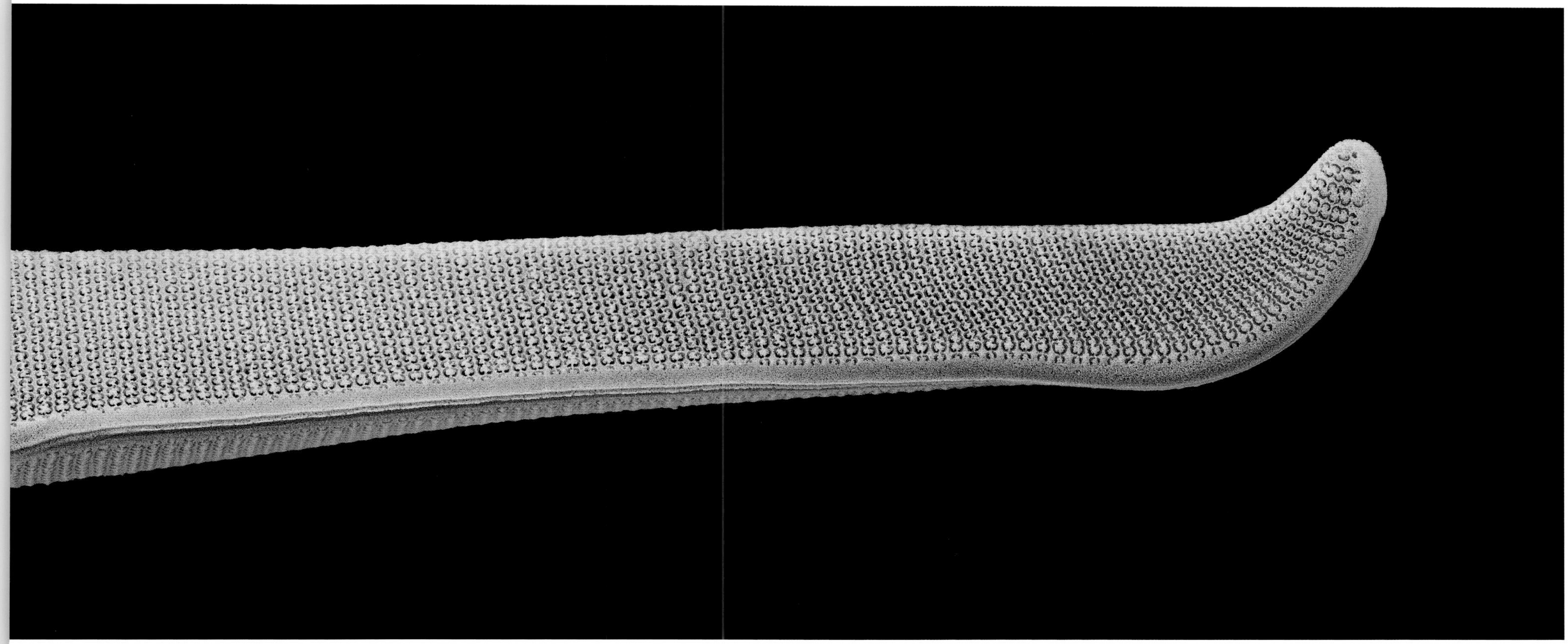

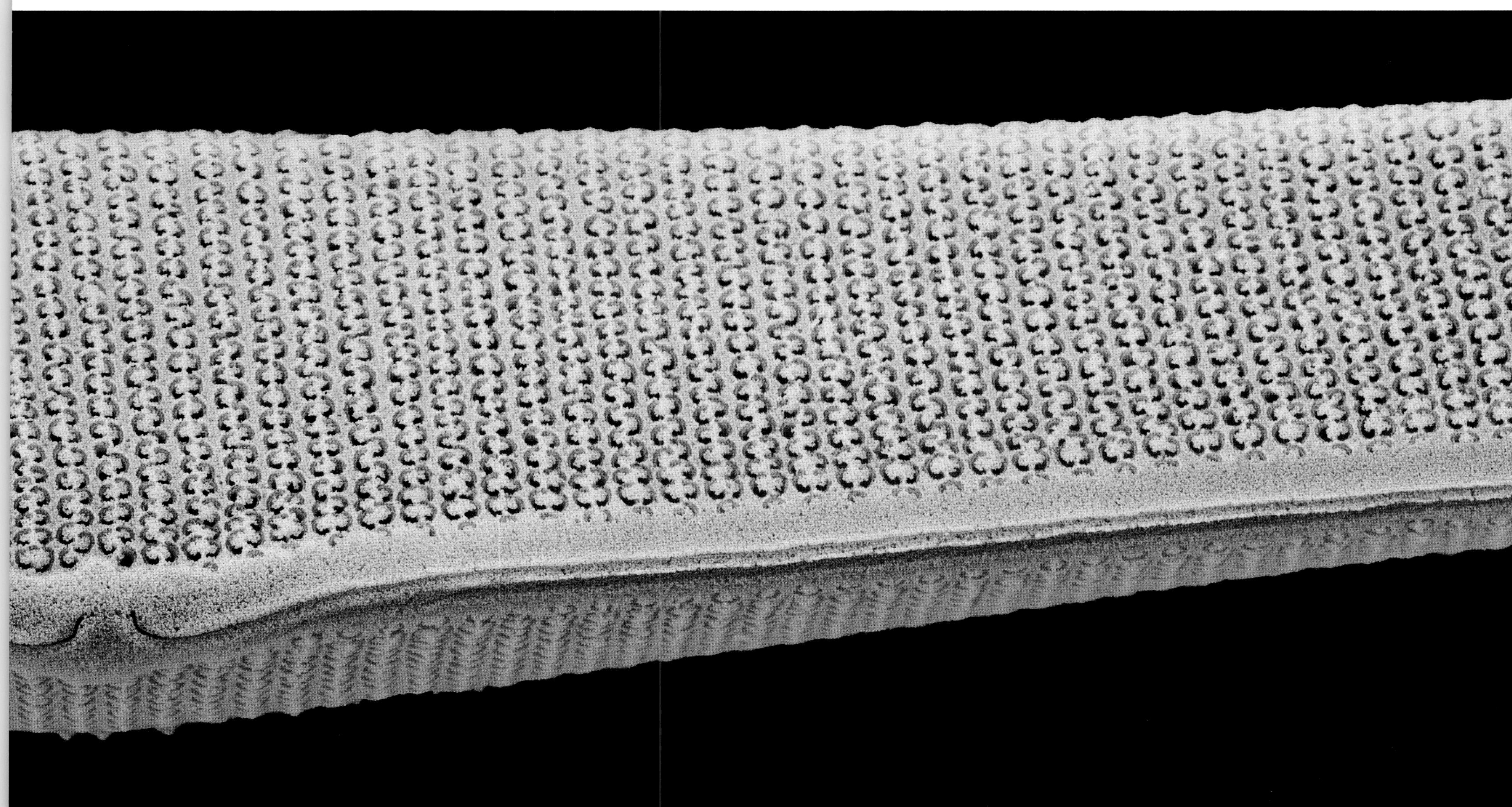

126

Facing Page, Pages 128–129 →

Marine diatom of the *Azpeitia* genus, Florida Gulf Stream.
3231 x. 65 microns wide, 0.06 millimeter

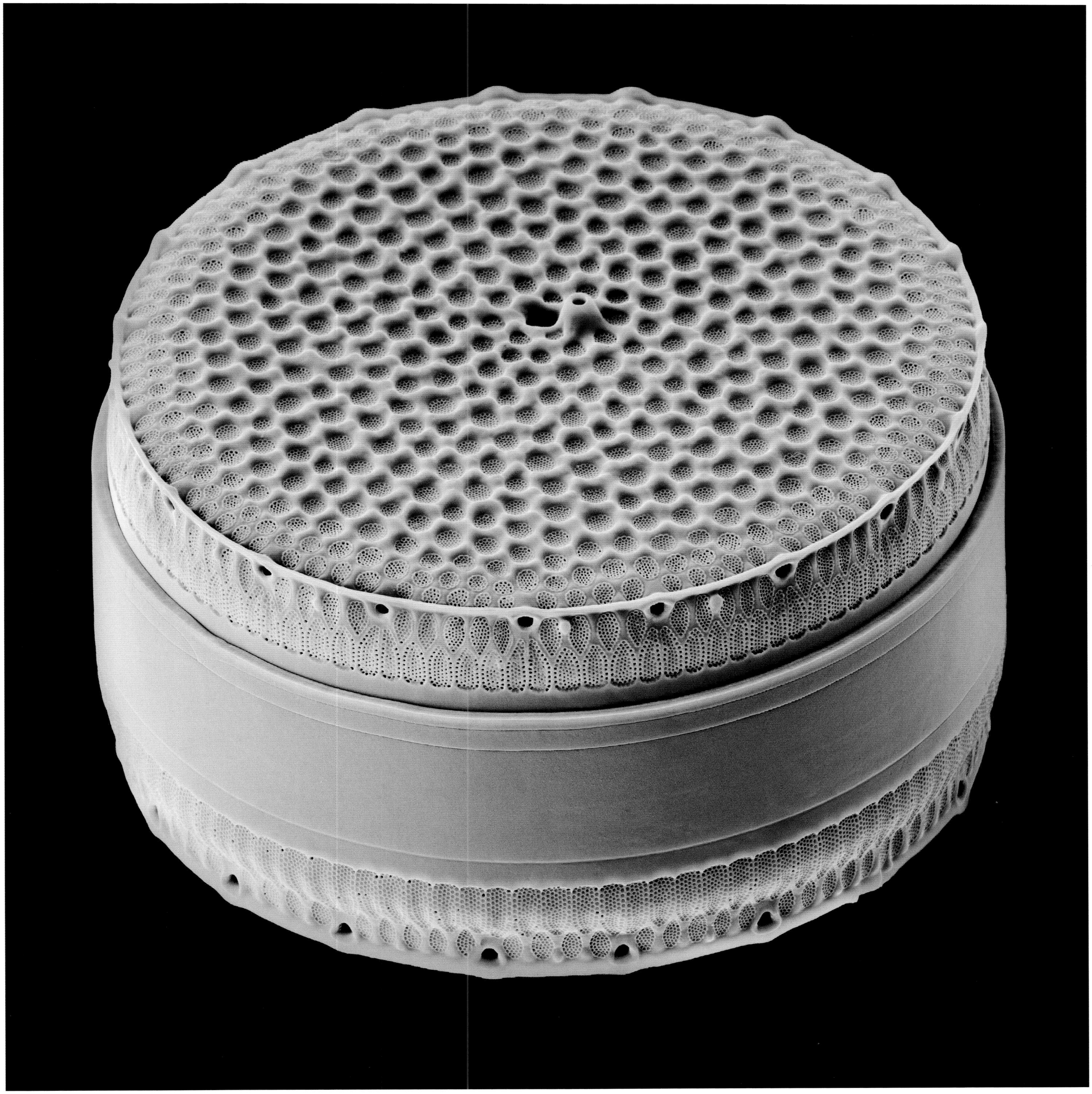

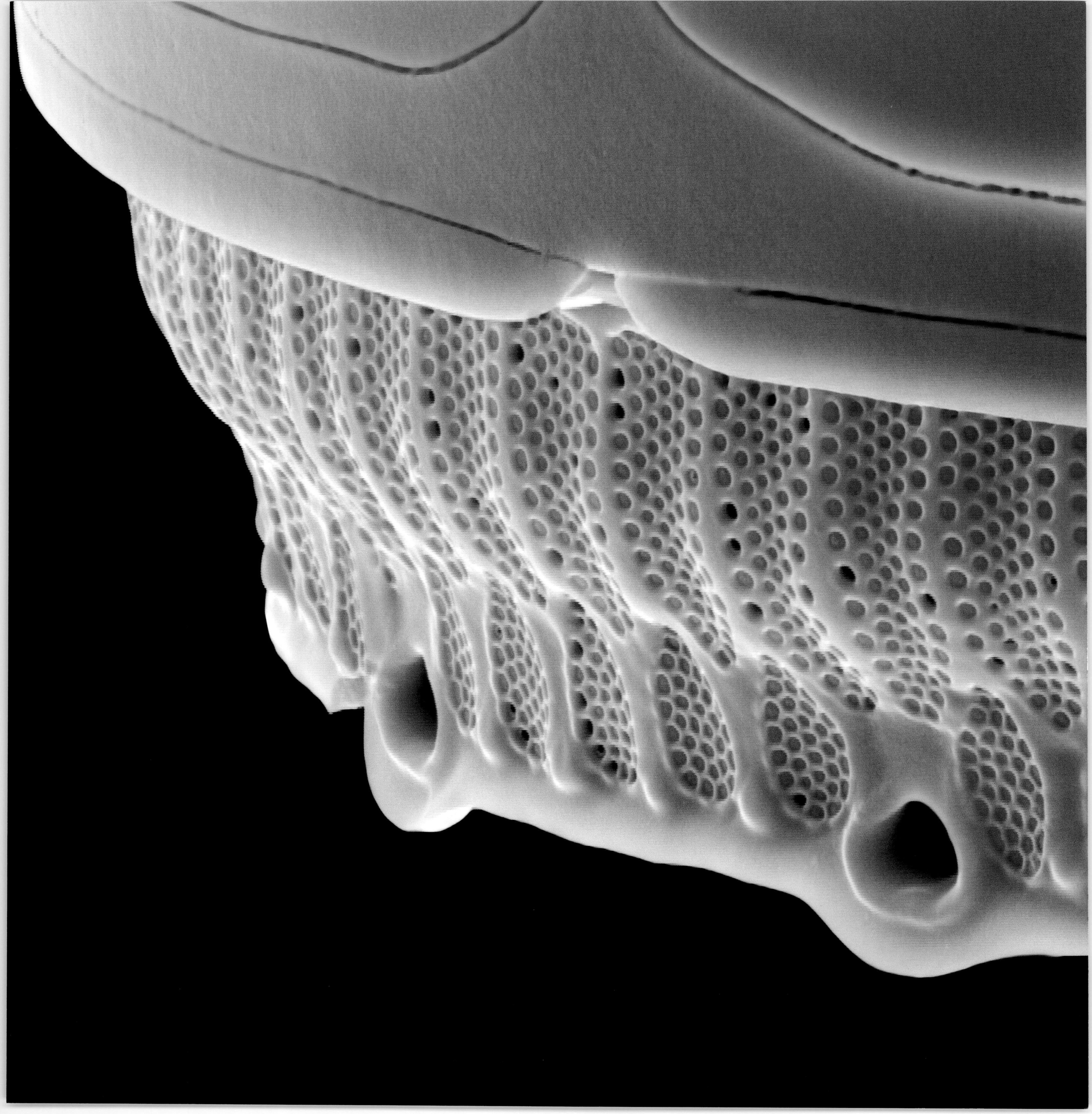

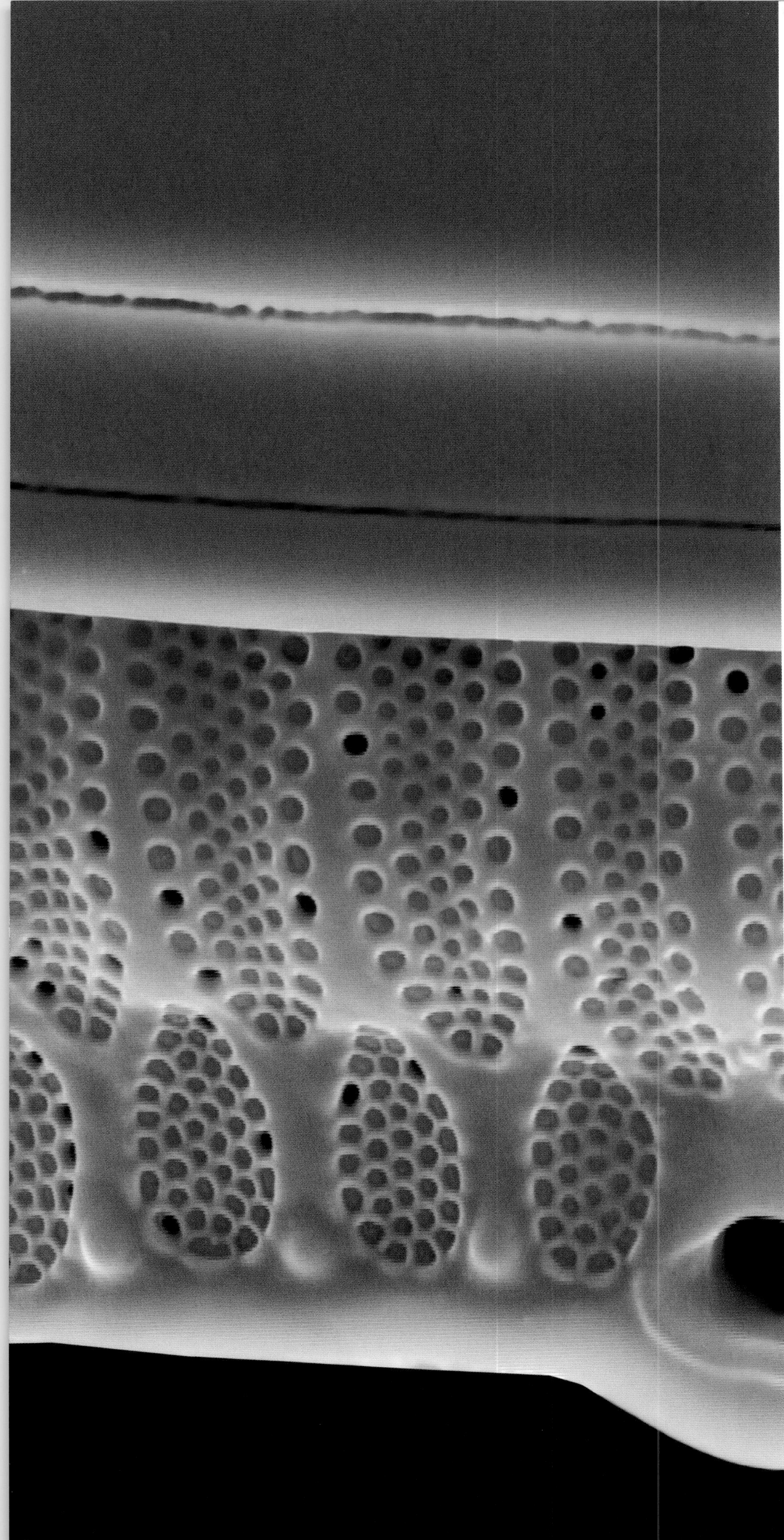



**Page 130 →**

*Cyclotella meneghiniana*, Wapusk National Park, Hudson Bay.
6571 x. 35 microns wide, 0.03 millimeter

**Page 131 →**

*Asterolampra marylandica*, Florida Gulf Stream.
3500 x. 60 microns wide, 0.06 millimeter

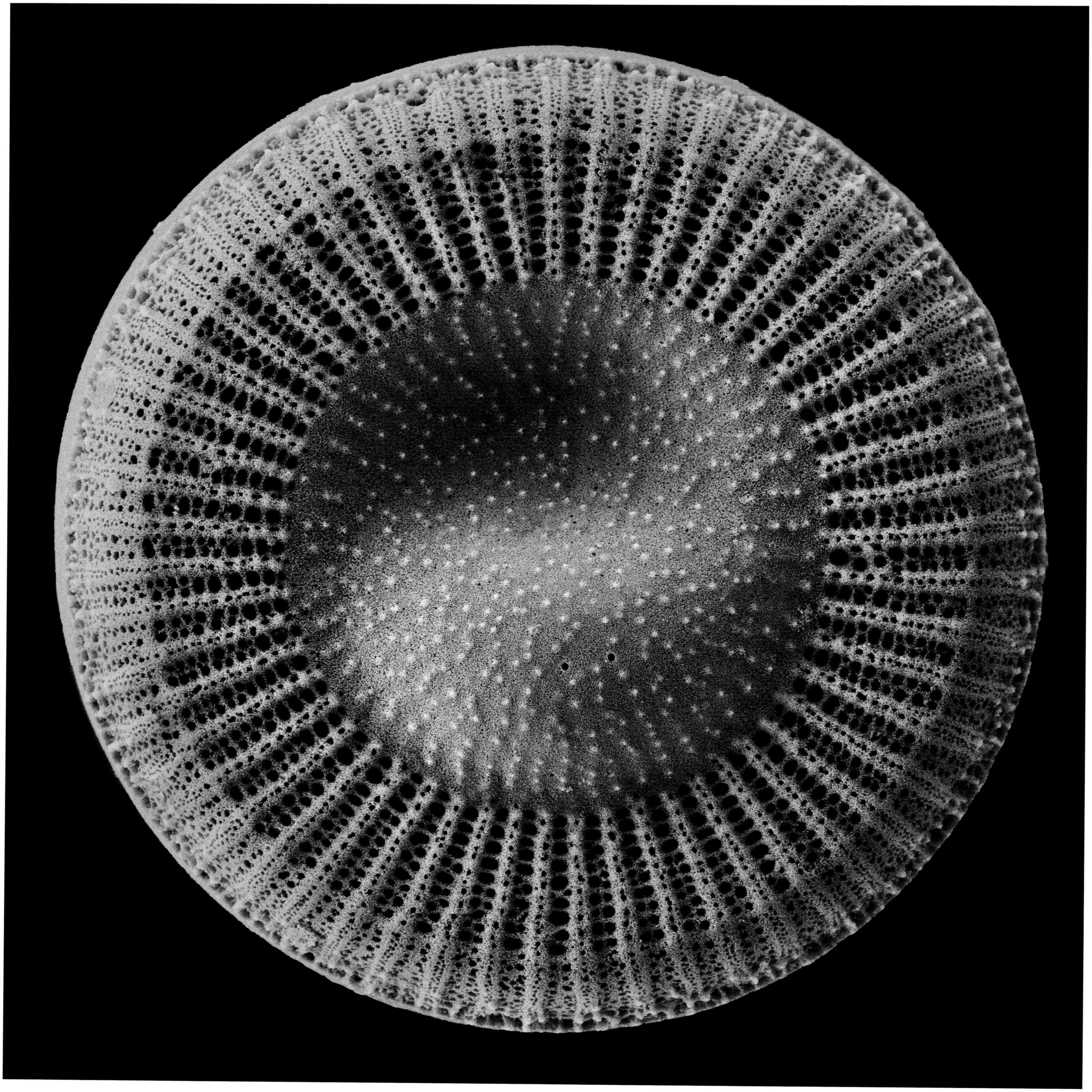

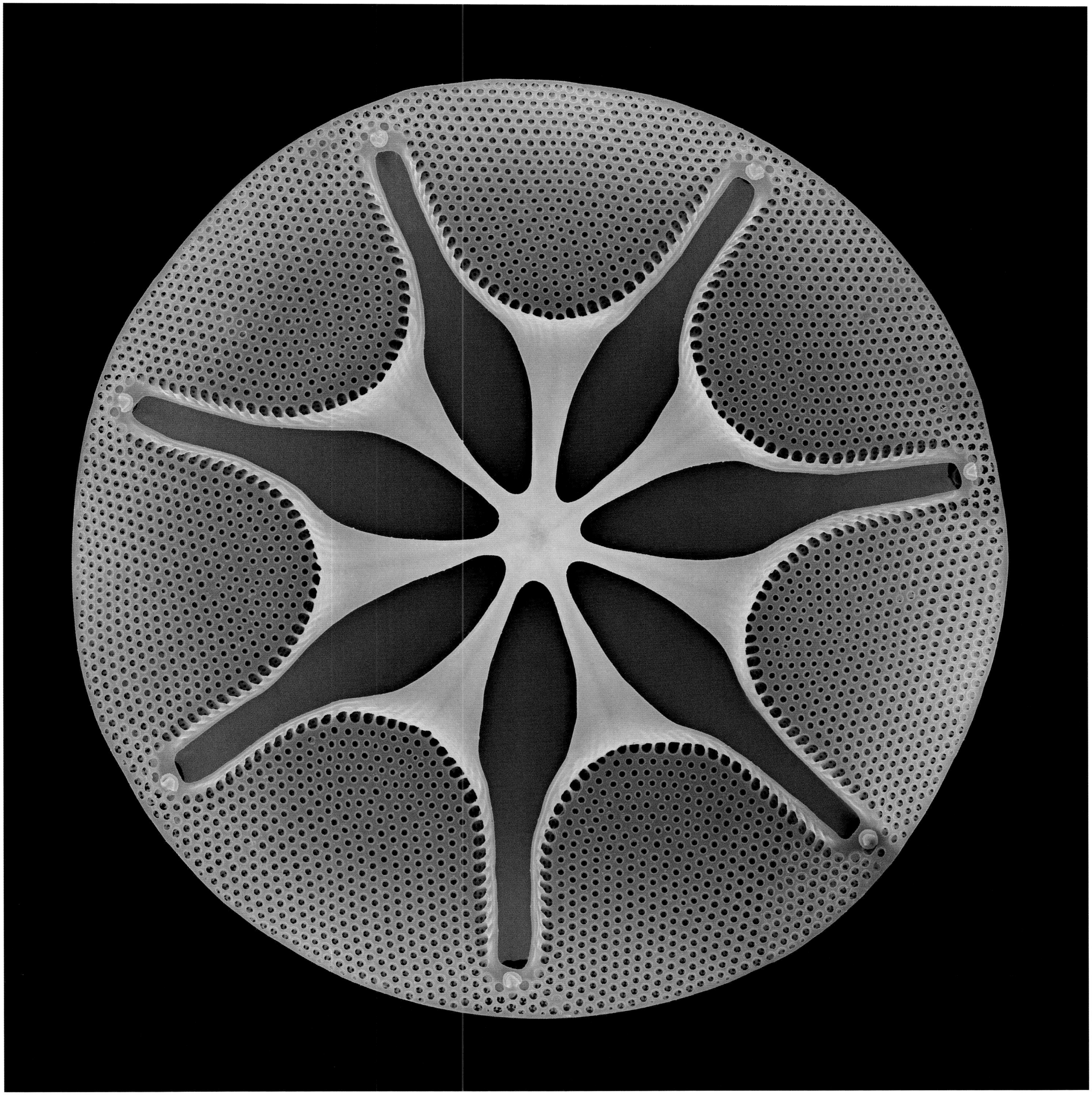

**Facing Page**

*Stauroneis amphicephala*, Wapusk National Park, Hudson Bay.
3182 x. 110 microns from end to end, 0.11 millimeter

**Pages 134–135 →**

*Eunotia exigua*, Wapusk National Park, Hudson Bay.
9200 x. 50 microns wide, 0.05 millimeter

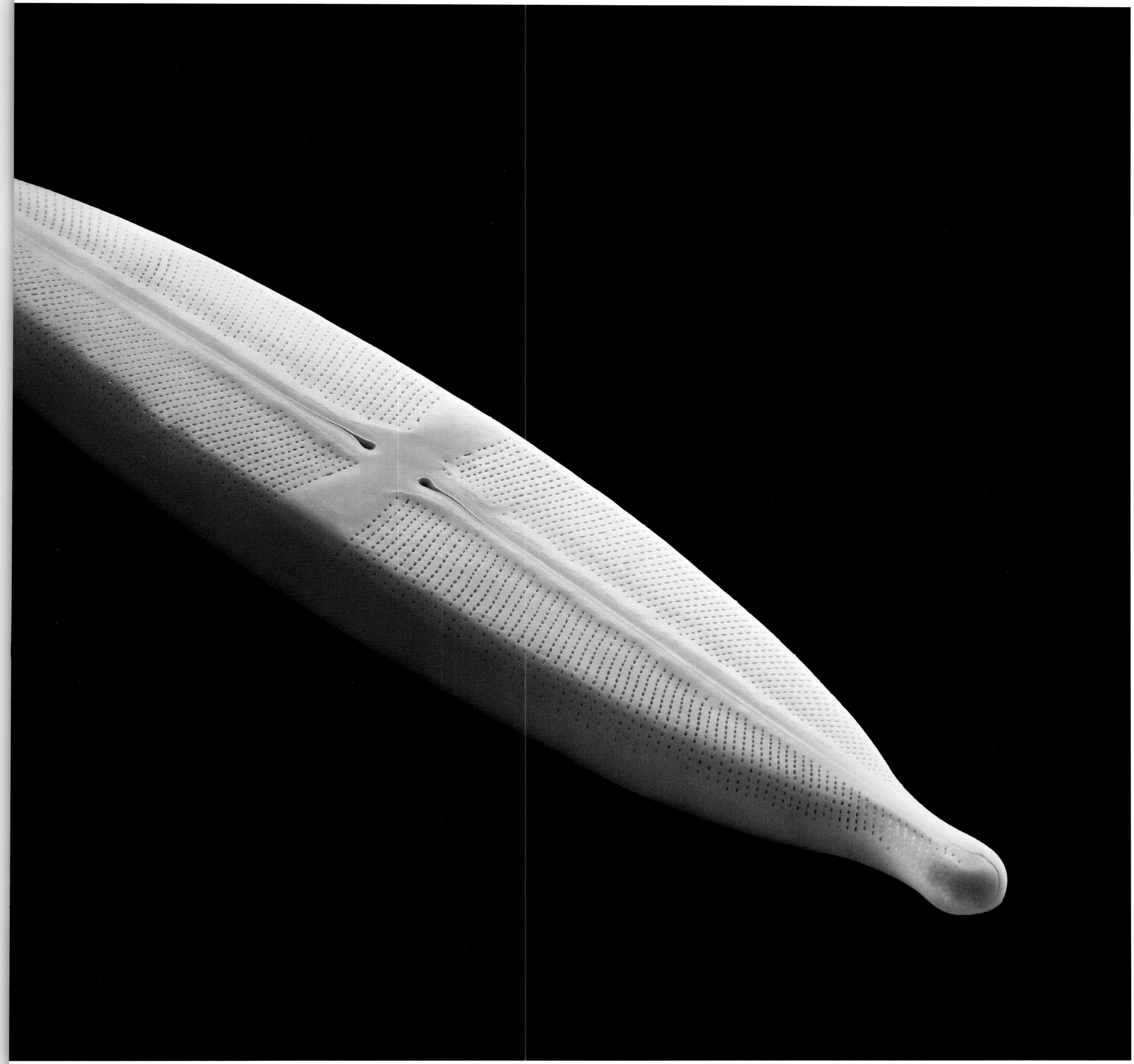

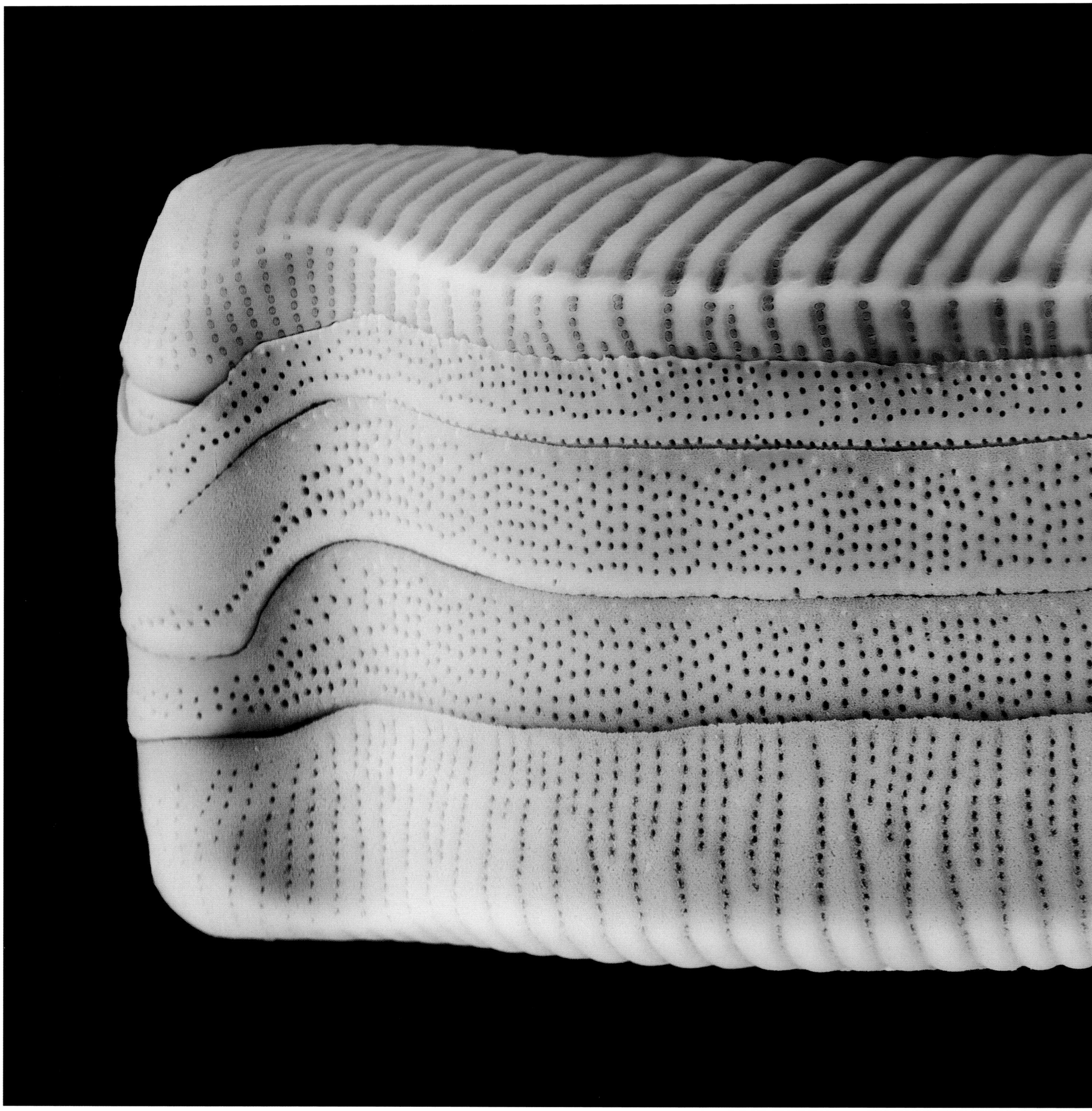

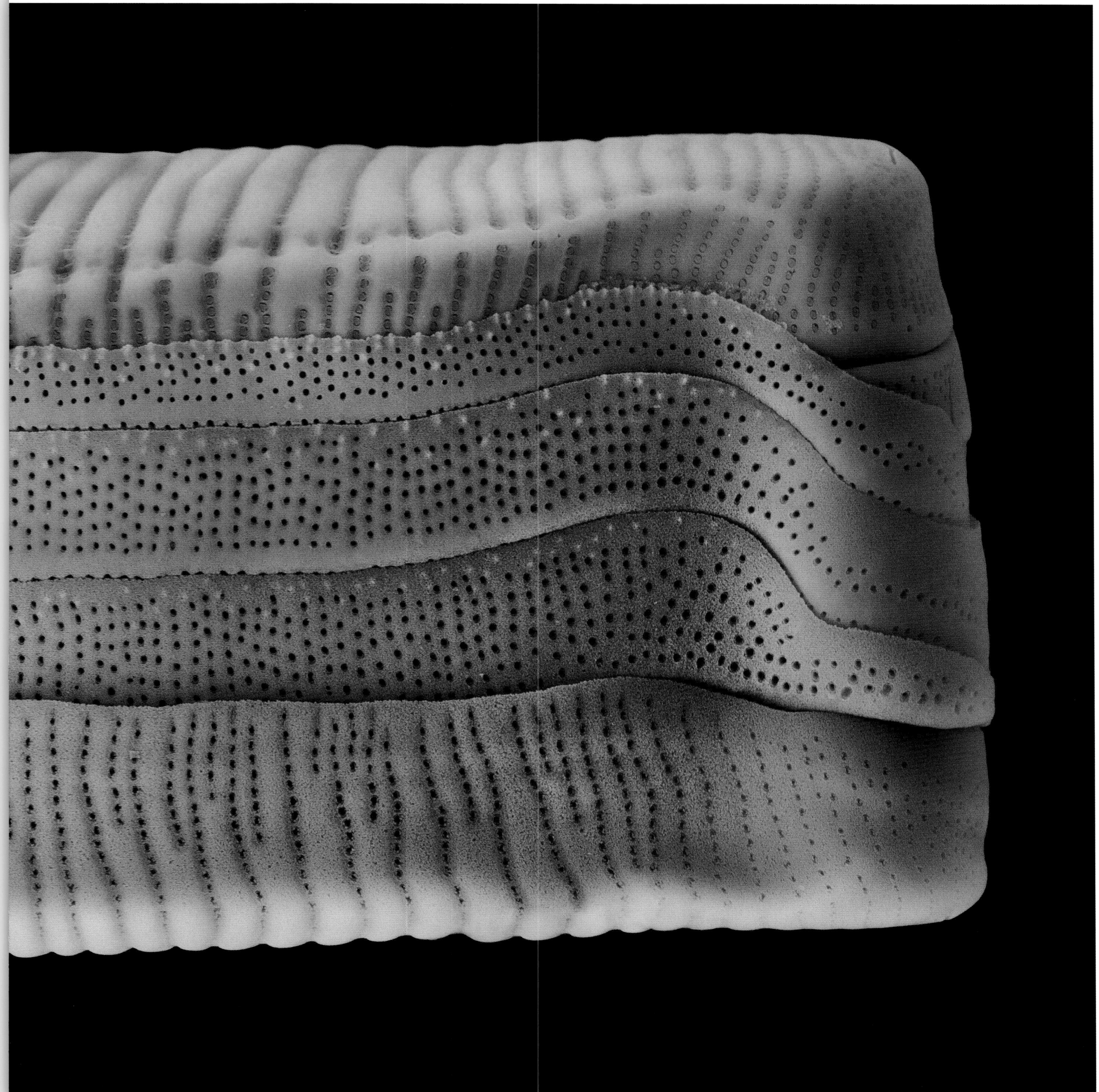

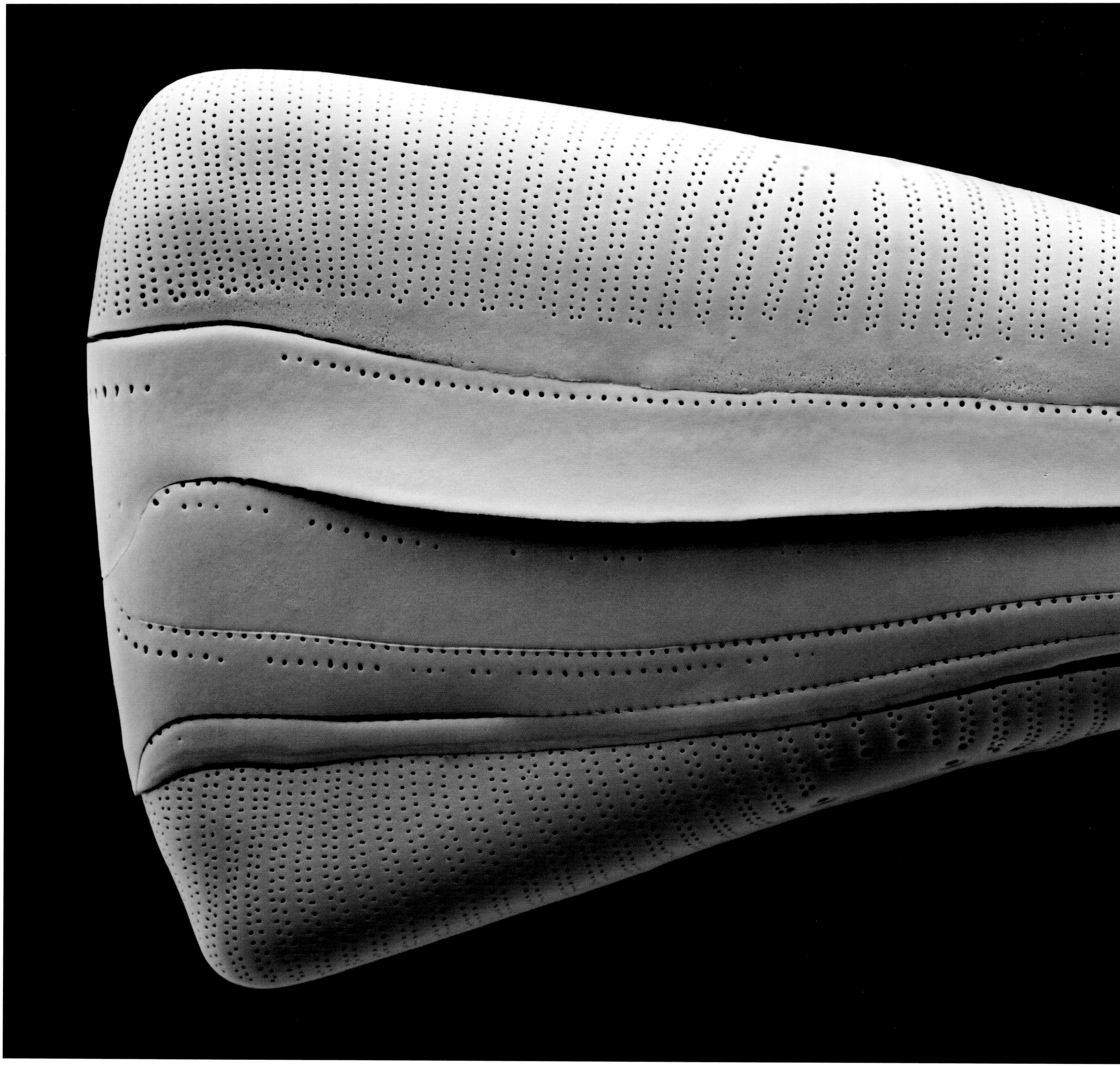

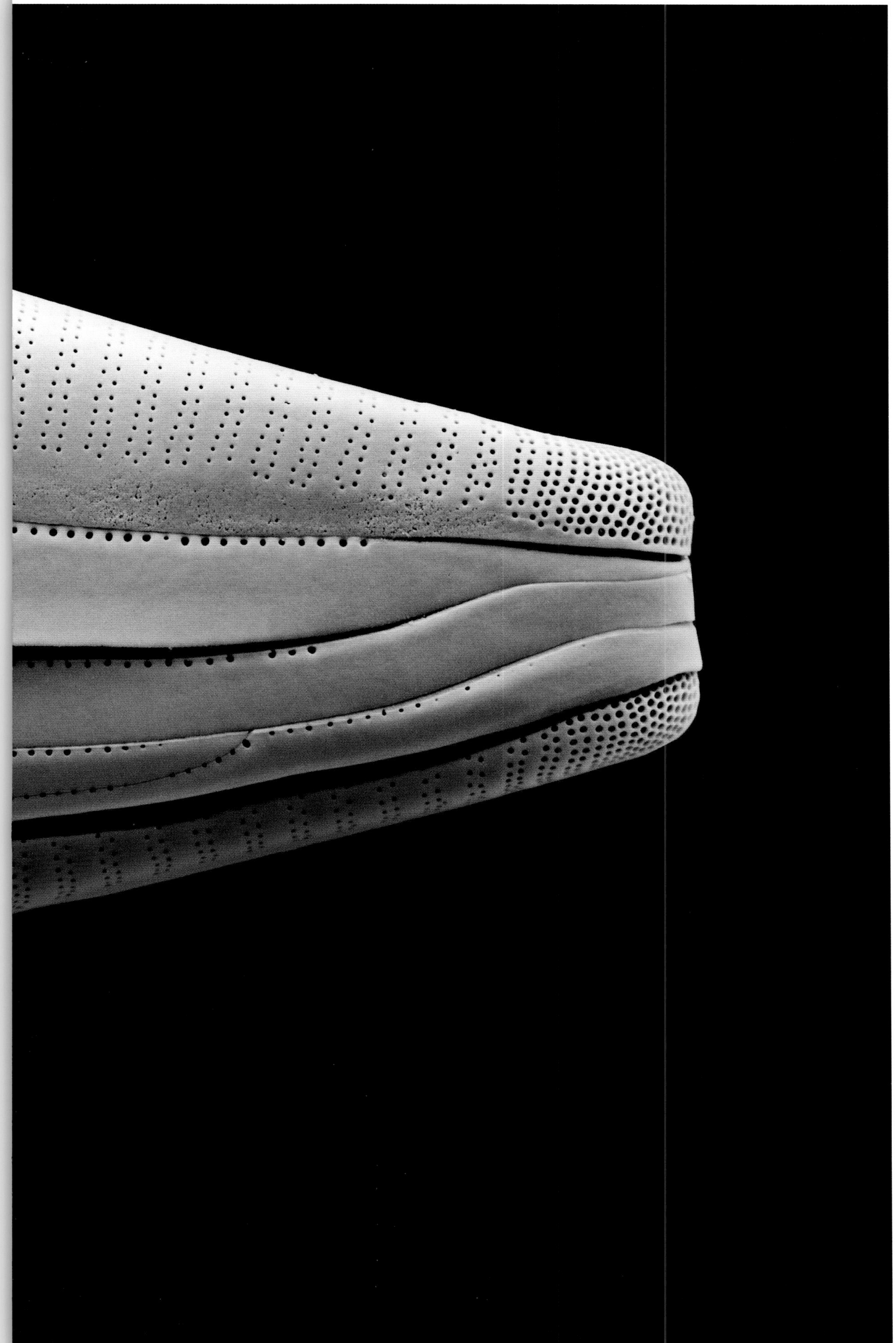

137

**Facing Page**

*Gomphonema* genus diatom, Lake Baikal, Siberia.
6909 x. 55 microns wide, 0.05 millimeter

Facing Page, Pages 140–141 →

*Epithemia adnata*, Wapusk National Park, Hudson Bay.
2250 x. 160 microns from end to end, 0.16 millimeter

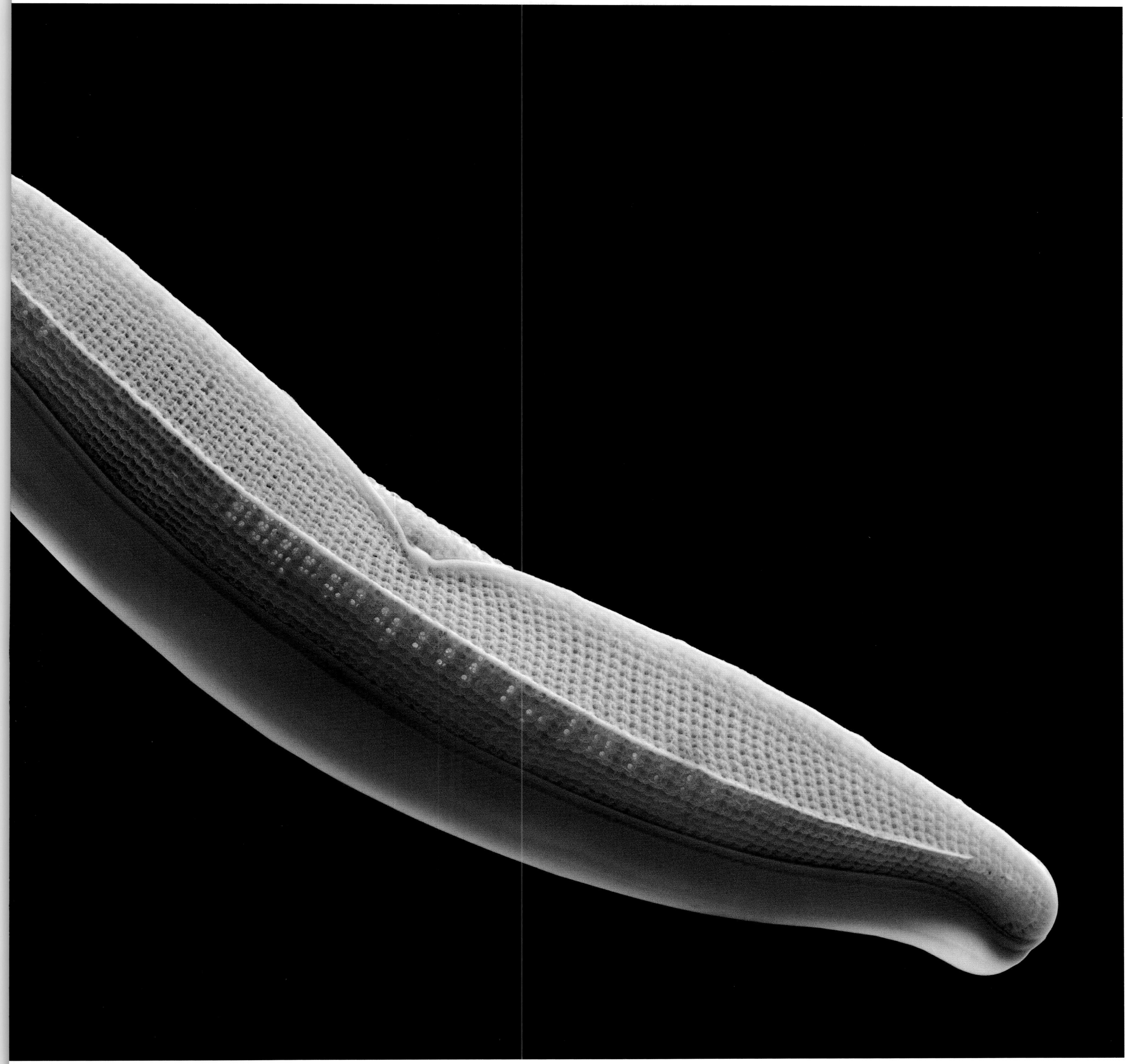

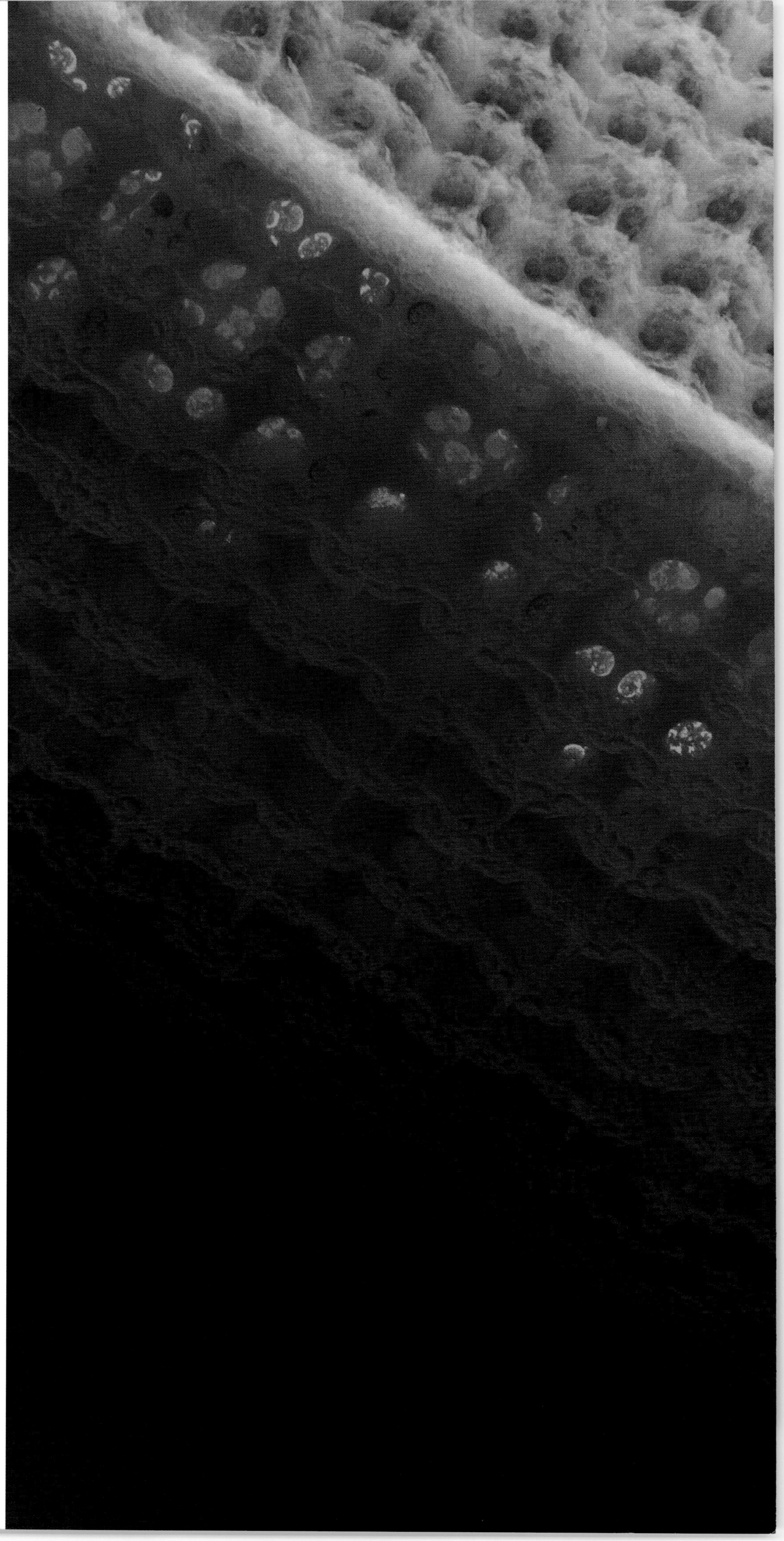

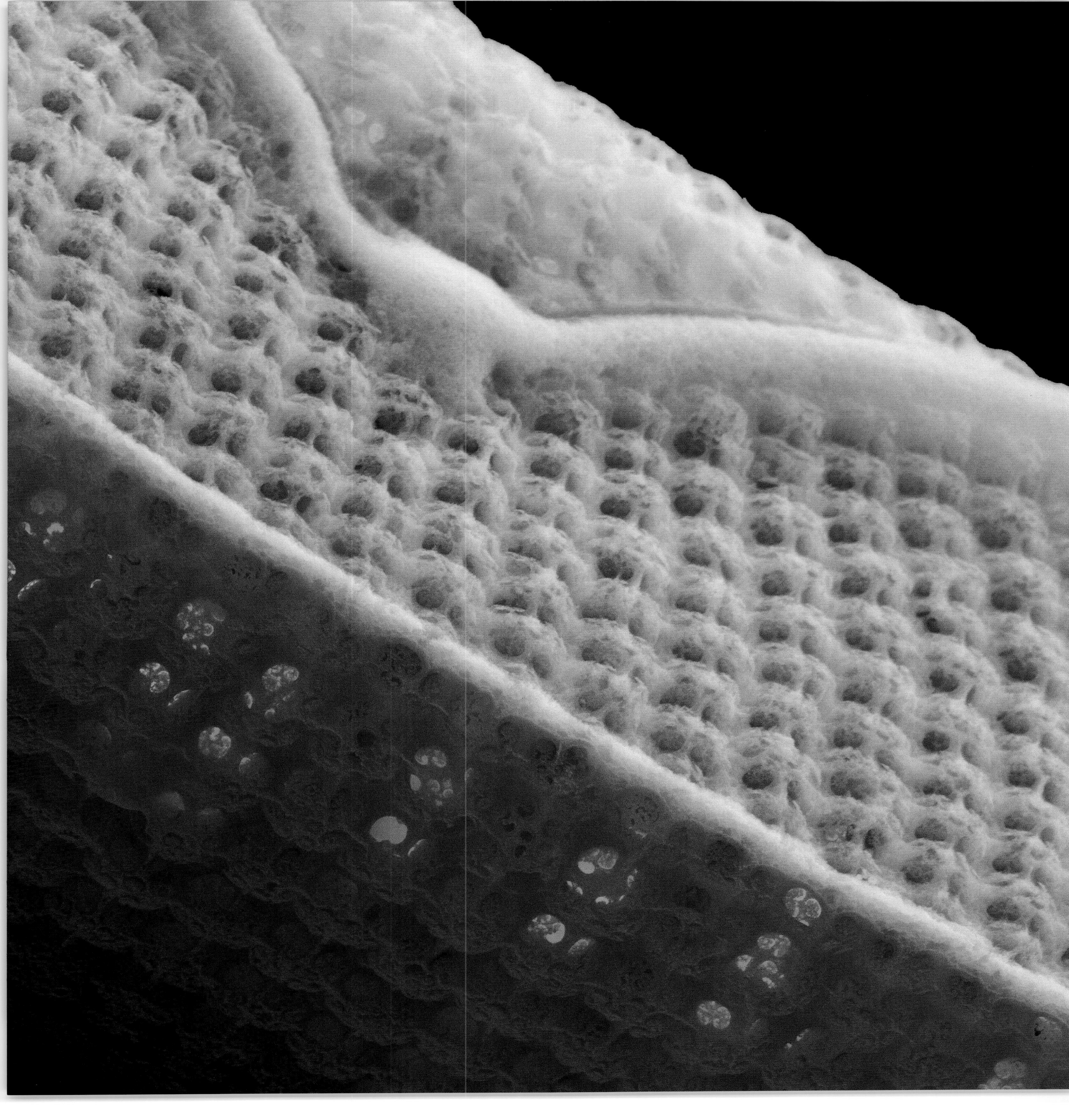

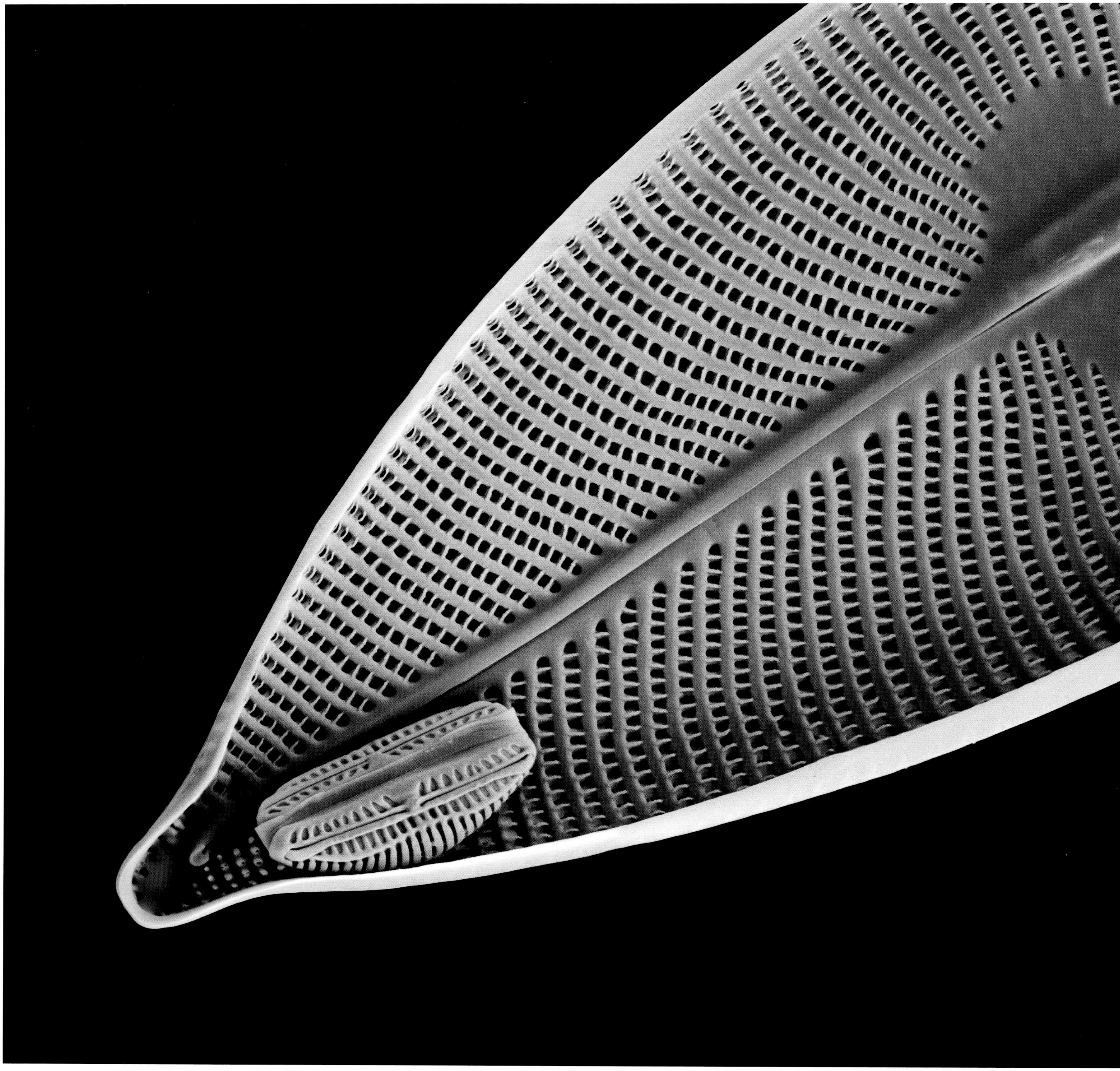

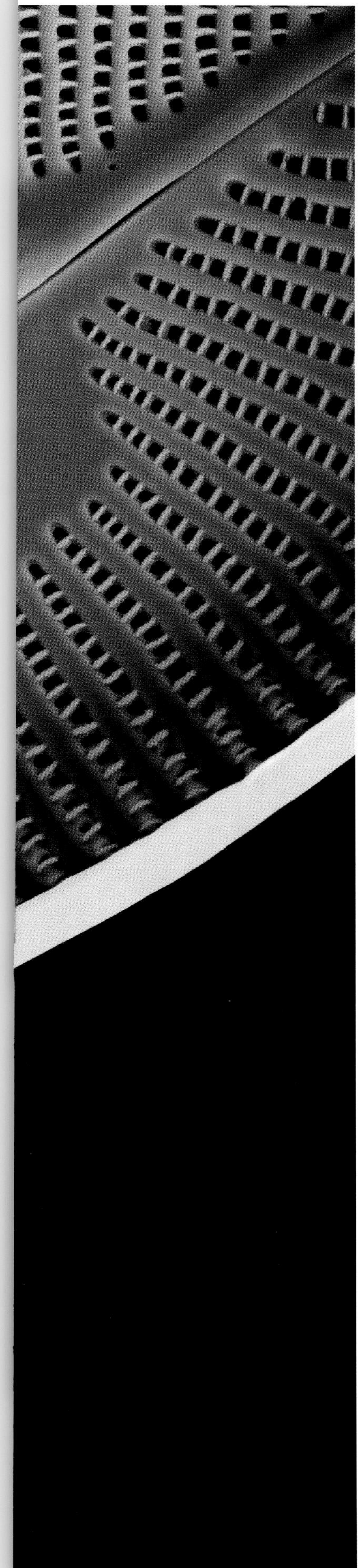



**Facing Page**

*Cymbopleura* genus half shell with *Amphora inariensis*, the smaller diatom, Lake Baikal, Siberia.
3111 x. 45 microns wide from tip to picture corner, 0.04 millimeter

**Pages 144–145 →**

*Denticula kuetzingii* half shell, Wapusk National Park, Hudson Bay.
19,200 x. 25 microns wide, 0.02 millimeter

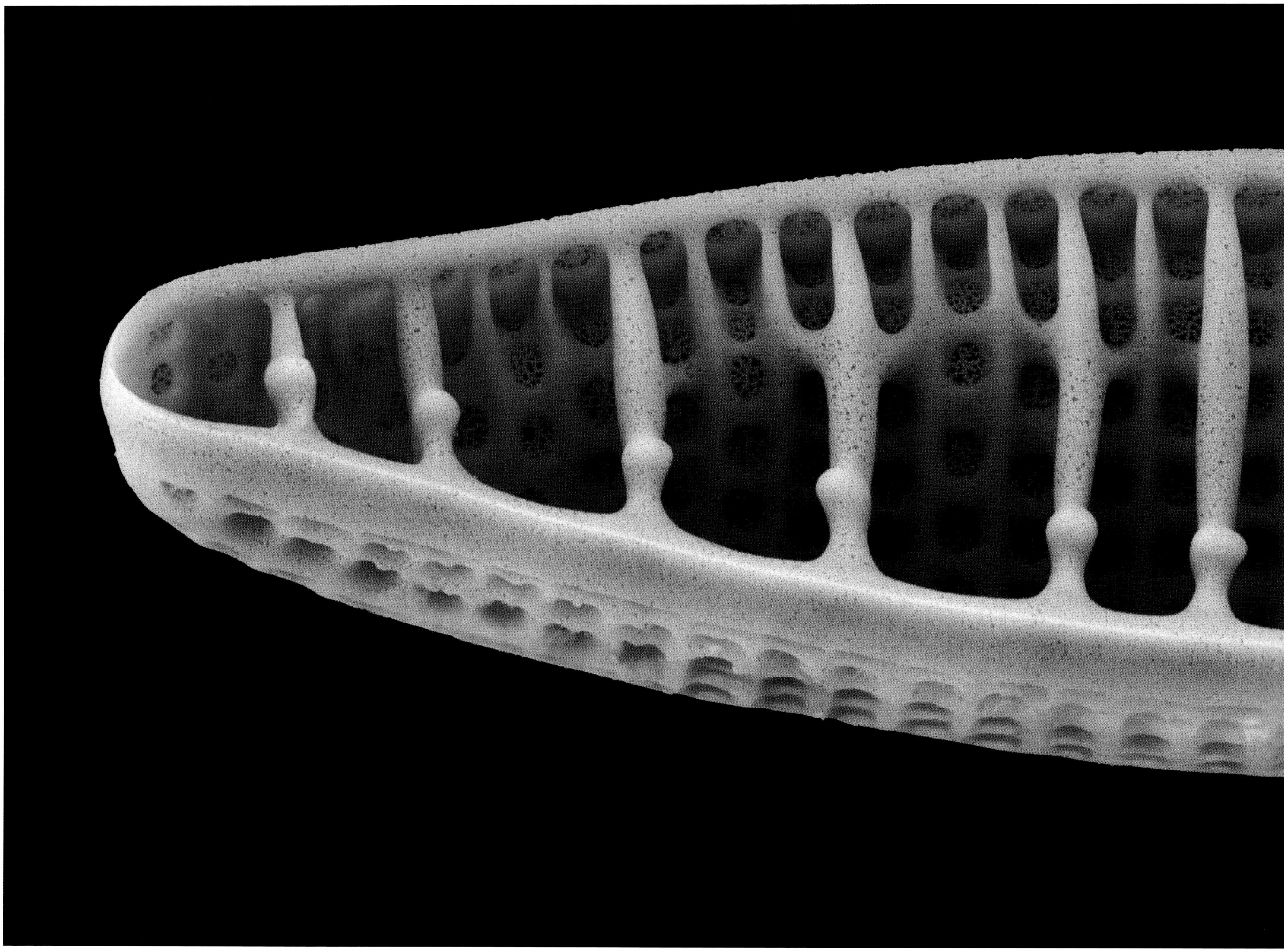

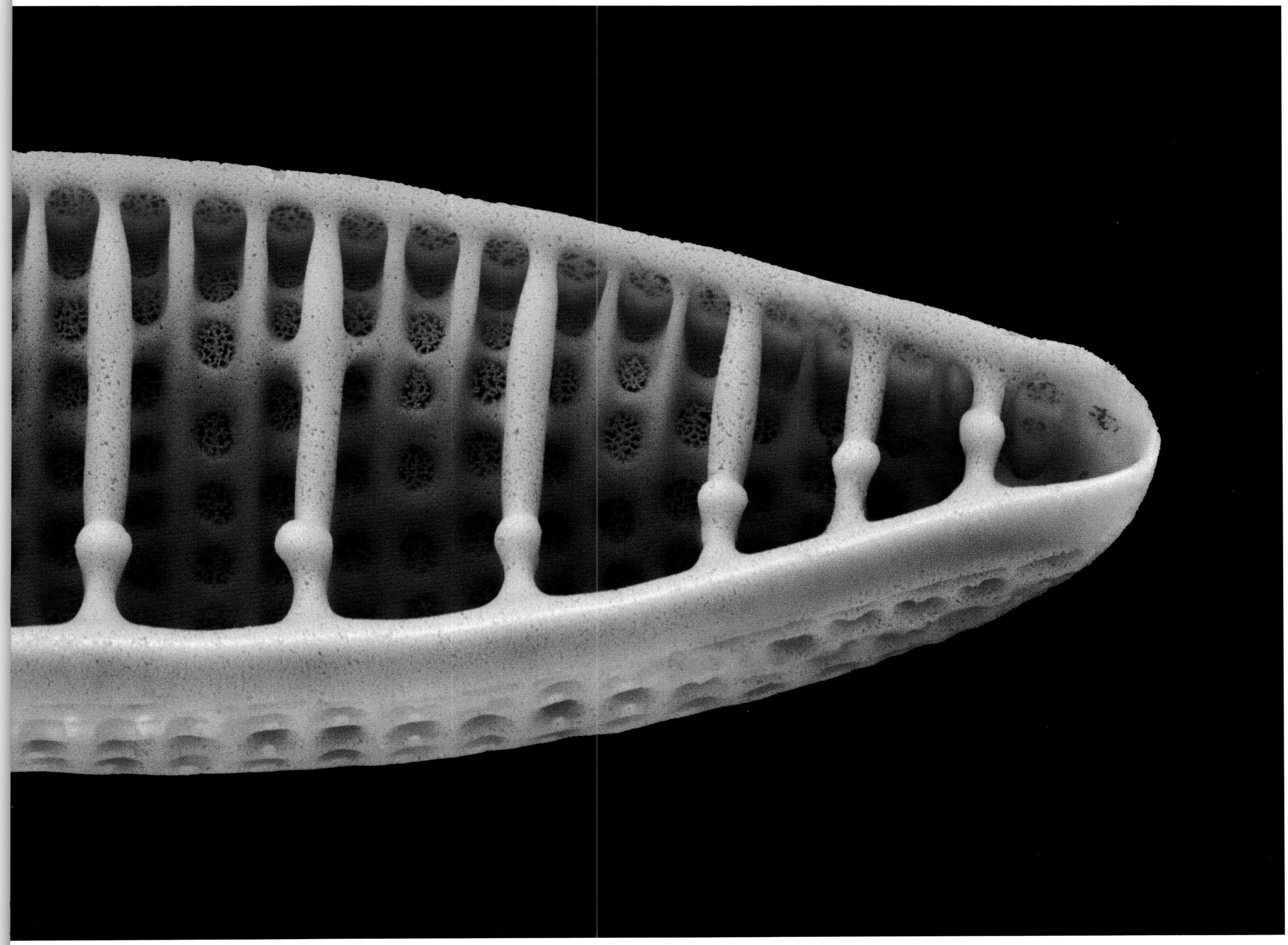

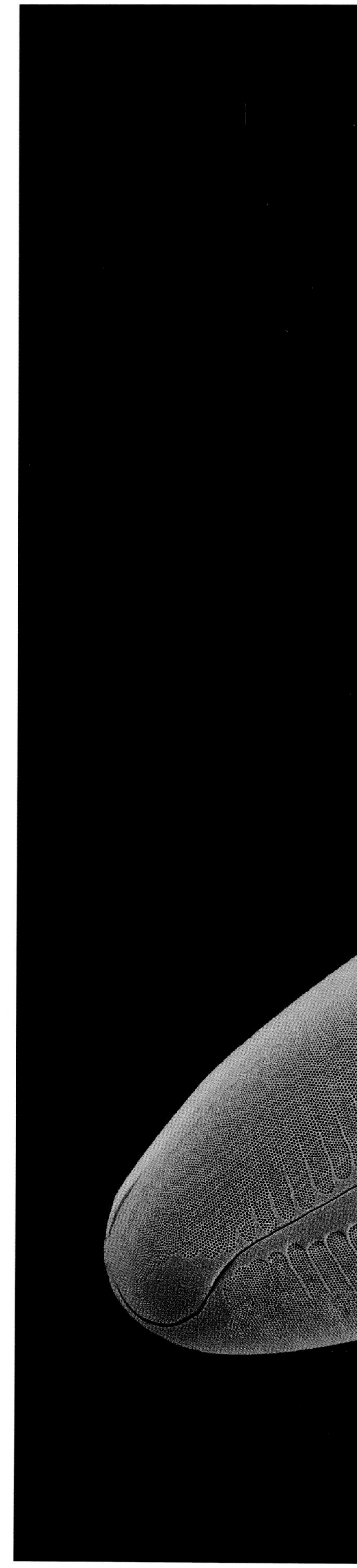

**Facing Page, page 148–149**

*Pinnularia neomajor*, Wapusk National Park, Hudson Bay.
1944 x. 180 microns long, 0.18 millimeter

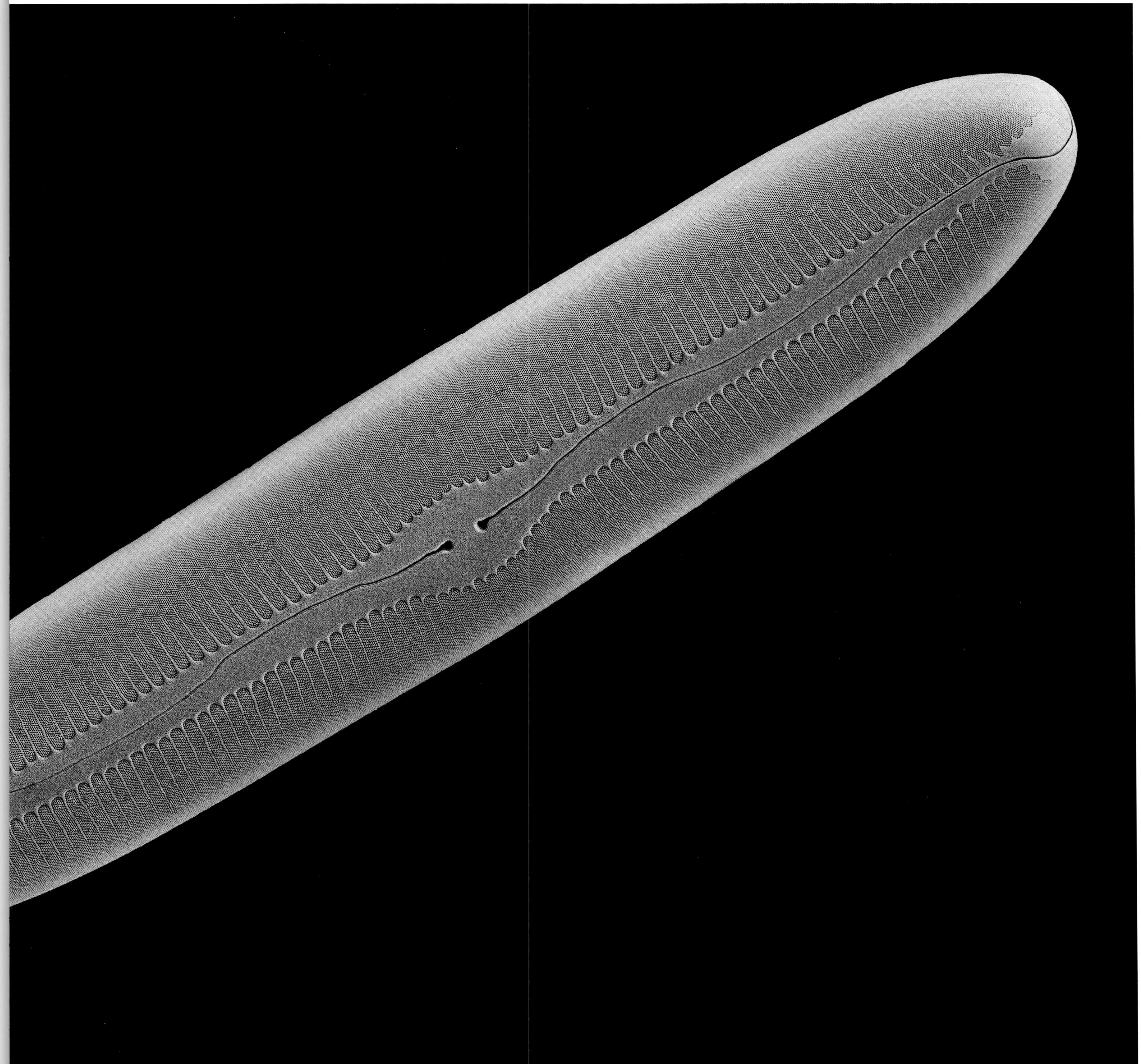

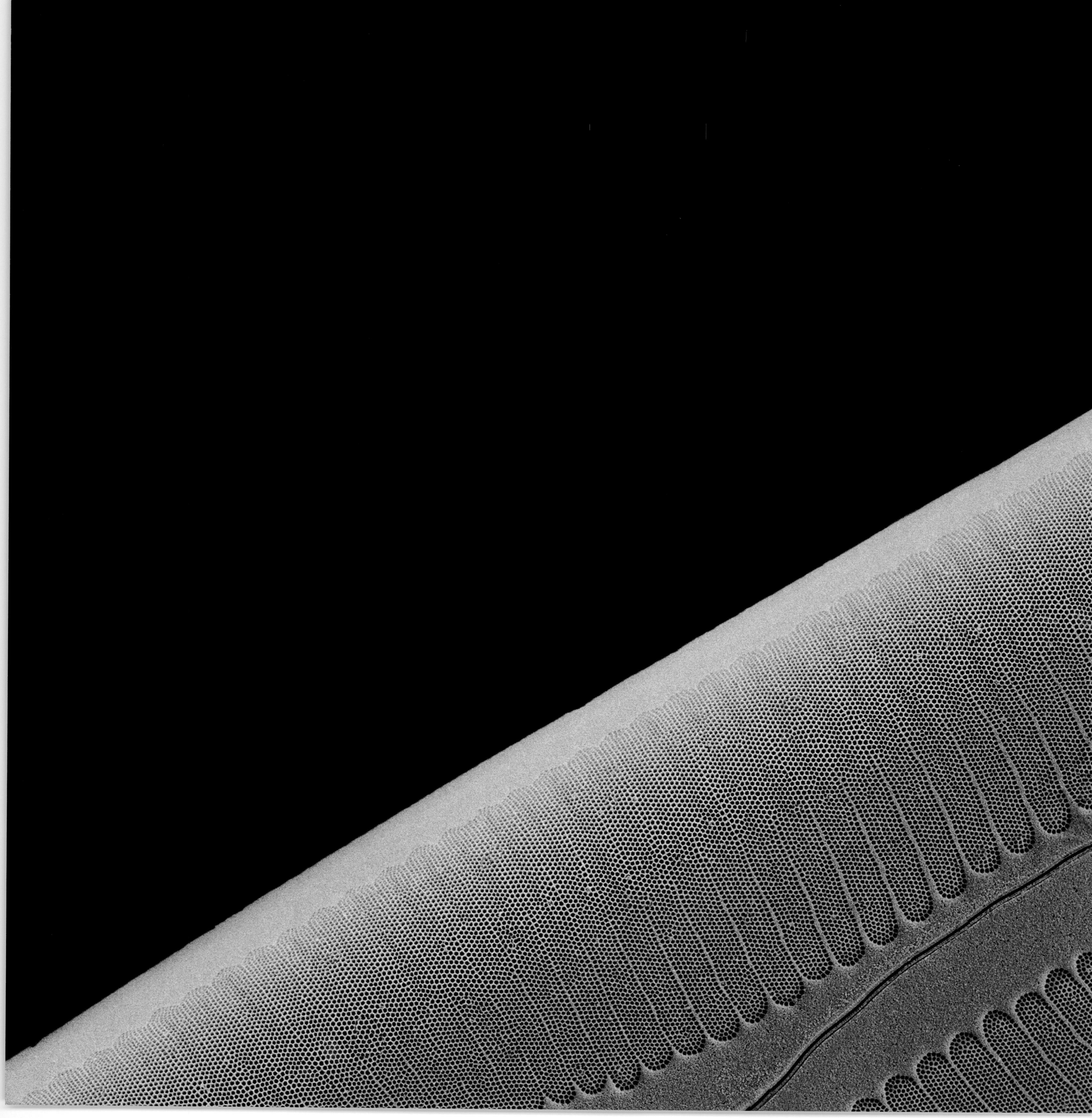

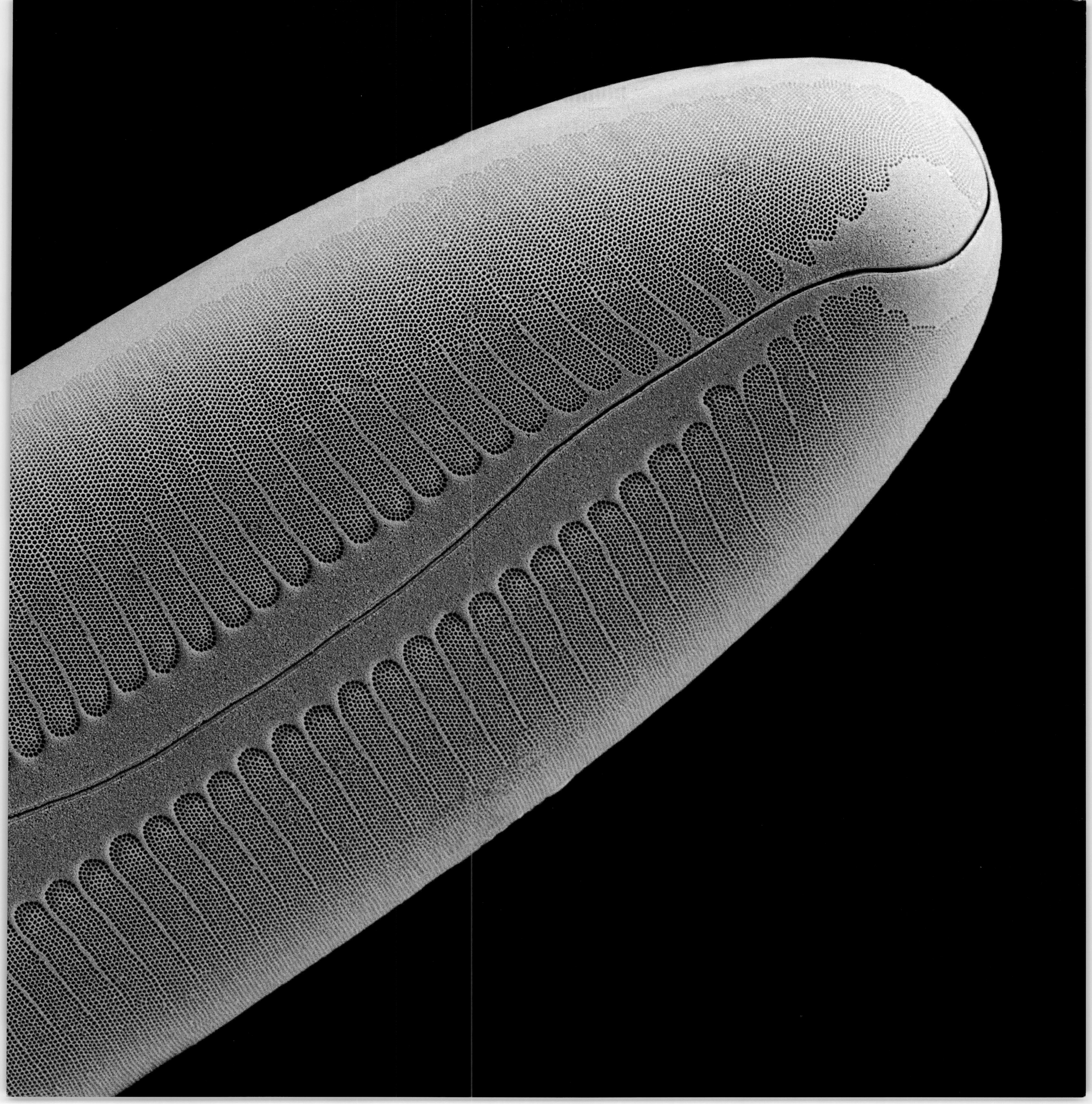

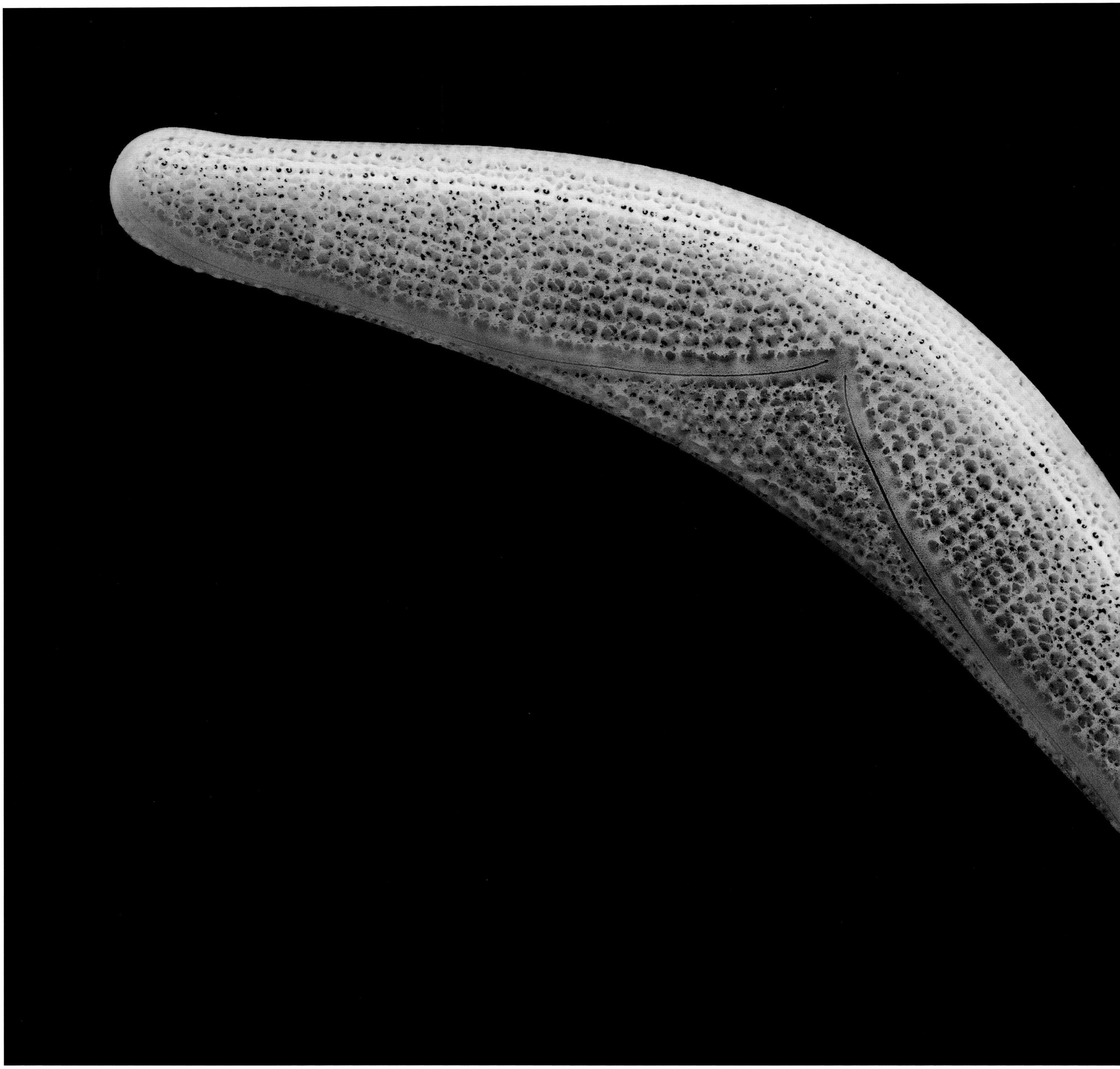

**Facing Page**

*Epithemia sorex var. gracilis*, Wapusk National Park, Hudson Bay.
4125 x. 80 microns from end to end, 0.08 millimeter

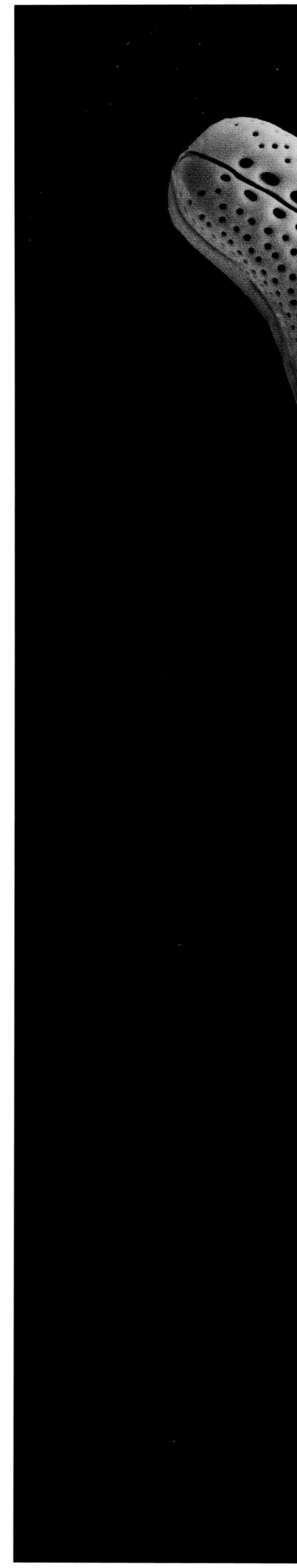

**Facing Page**

*Aneumastus tusculus*, Wapusk National Park, Hudson Bay.
4124 x. 80 microns long, 0.08 millimeter

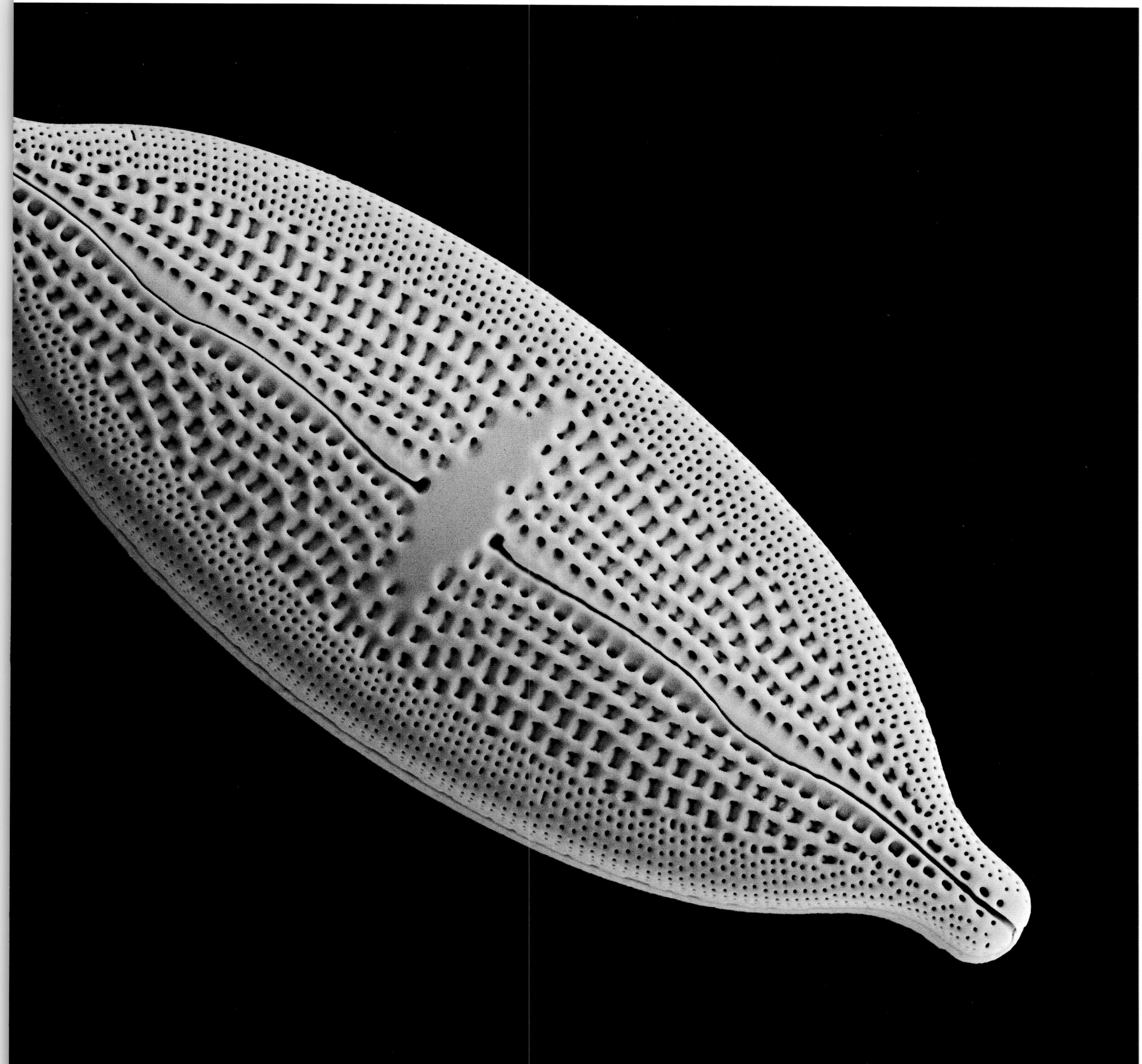

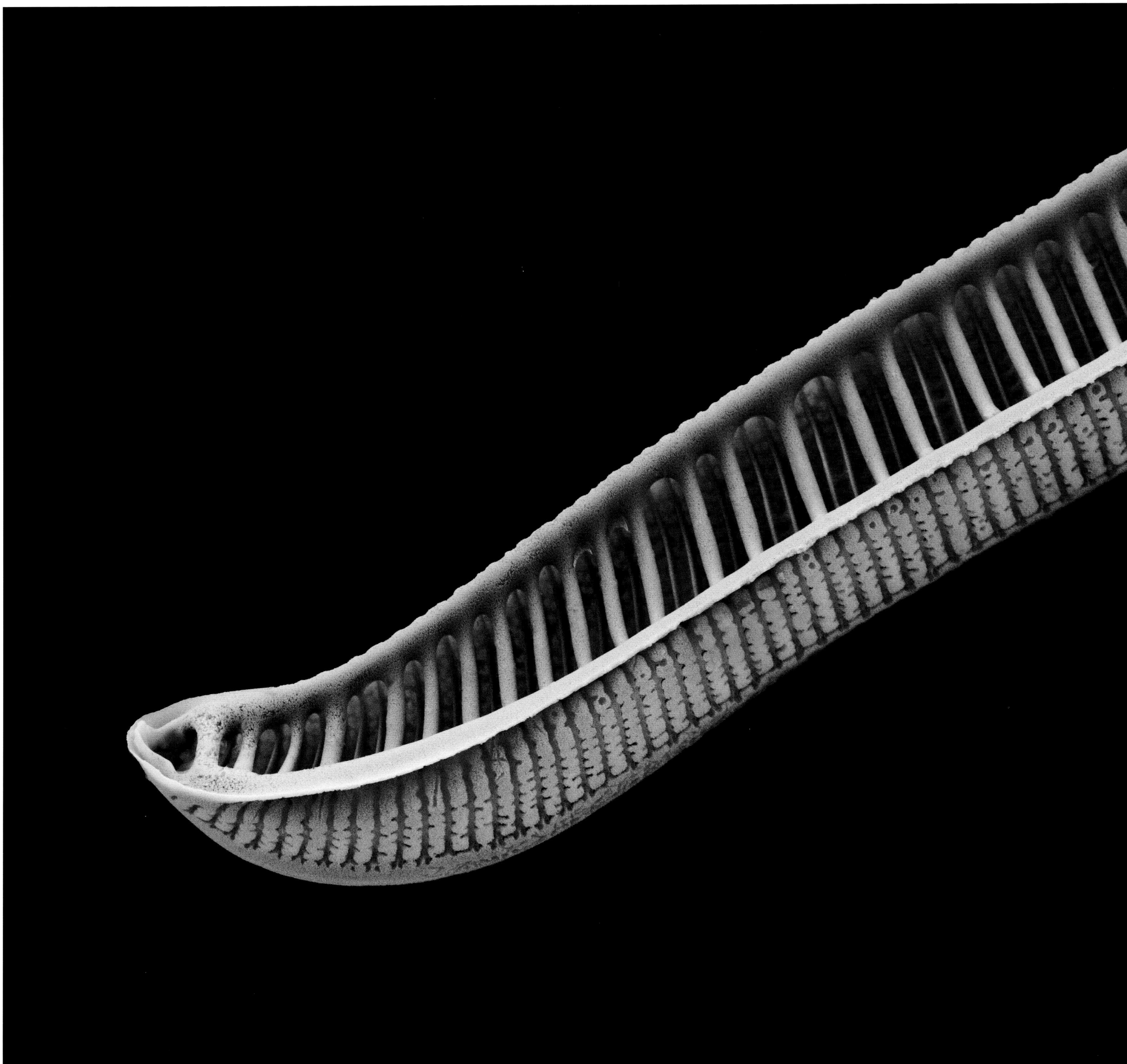

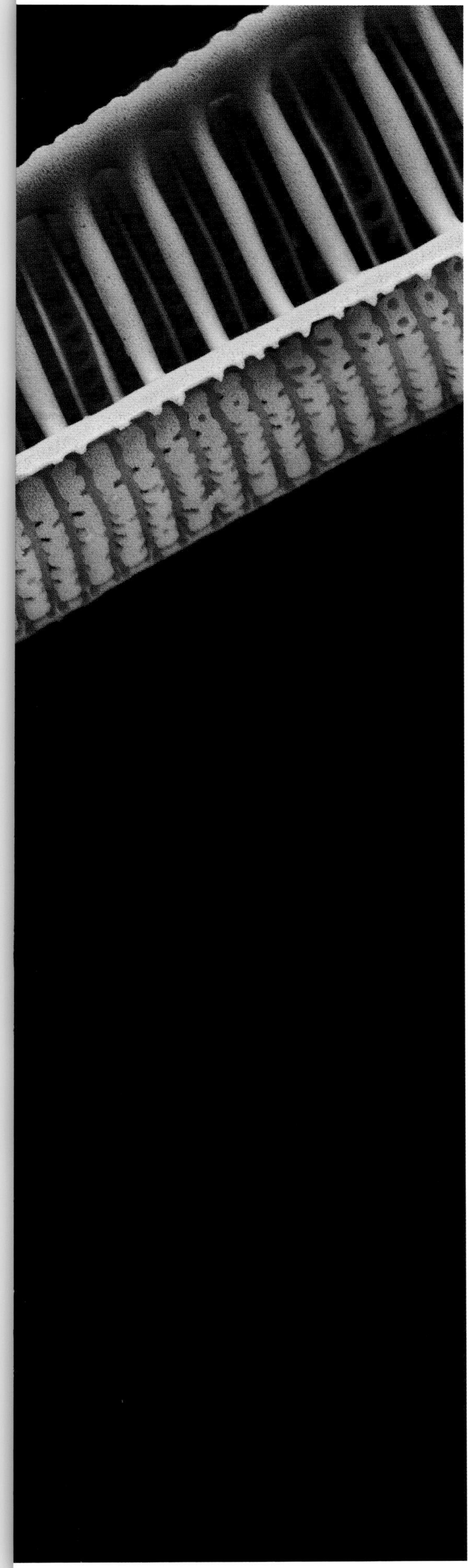



**Facing Page**

*Rhopalodia gibba* half shell, Wapusk National Park, Hudson Bay. Also see pages 124–125.
8000 x. About 50 microns from tip to picture corner, 0.05 millimeter

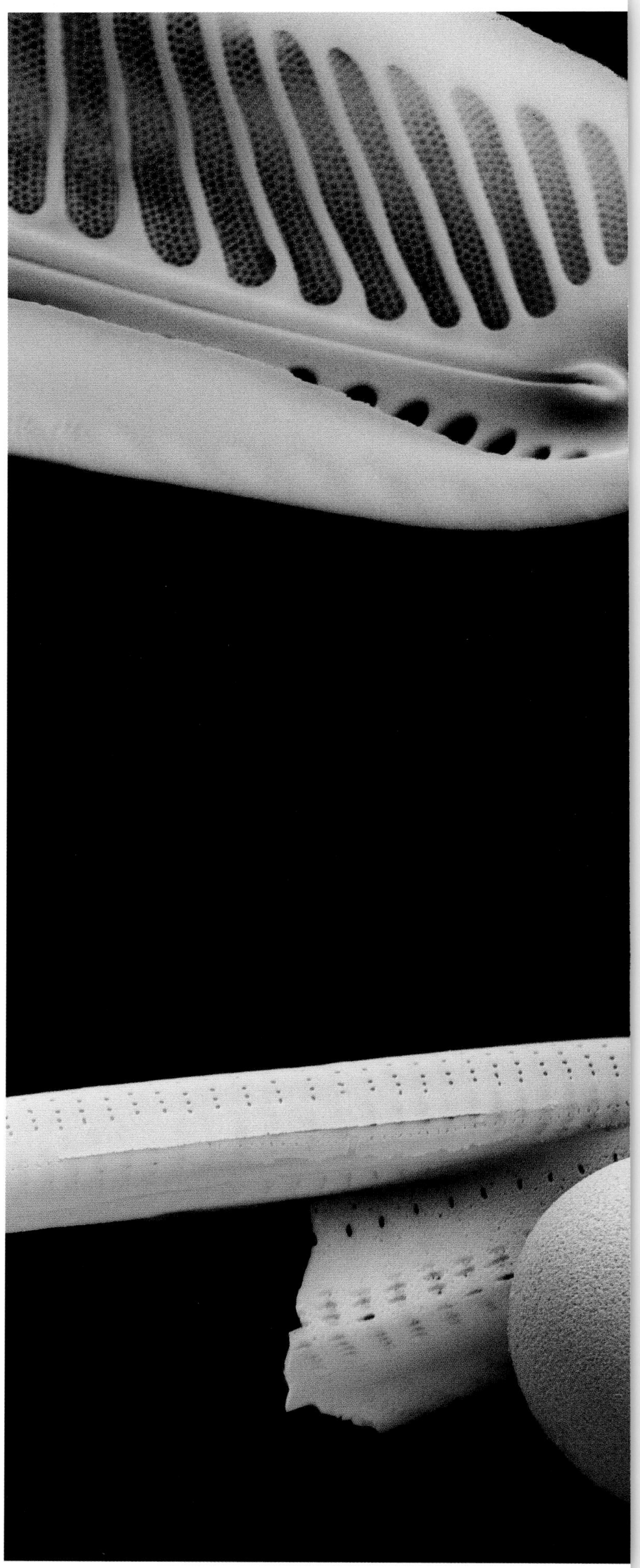

**Facing Page**

Jumbled diatoms and a single chrysophyte cyst, Wapusk National Park, Hudson Bay, including fragments of the genera *Cymbella*, *Pinnularia*, *Navicula*, and *Kobayasiella*. 8222 x. About 45 microns wide, 0.04 millimeter

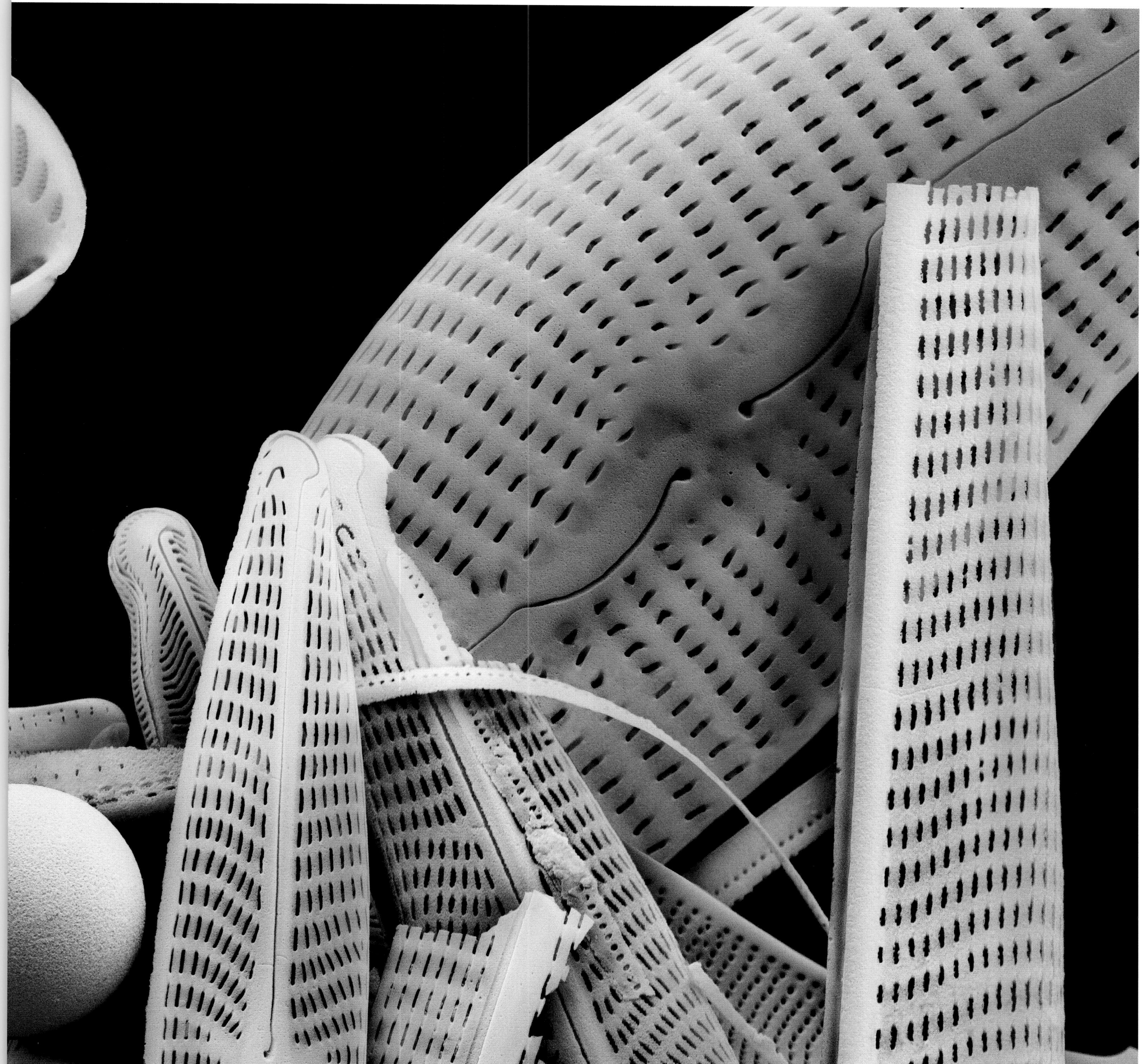

AQUA

# DINOFLAGELLATES

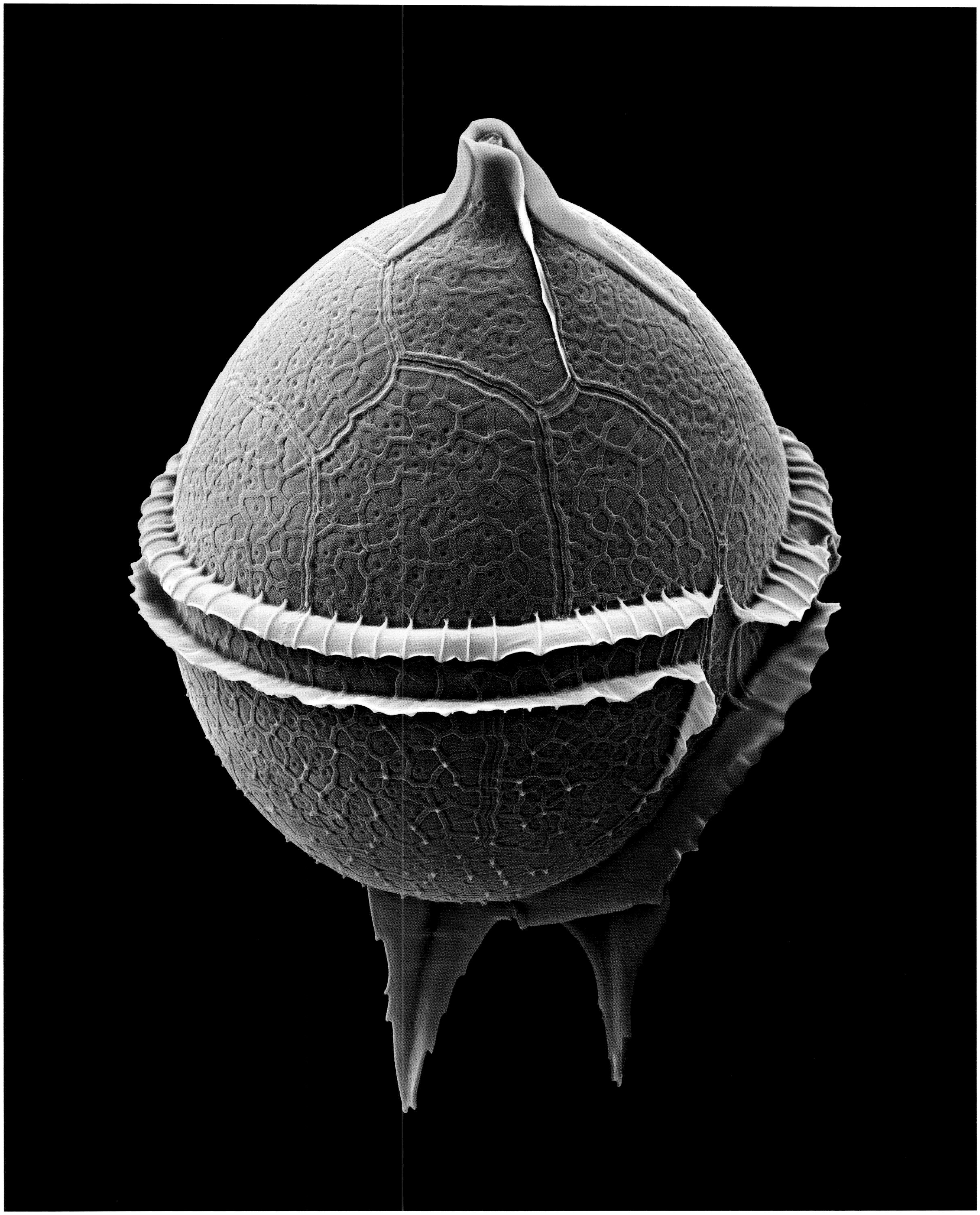

← Page 159

*Protoperidinium michaelis*, Carrie Bow Cay, Belize.
4844 x. About 45 microns from top to bottom, 0.04 millimeter

Facing Page, Pages 162–163 →

*Ceratocorys horrida*, Carrie Bow Cay, Belize.
1558 x. 120 microns from top to bottom, 0.12 millimeter

Page 164 →

*Ceratocorys horrida*, Florida Gulf Stream.
1667 X. 120 microns from top to bottom, 0.12 millimeter

Page 165 →

*Dinophysis cuneus*, Florida Gulf Stream.
1367 x. About 150 microns across the top, 0.15 millimeter

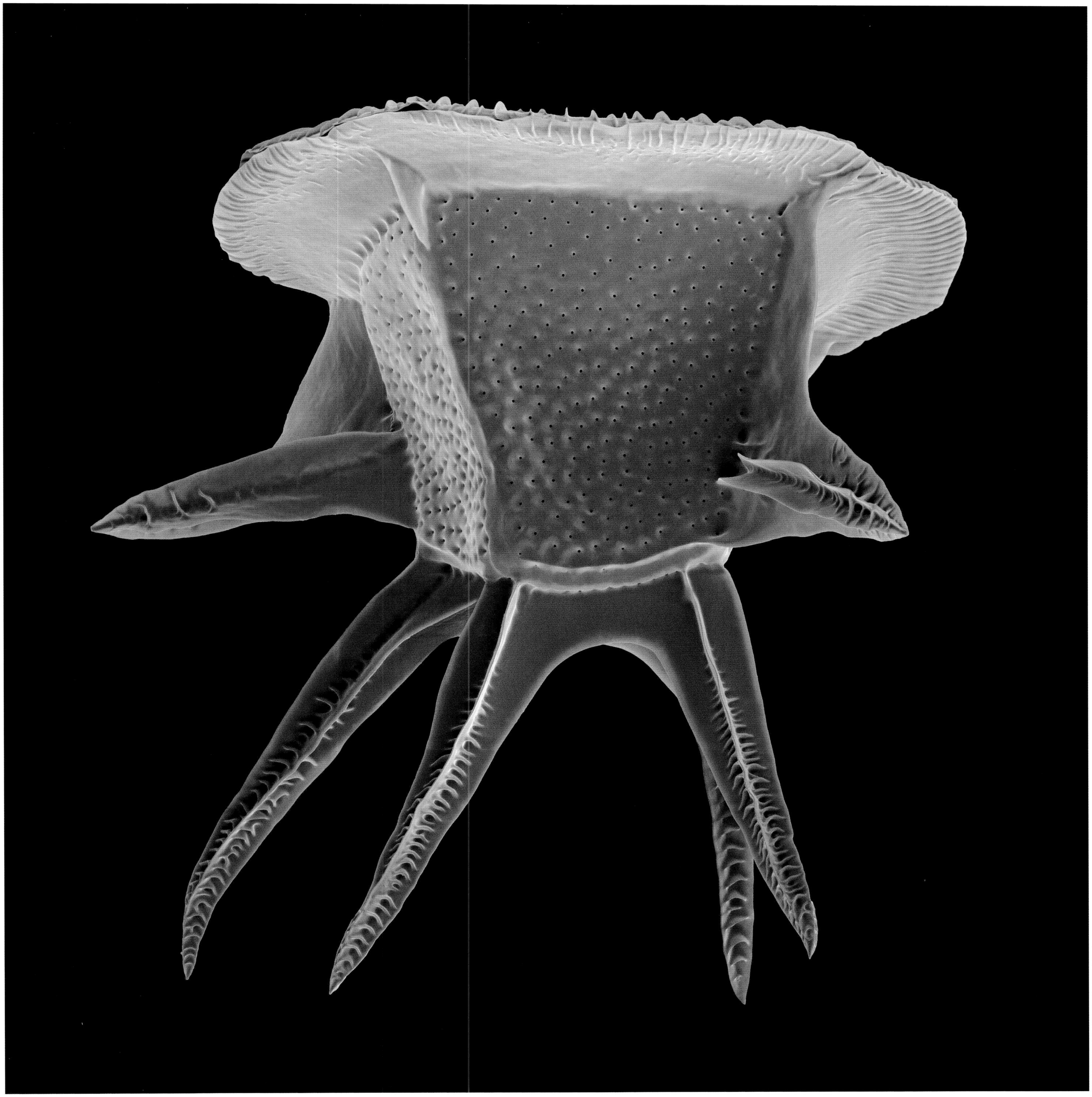

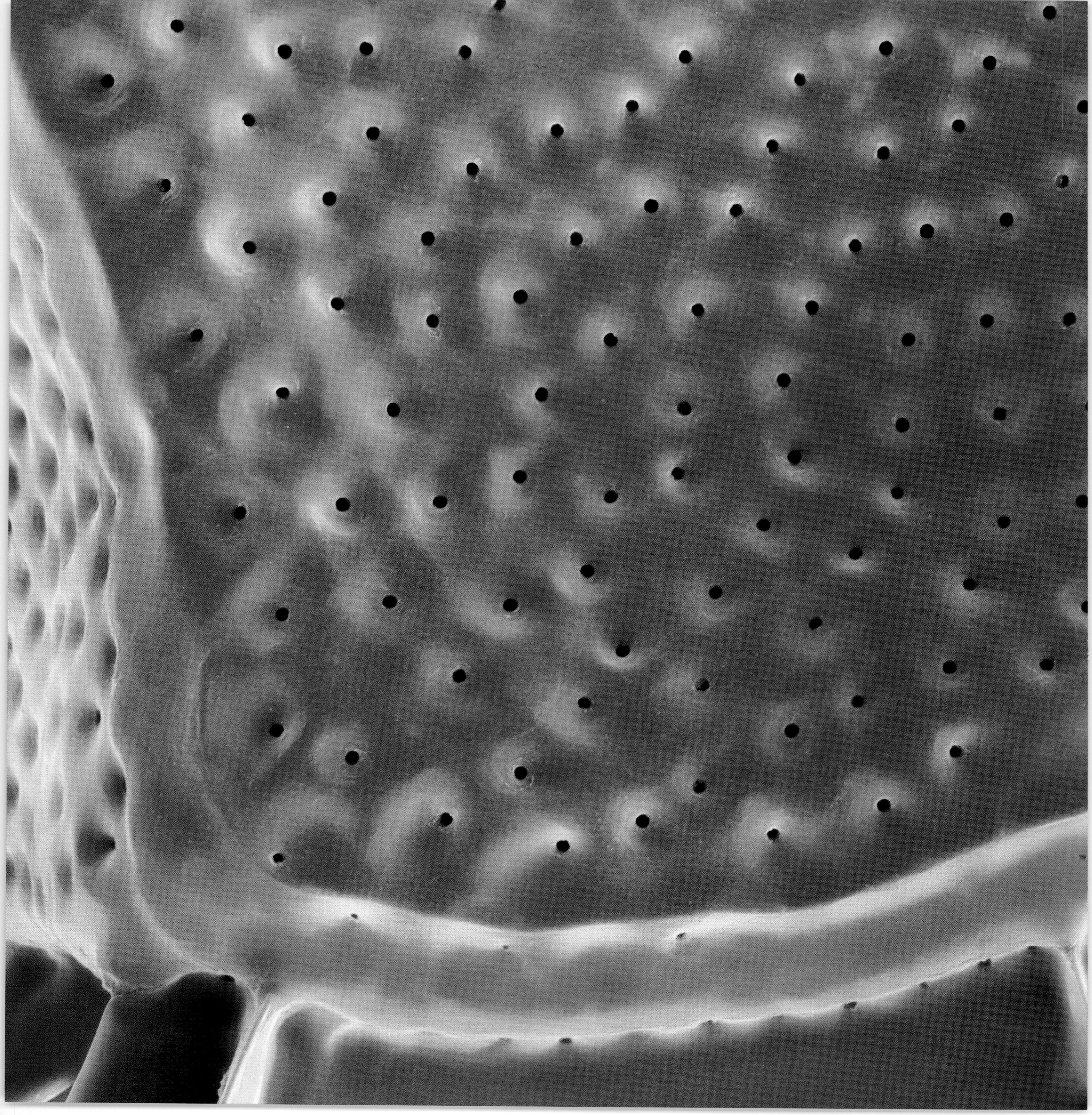

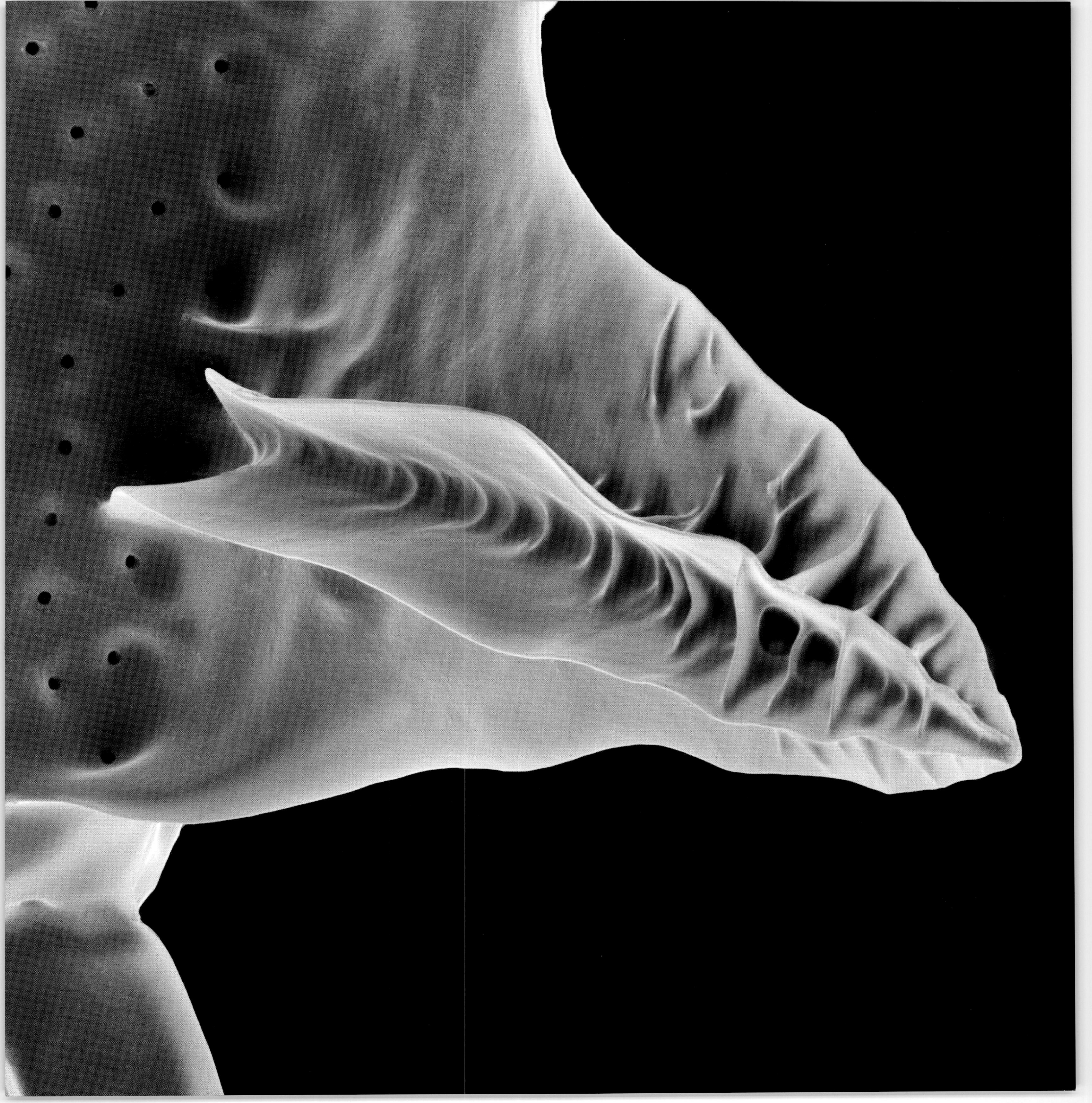

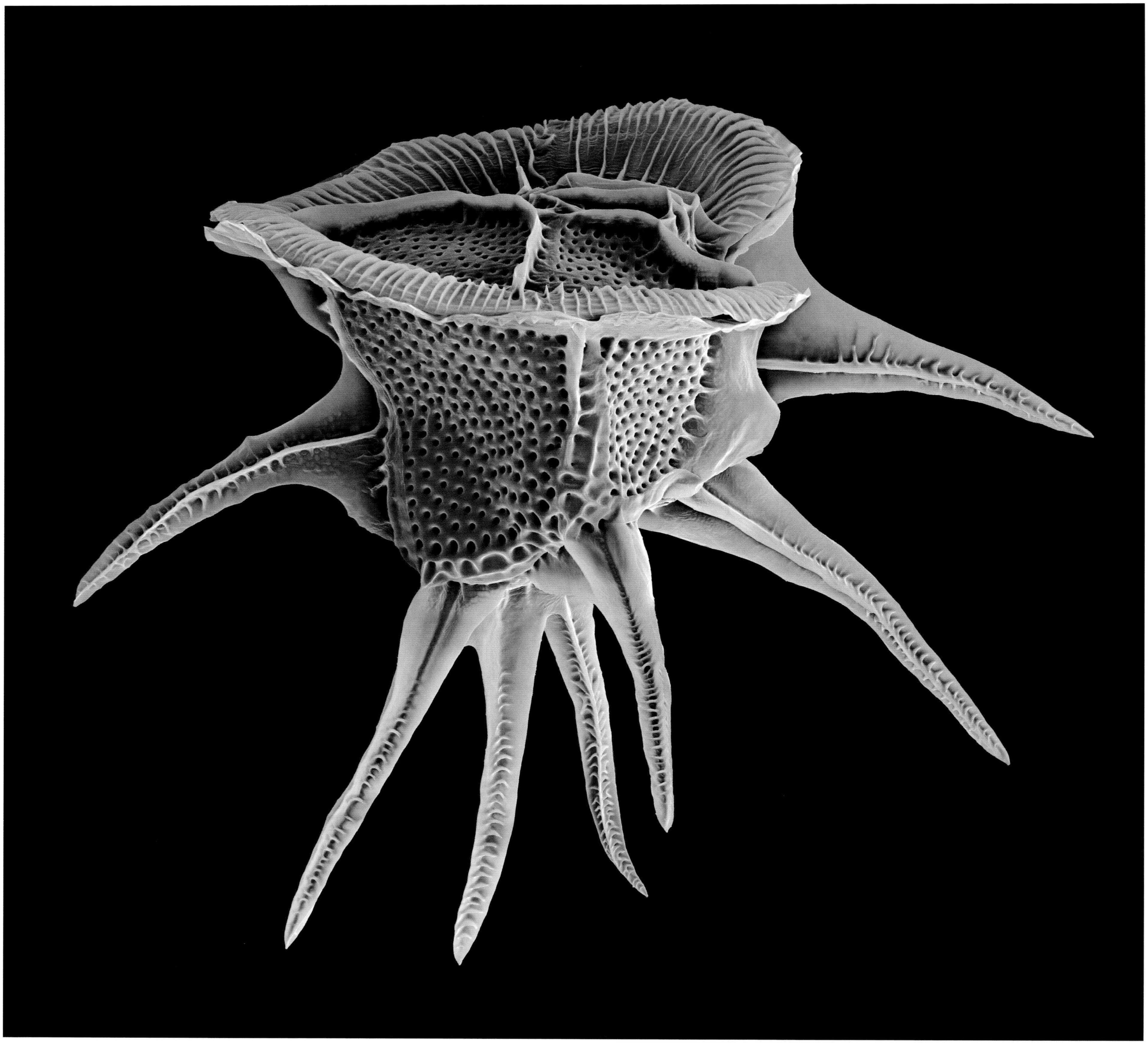

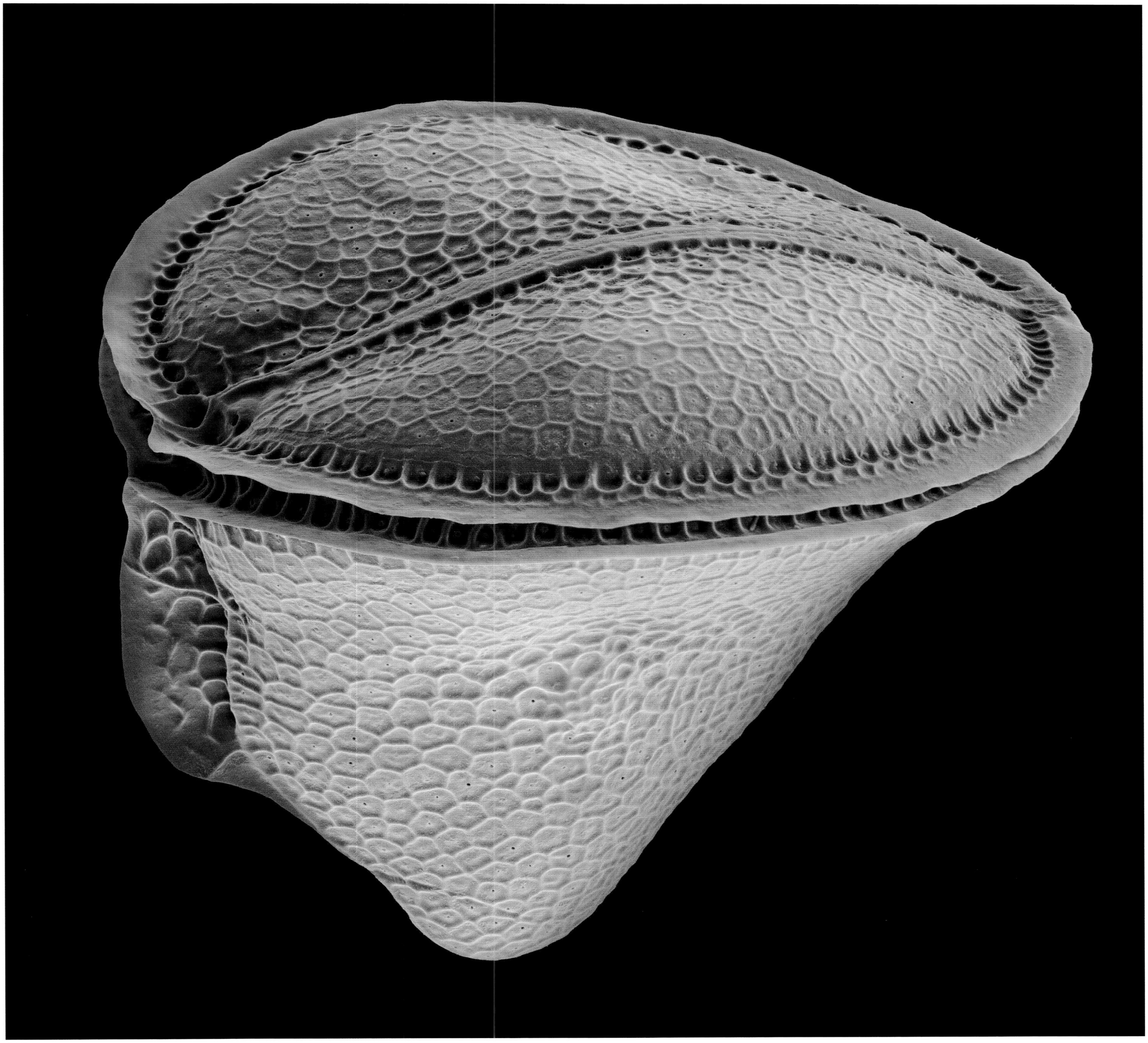

**Facing Page, Pages 168–169 →**

*Protoperidinium brochii*, Carrie Bow Cay, Belize.
2444 x. 90 microns from top to bottom, 0.09 millimeter

**Page 170 →**

*Lophodinium* genus dinoflagellate, North Carolina Gulf Stream.
3143 x. 70 microns from top to bottom, 0.07 millimeter

**Page 171 →**

*Oxytoxum* genus dinoflagellate, Florida Gulf Stream.
2555 x. About 90 microns from top to bottom, 0.09 millimeter

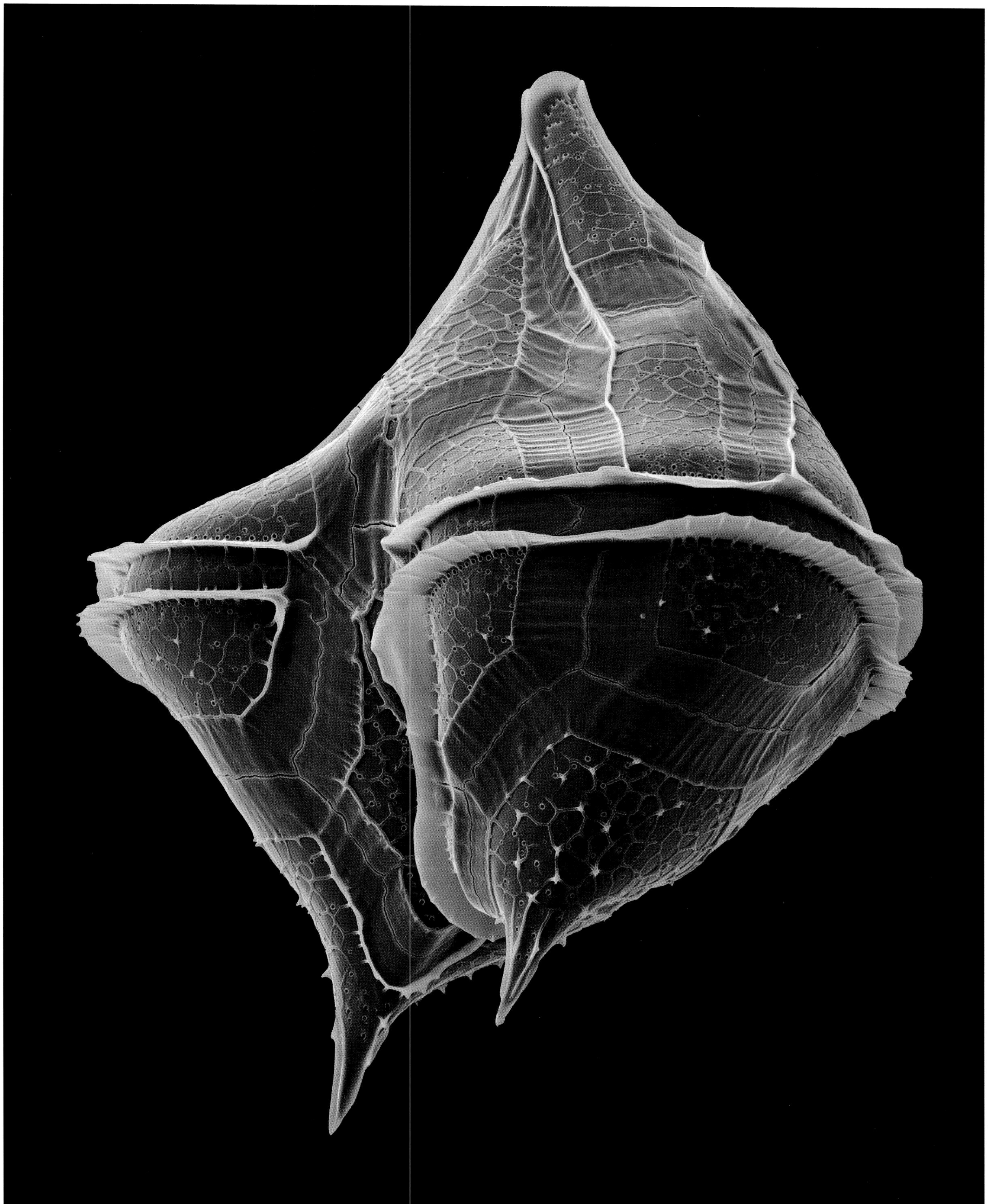

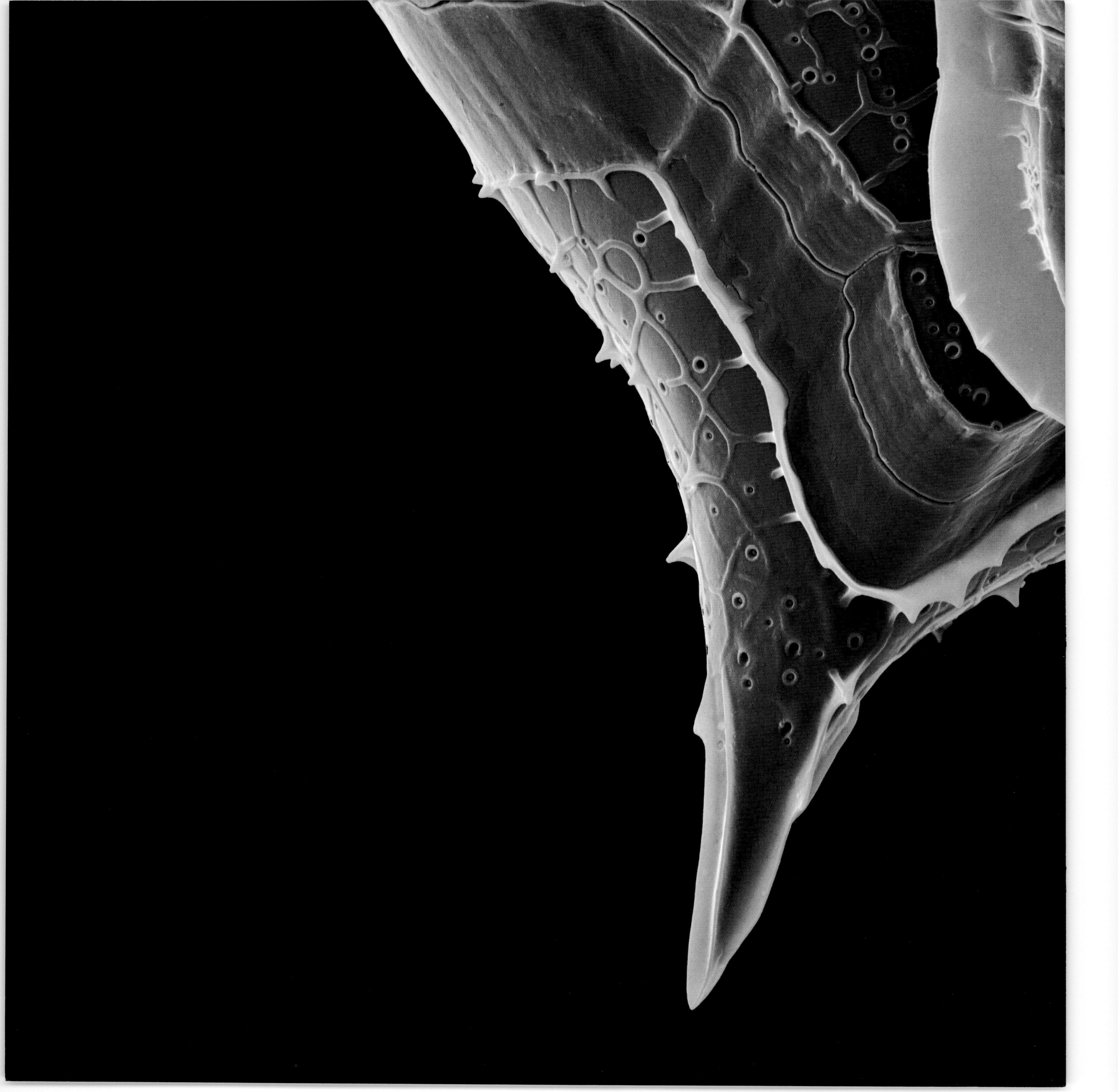

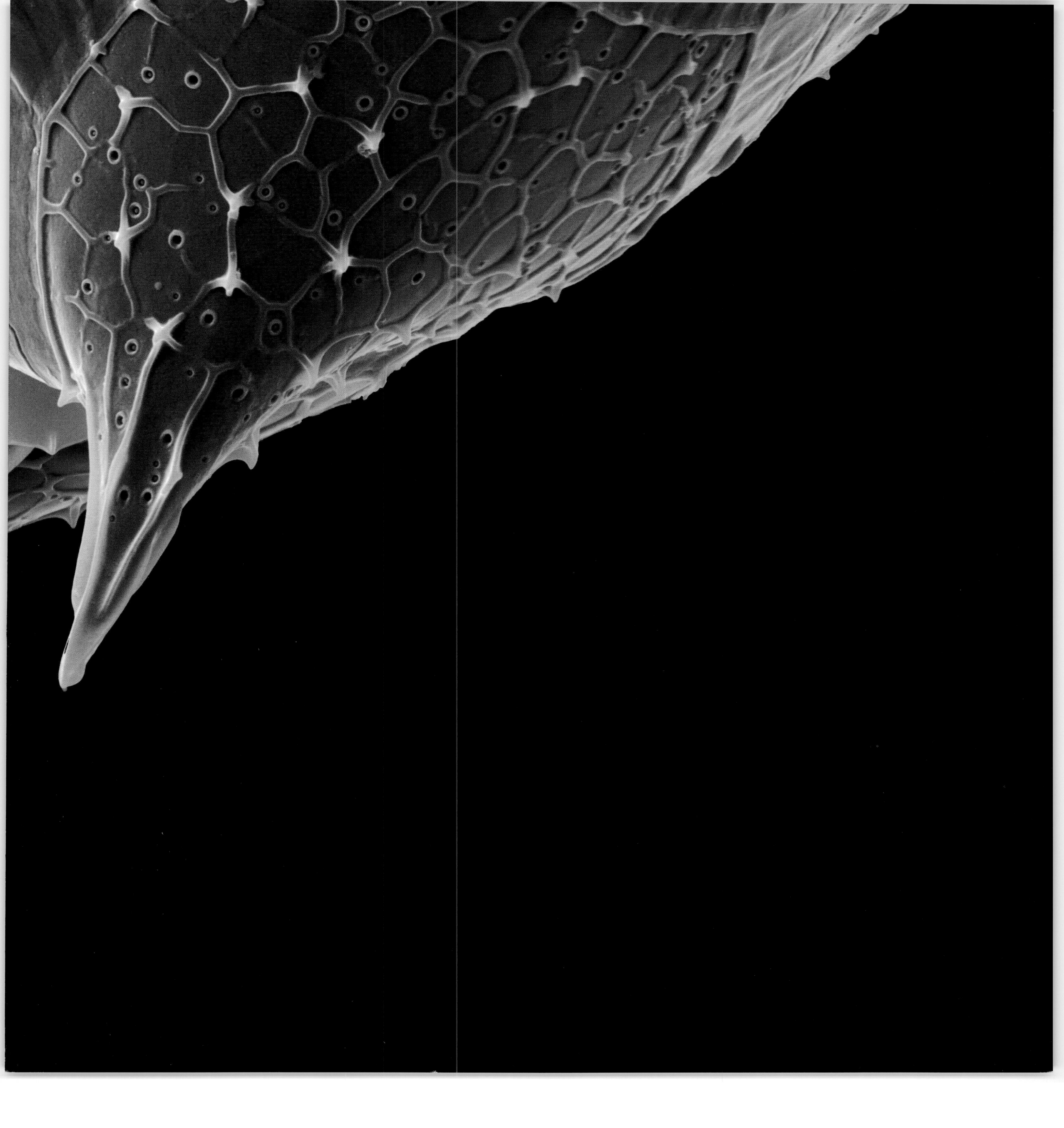

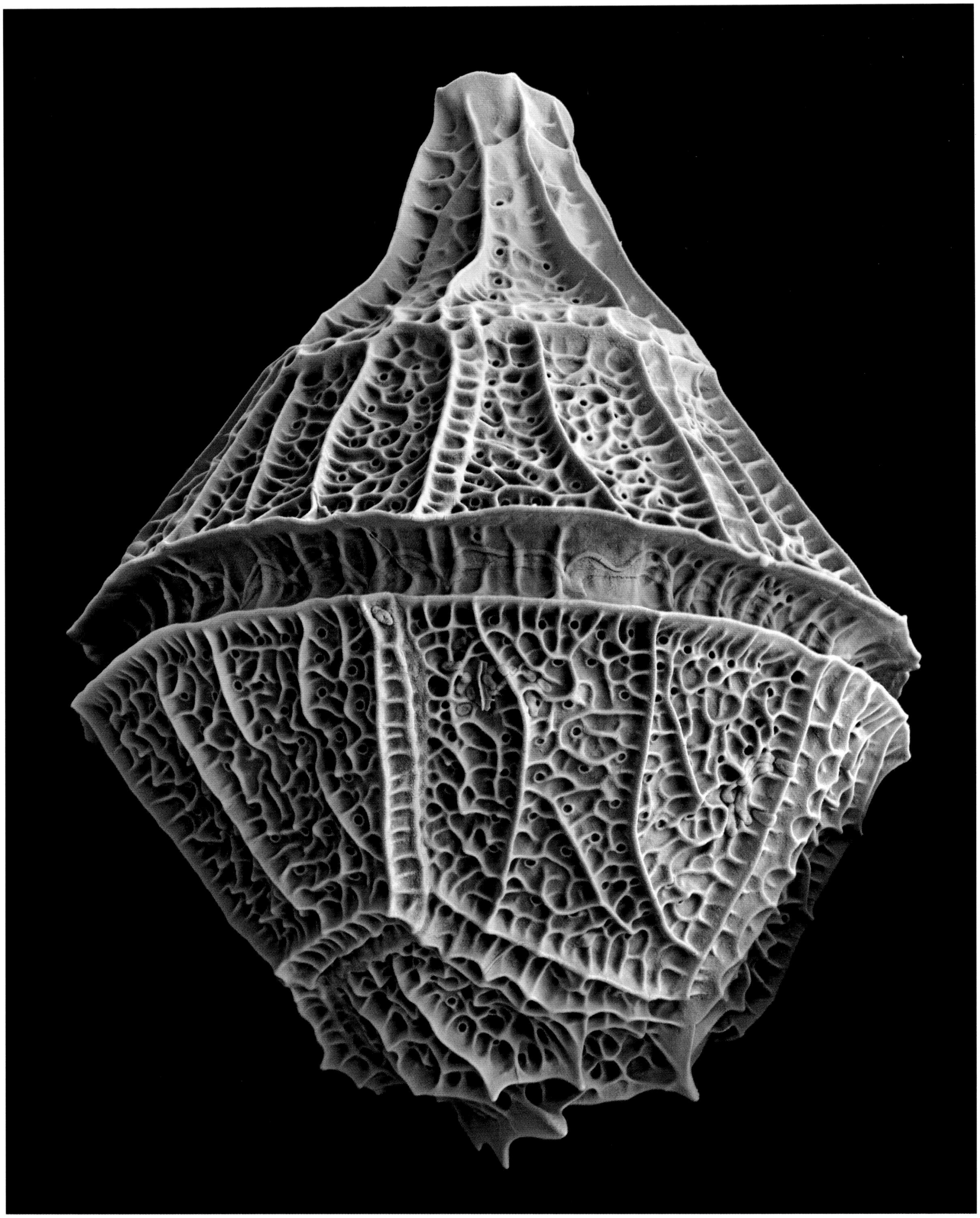

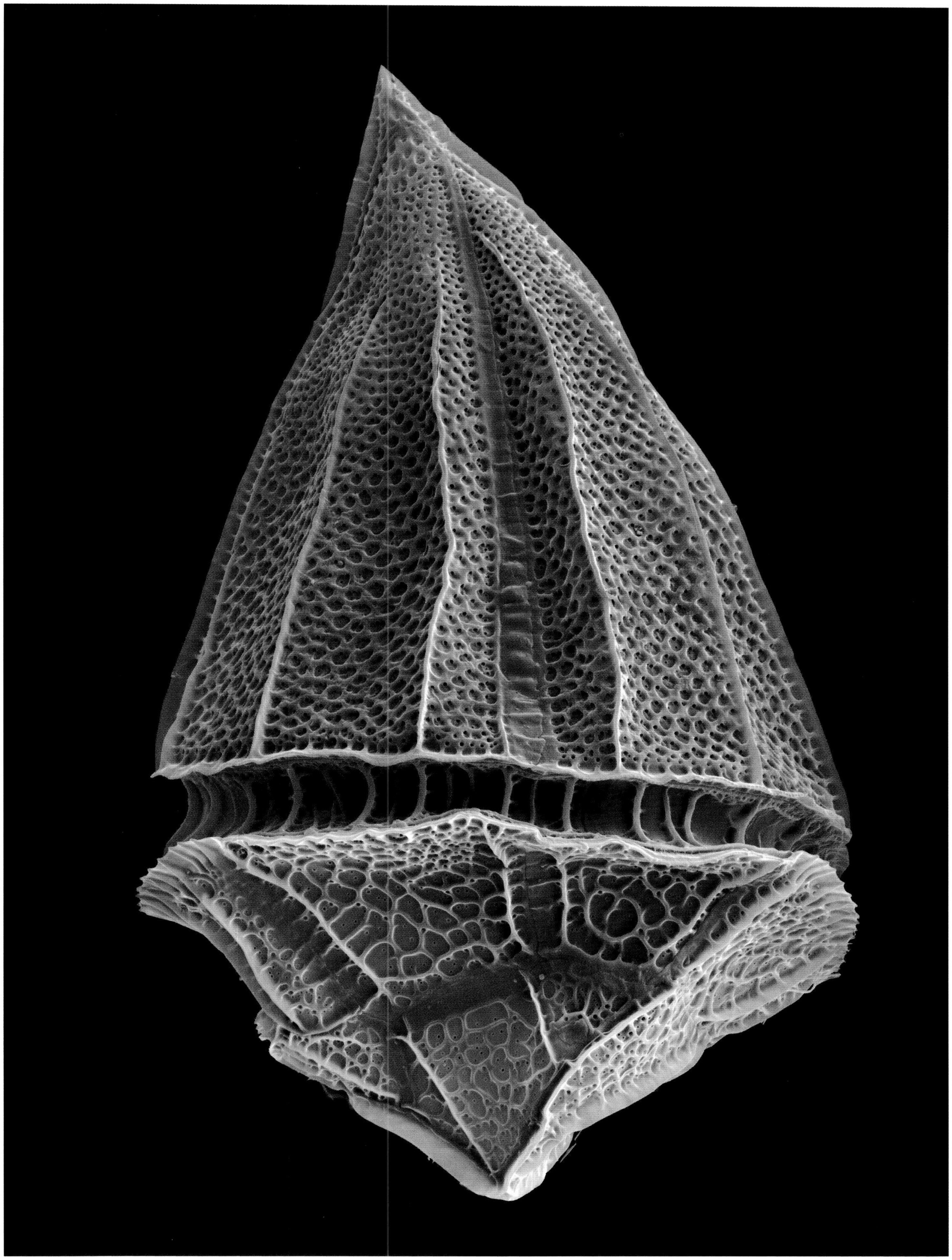

172

Facing Page, Pages 174–175 →

*Protoperidinium grande*, North Carolina Gulf Stream.
1800 x. About 100 microns wide, 0.1 millimeter

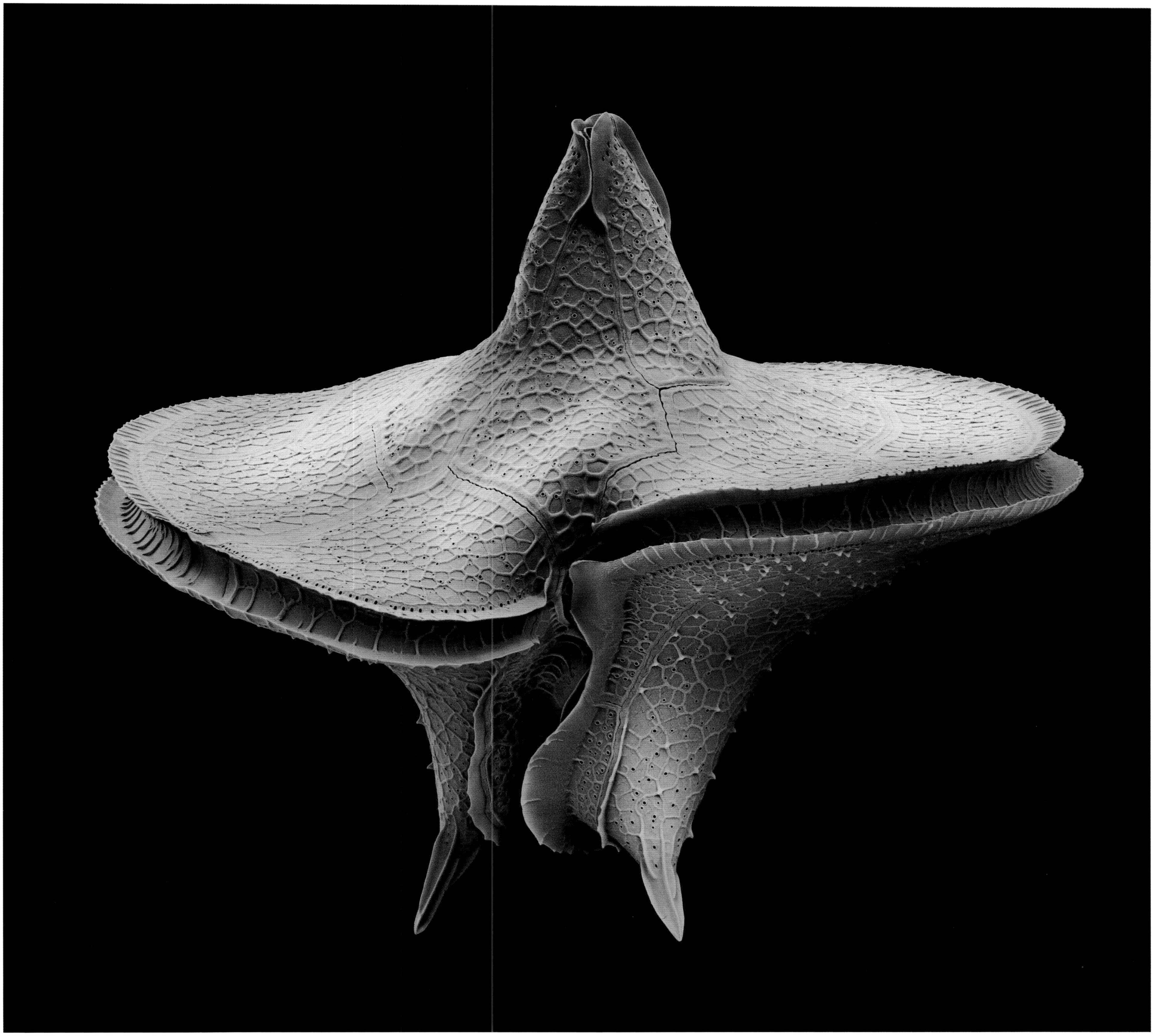

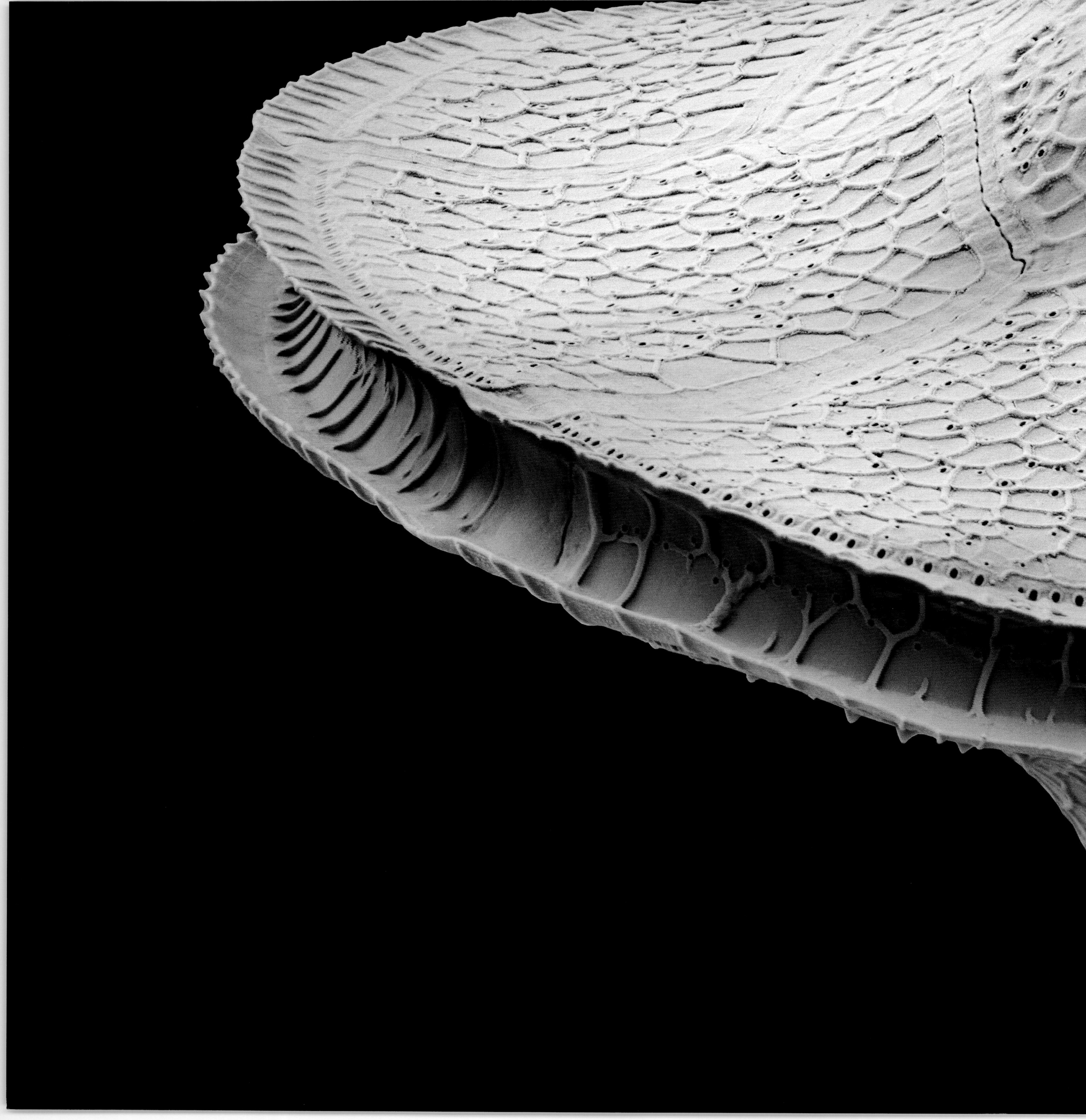

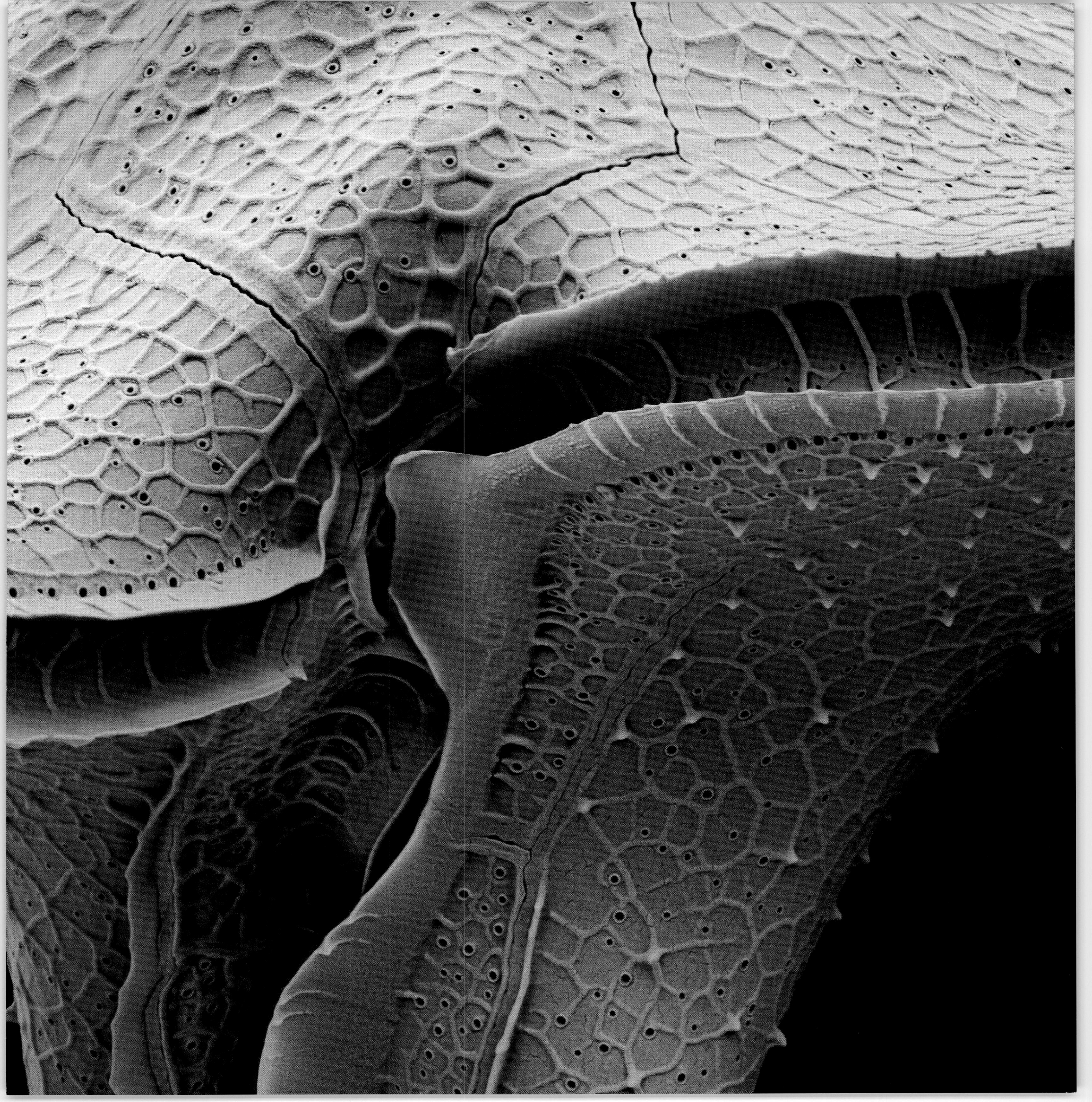

**Facing Page**

*Protoperidinium grande*, North Carolina Gulf Stream.
2200 x. About 100 microns wide, 0.1 millimeter

**Page 178, Left →**

*Ceratium* genus dinoflagellate, North Carolina Gulf Stream.
896.6 x. About 240 microns from top to bottom, 0.24 millimeter

**Page 178, Right →**

*Tripos symmetricus*, Florida Gulf Stream.
916.6 x. About 240 microns from top to bottom, 0.24 millimeter

**Page 179 →**

*Tripos symmetricus*, Carrie Bow Cay, Belize.
683 x. About 300 microns from top to bottom, 0.3 millimeter

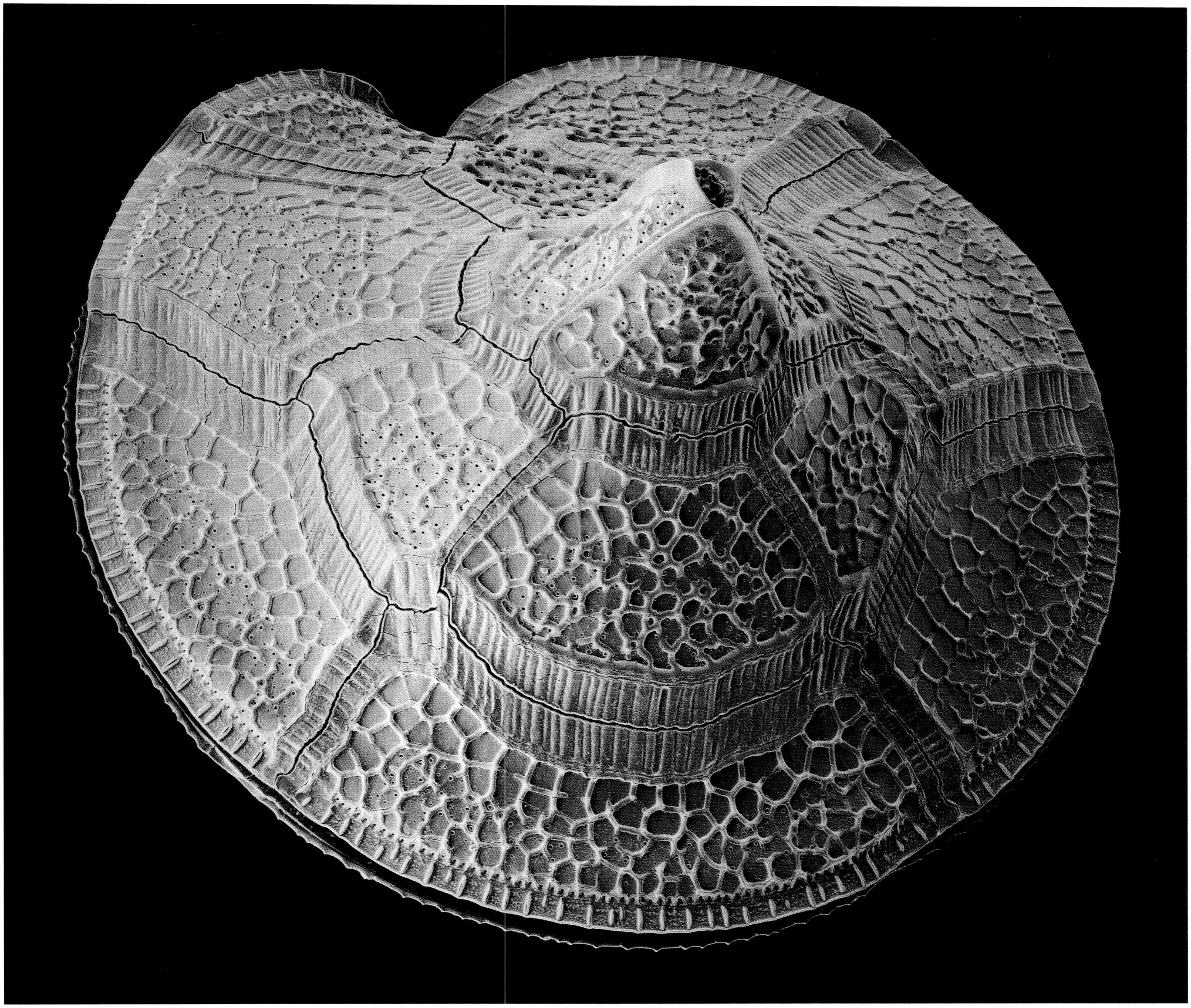

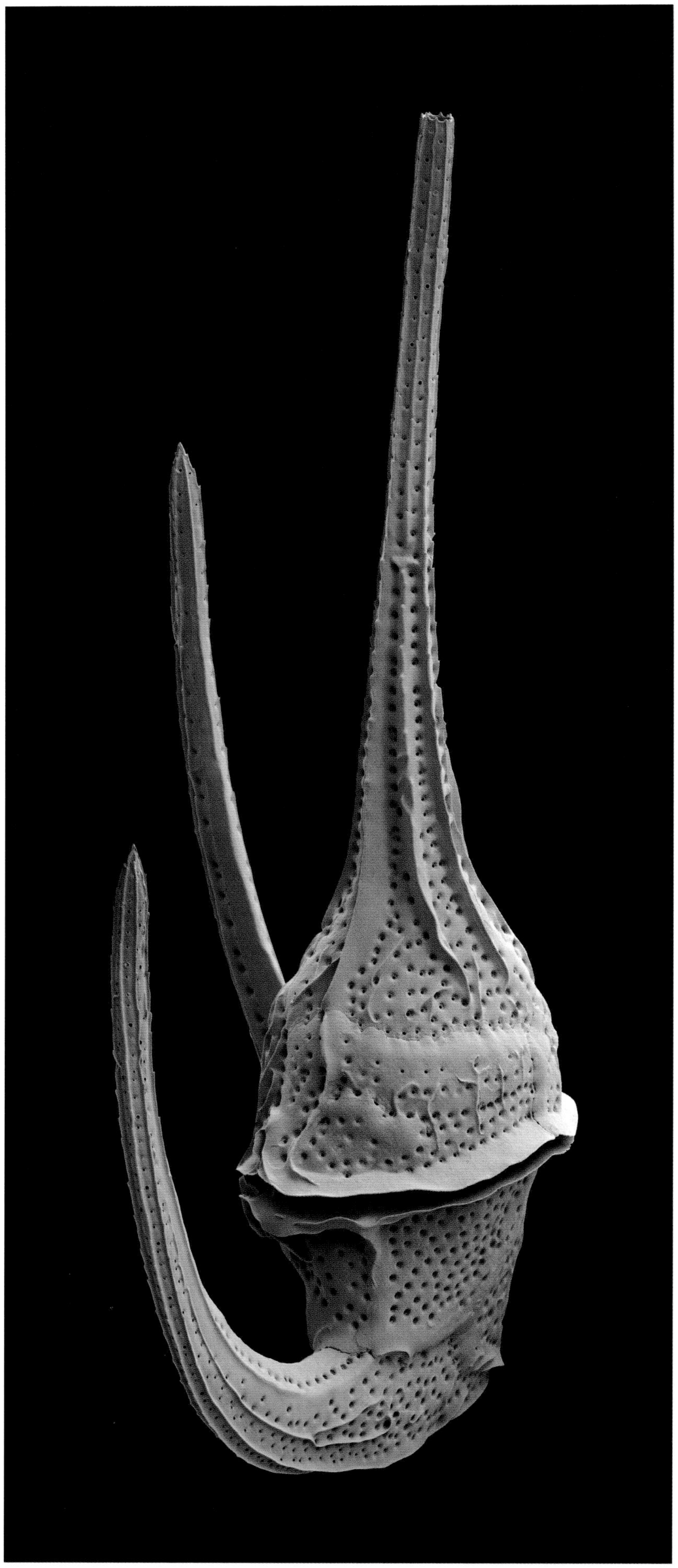
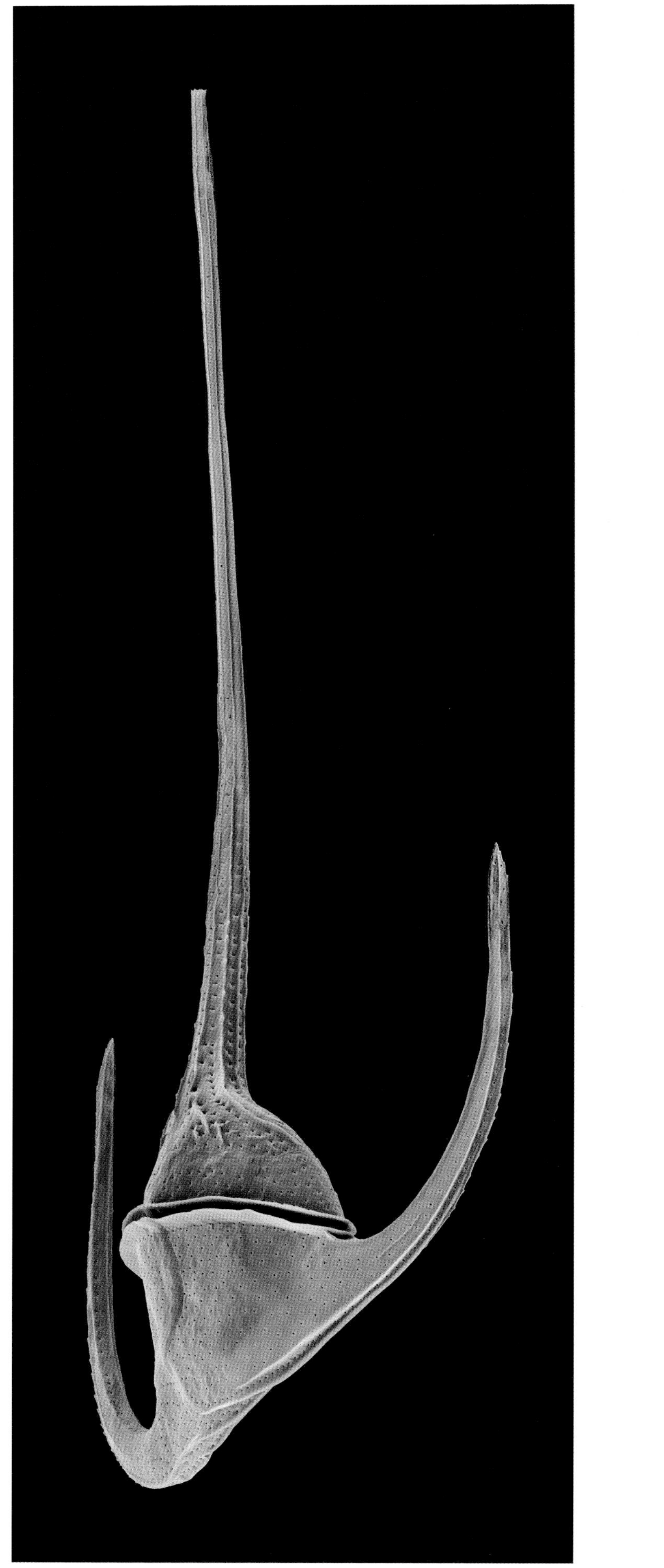

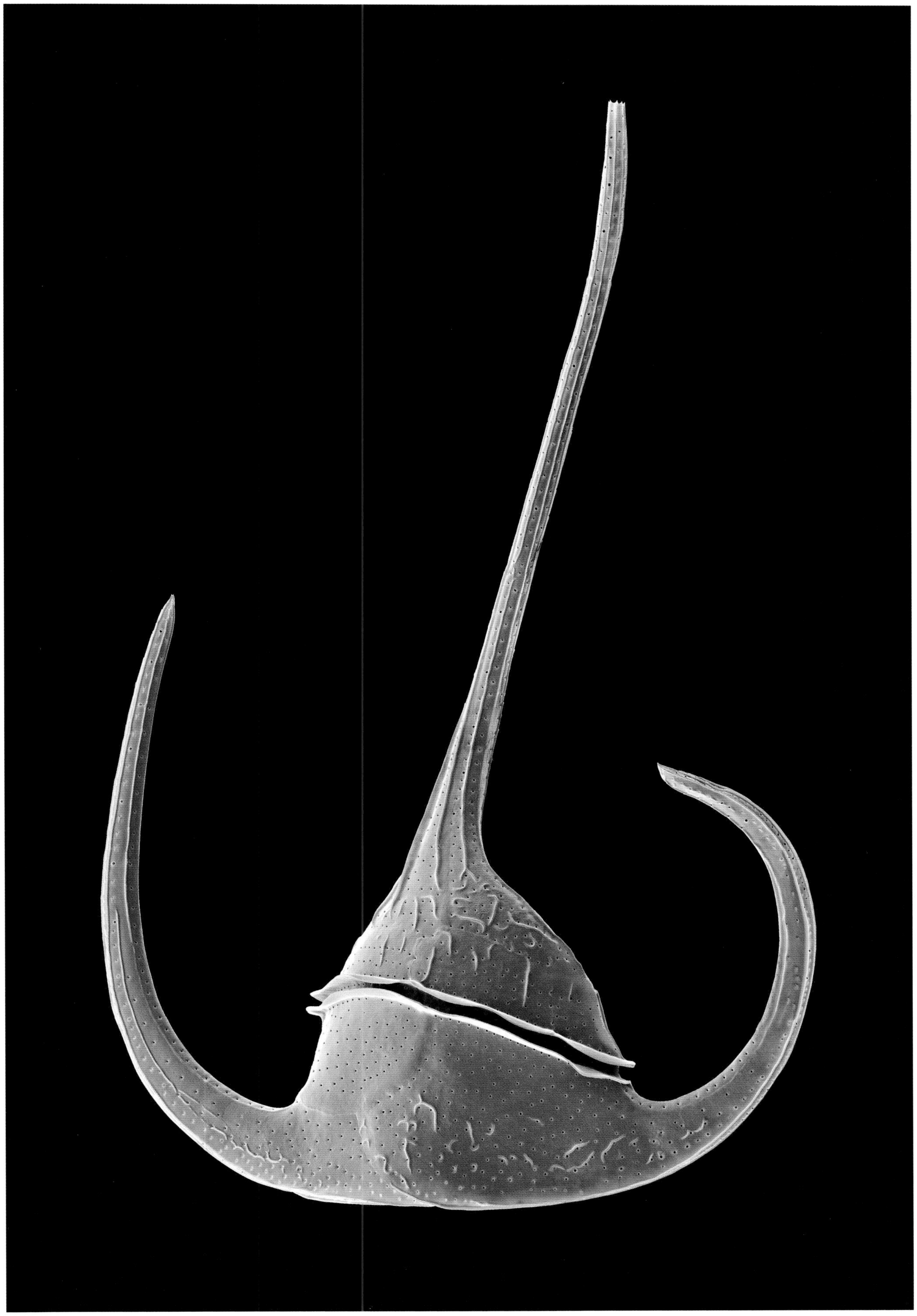

**Facing Page, Pages 182–183 →**

*Ceratium* genus dinoflagellates, possibly *Ceratum karstenii*, North Carolina Gulf Stream. 607 x. About 280 microns from top to bottom, 0.28 millimeter

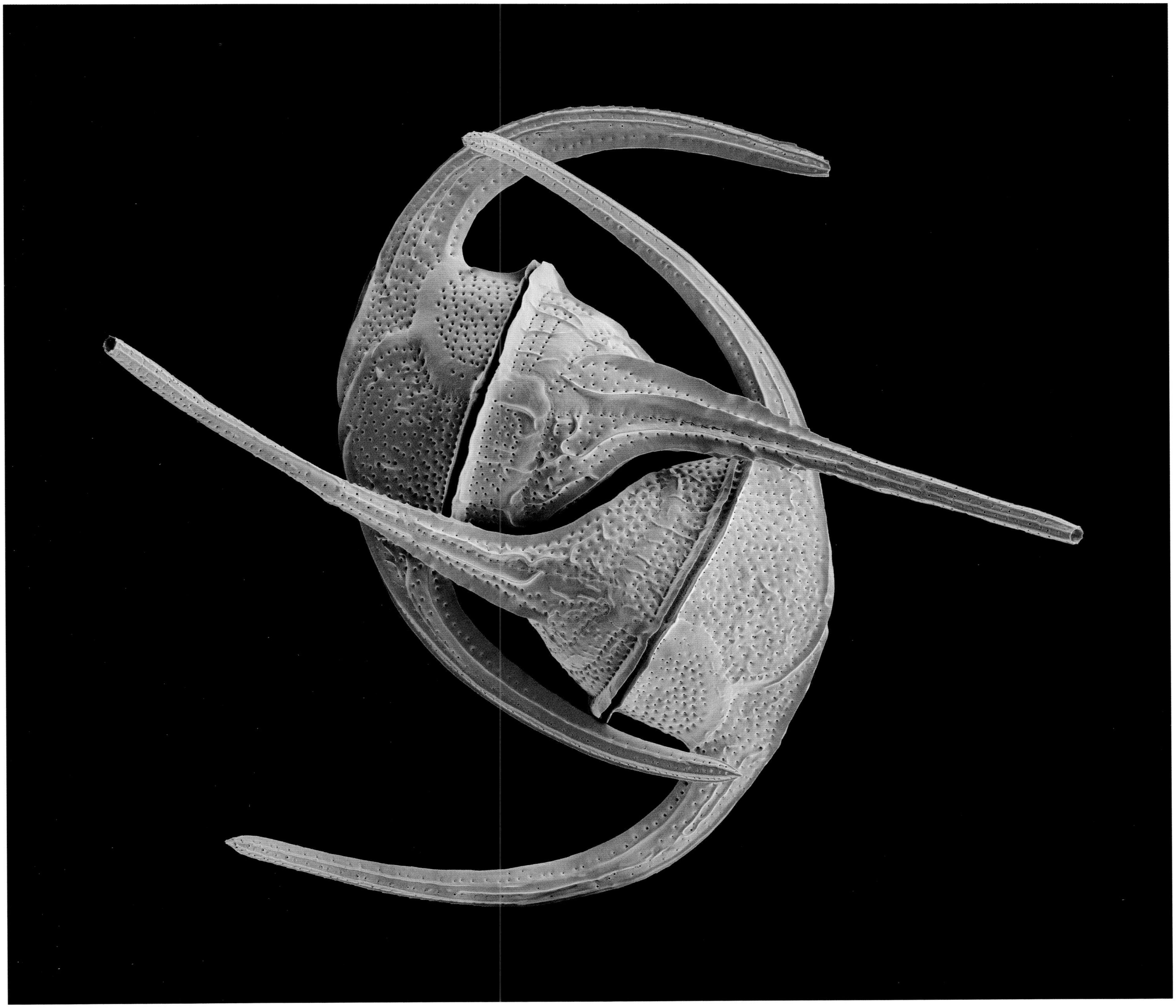

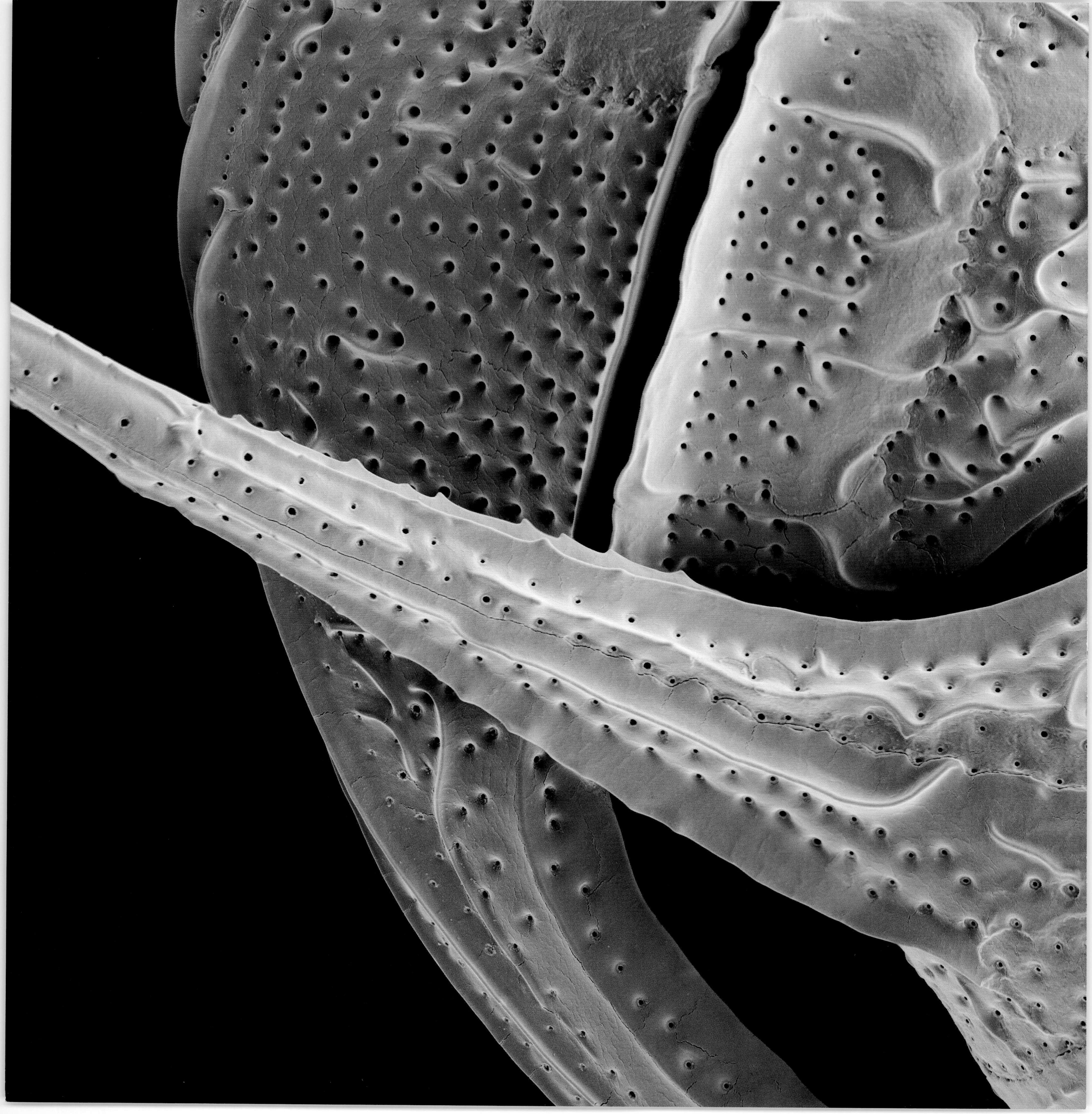

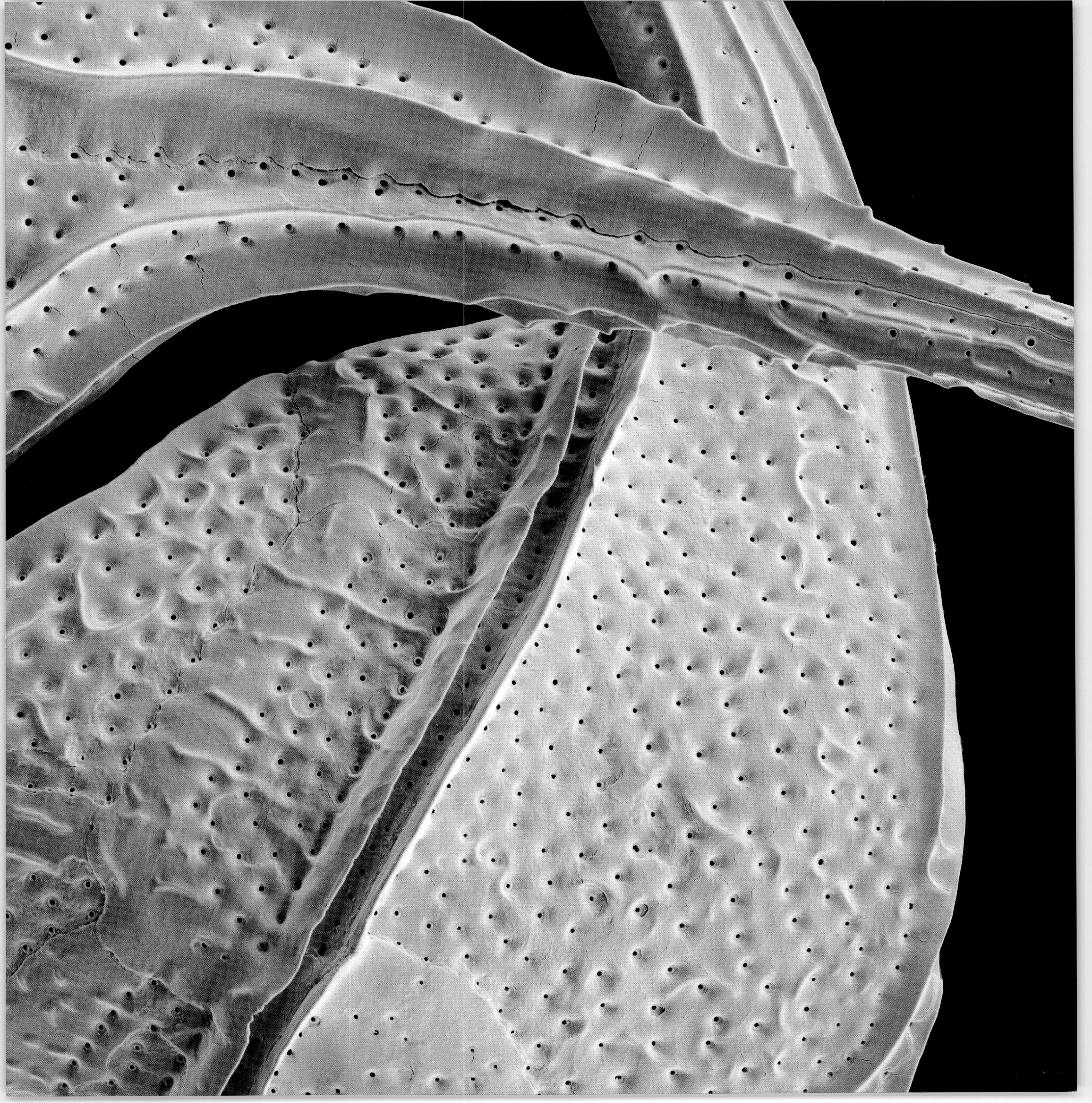

**Facing Page**

*Protoperidinium* genus dinoflagellate, Carrie Bow Cay, Belize.
3333 x. About 60 microns from top to bottom, 0.06 millimeter

**Page 186 →**

*Diplopsalopsis* genus dinoflagellate, North Carolina Gulf Stream.
3750 x. About 60 microns across, 0.06 millimeter

**Page 187 →**

*Protoperidinium globulus*, North Carolina Gulf Stream.
3500 x. About 60 microns wide, 0.06 millimeter

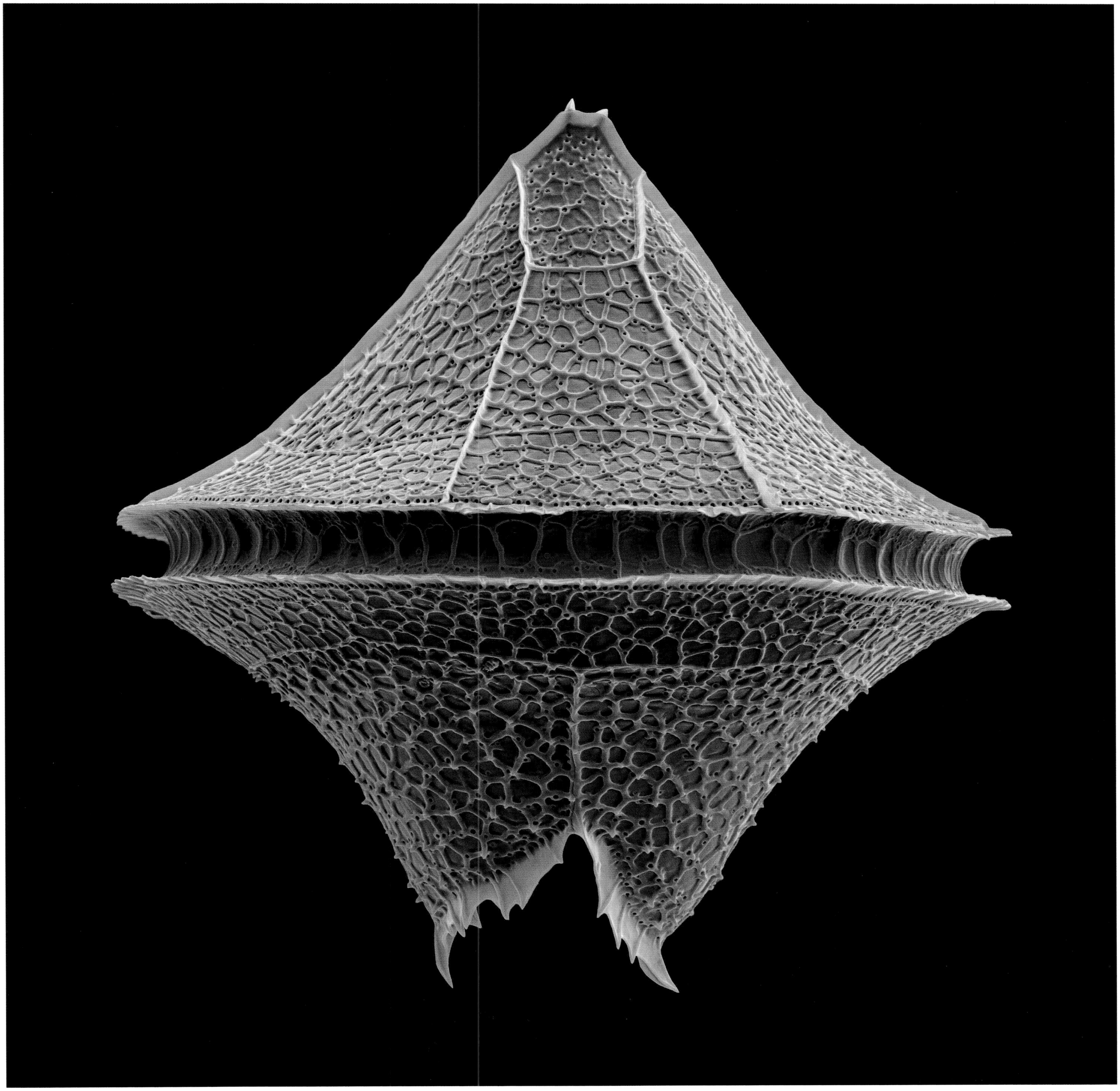

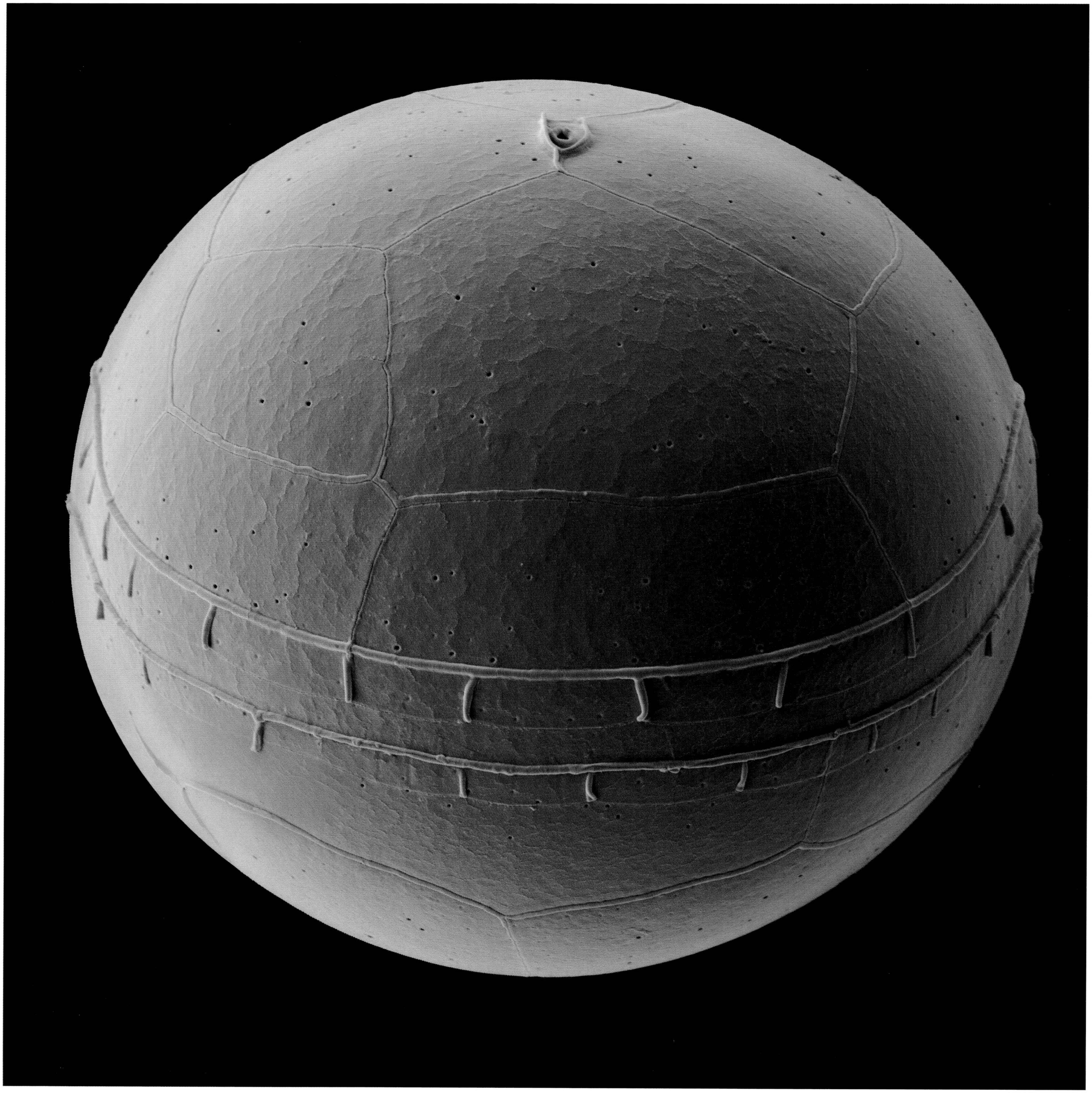

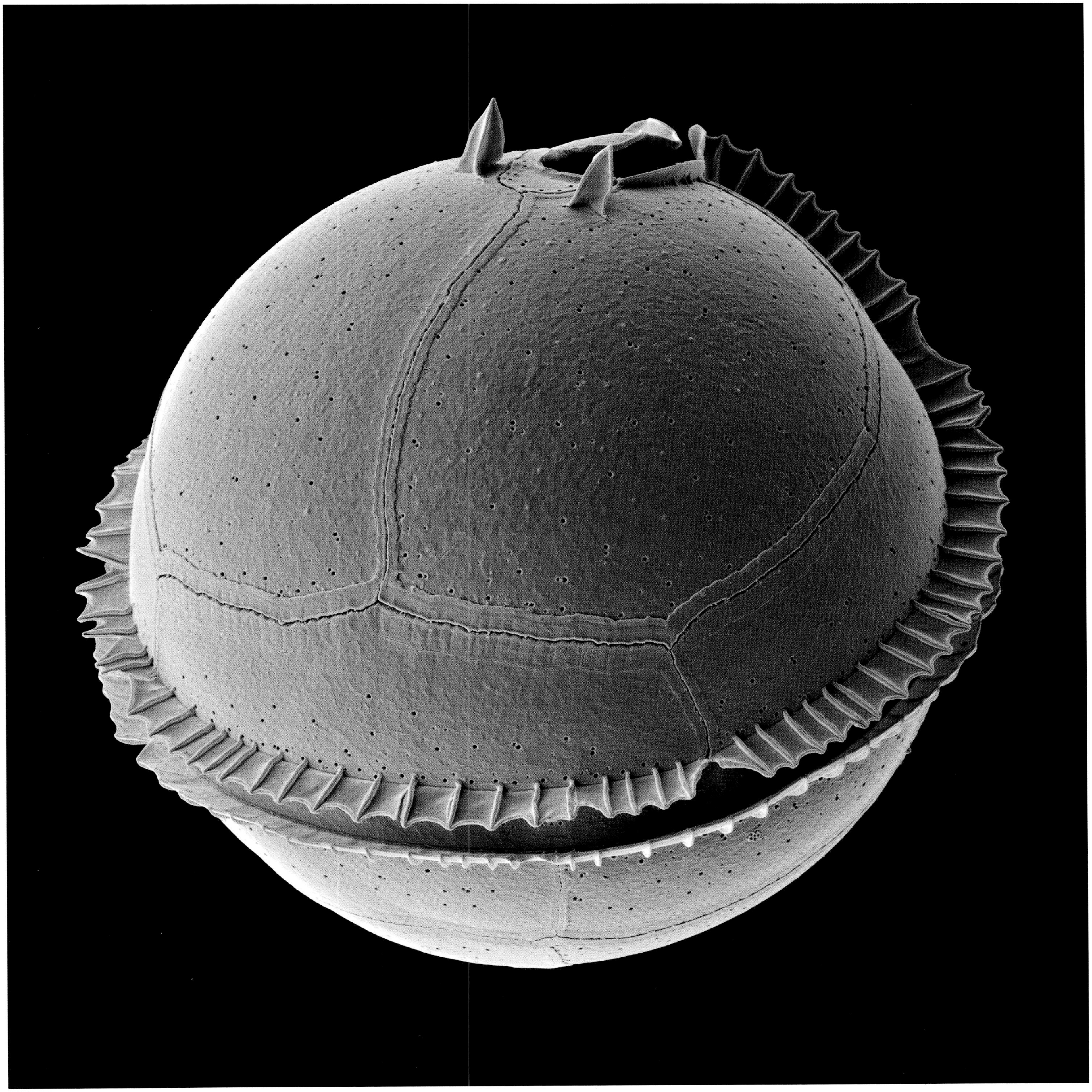

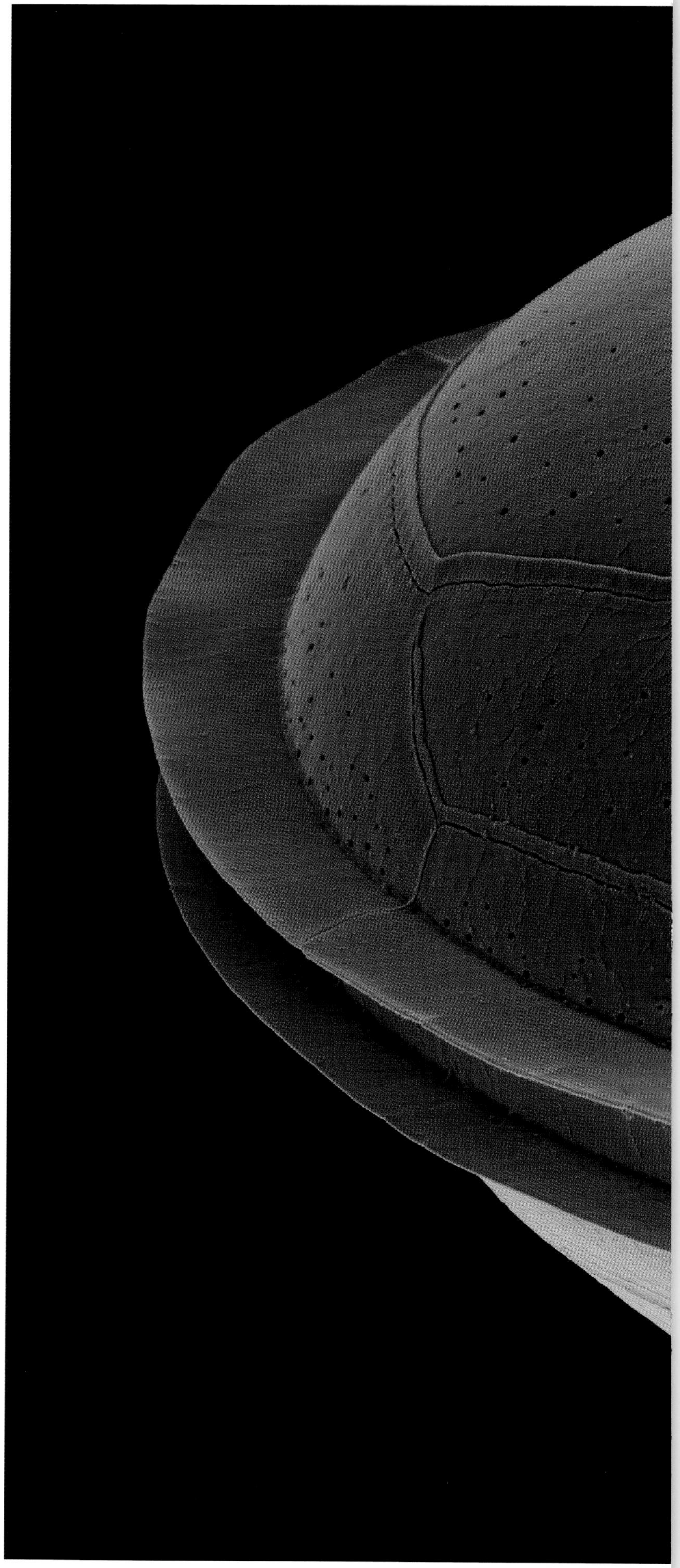

**Facing Page**

*Diplopsalopsis orbicularis*, North Carolina Gulf Stream.
3400 x. 100 microns across, 0.1 millimeter

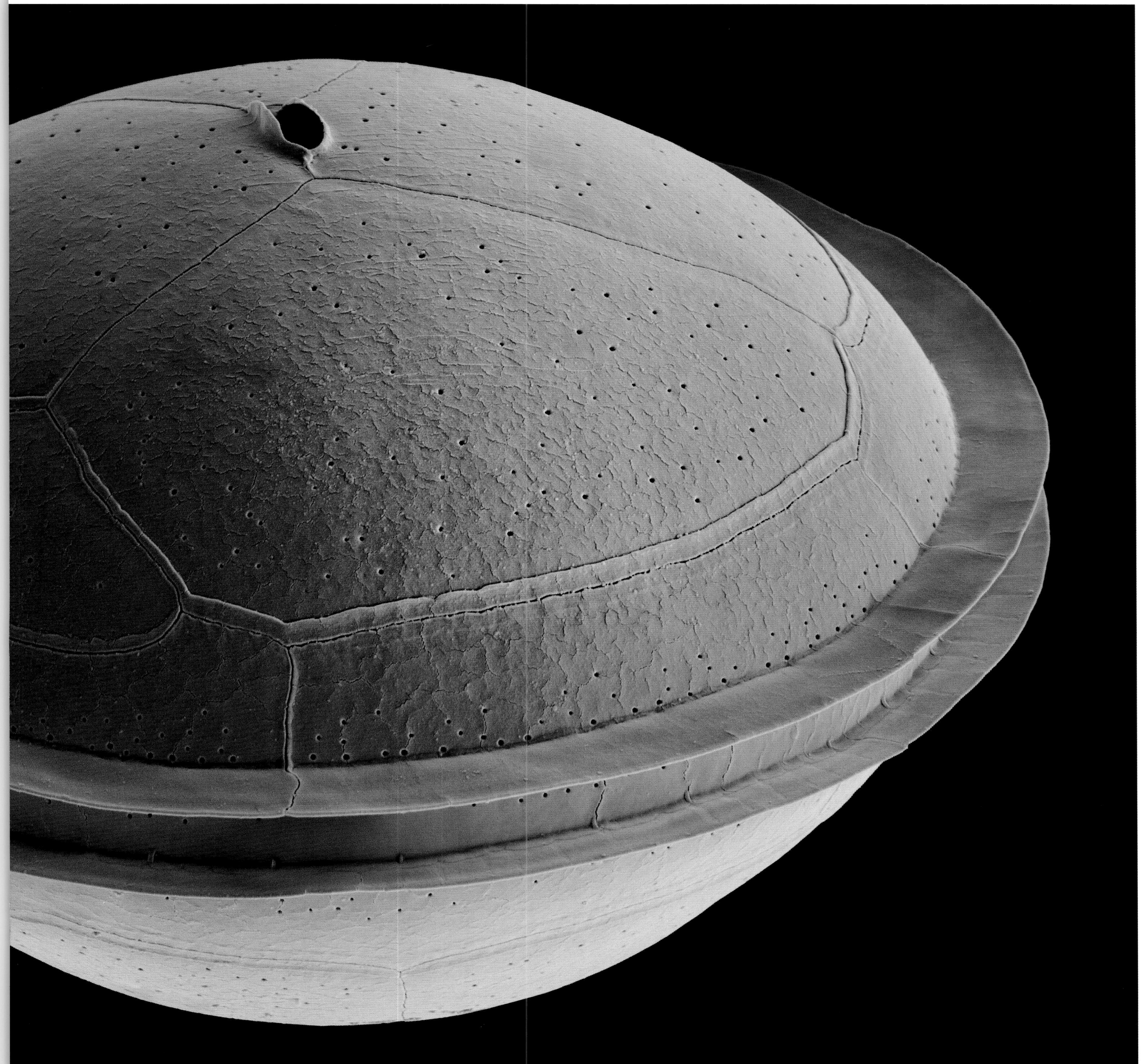

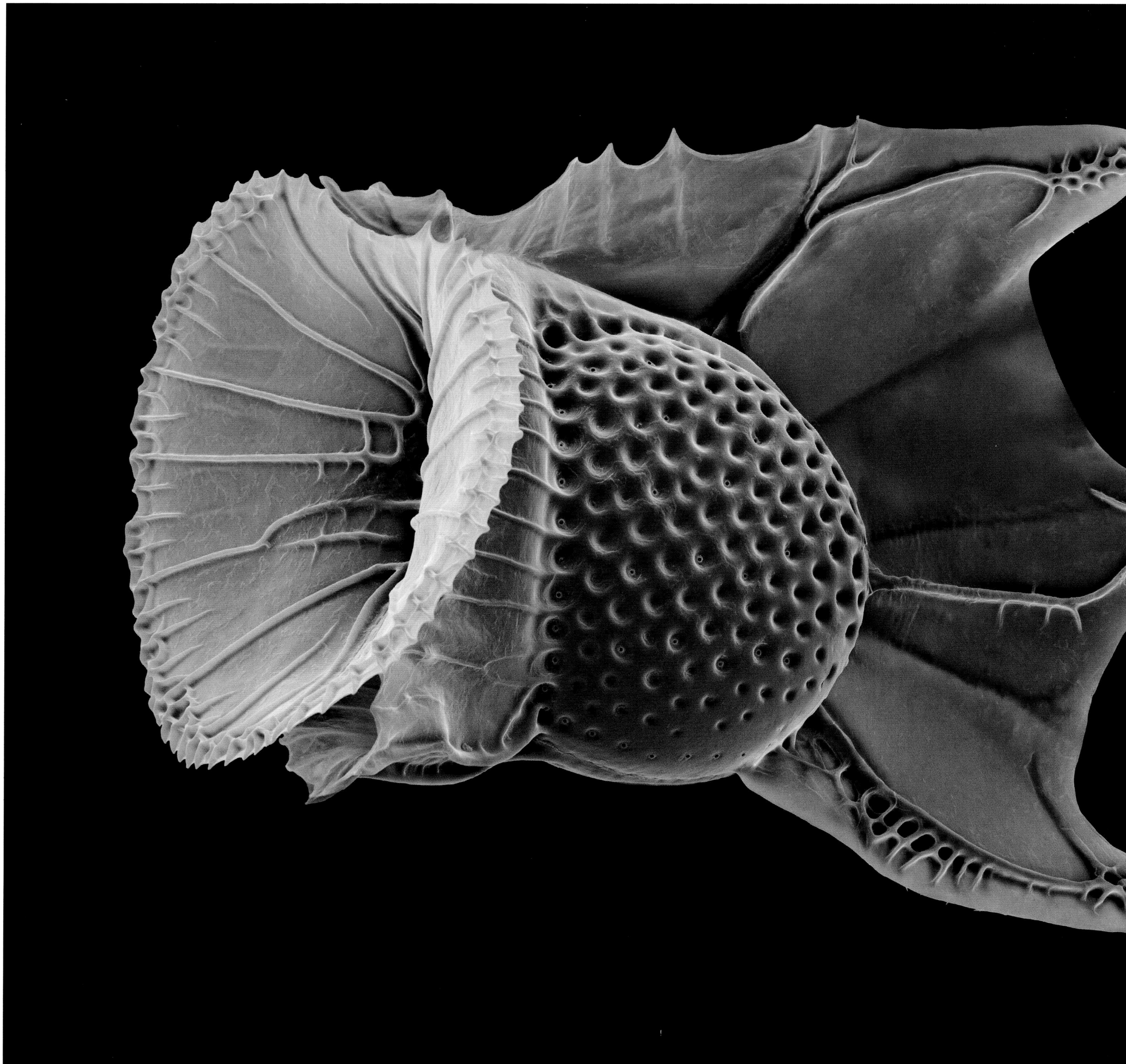

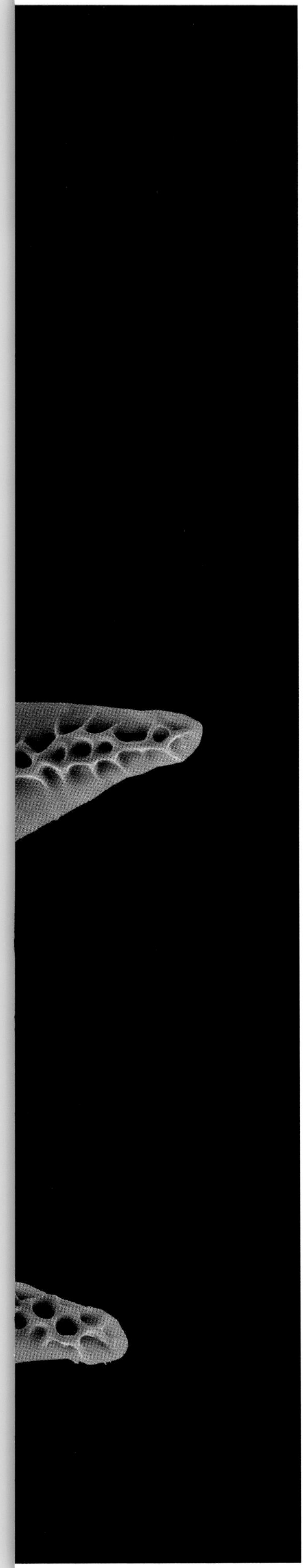

191

**Facing Page**

*Ornithocercus magnificus*, Florida Gulf Stream.
2800 x. About 100 microns wide, 0.1 millimeter

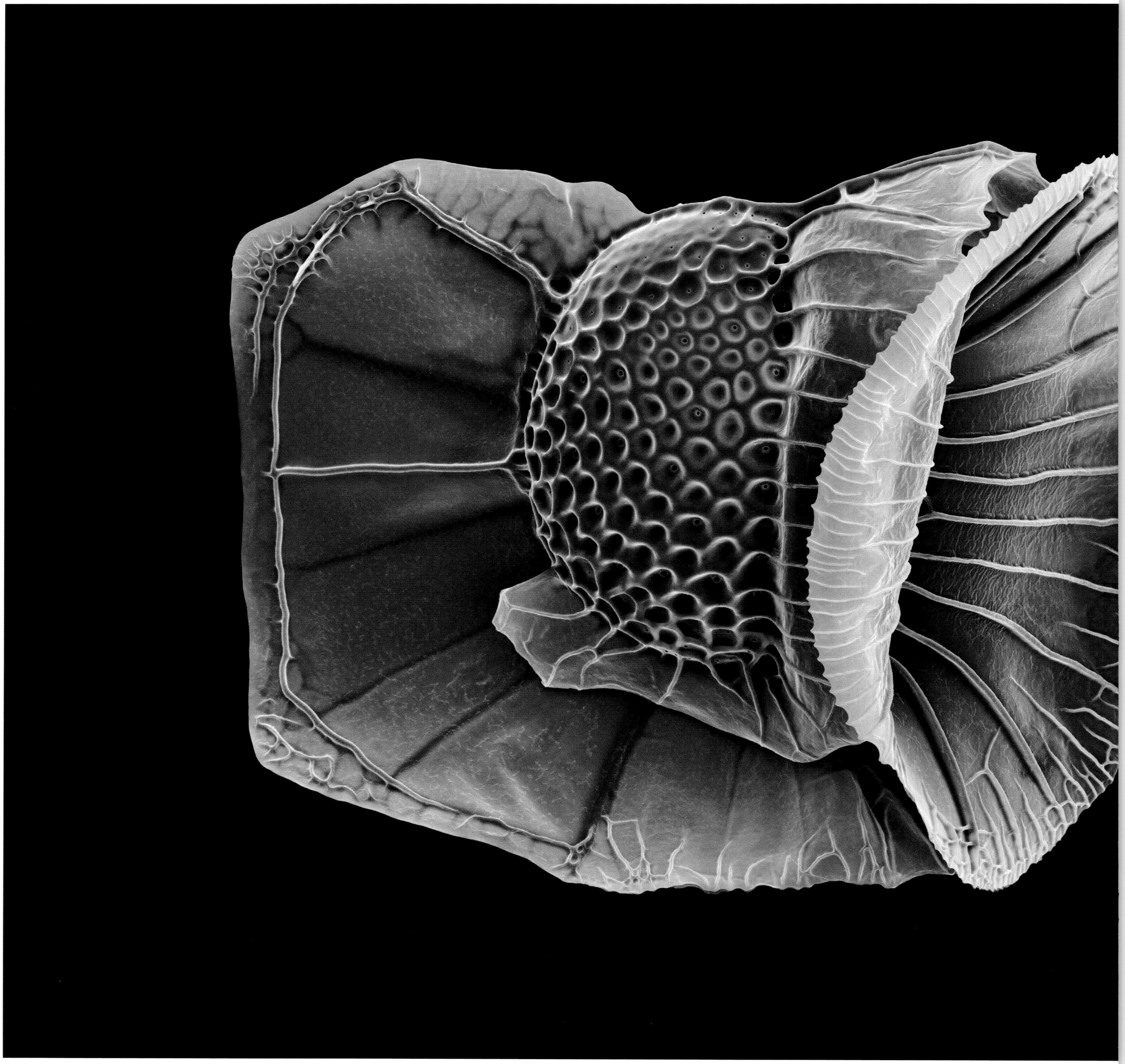

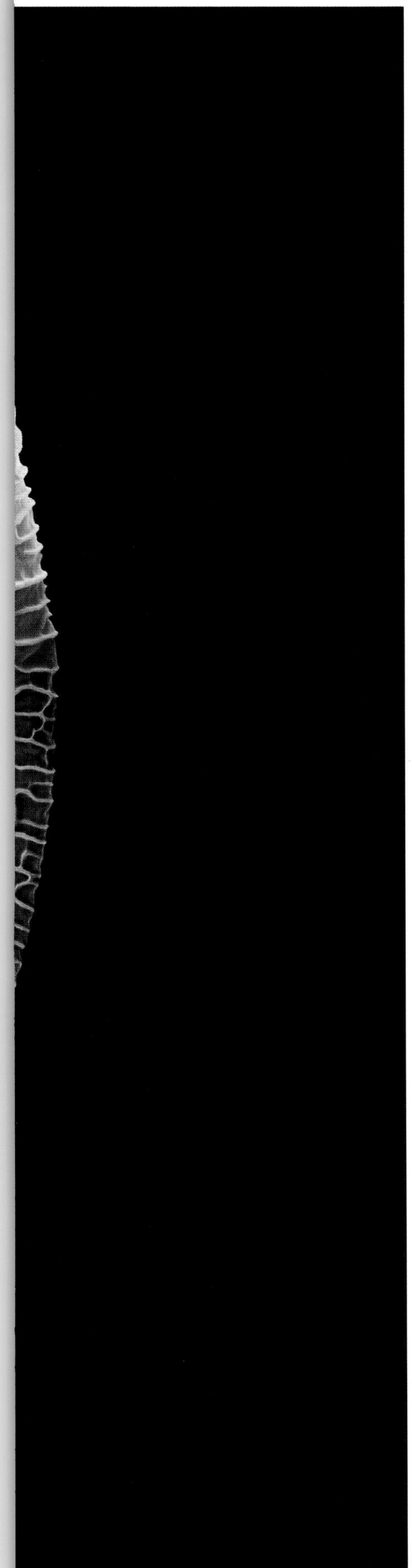



**Facing Page, Pages 194–195 →**

*Ornithocercus quadratus*, Florida Gulf Stream.
1769 x. About 130 microns wide, 0.13 millimeter

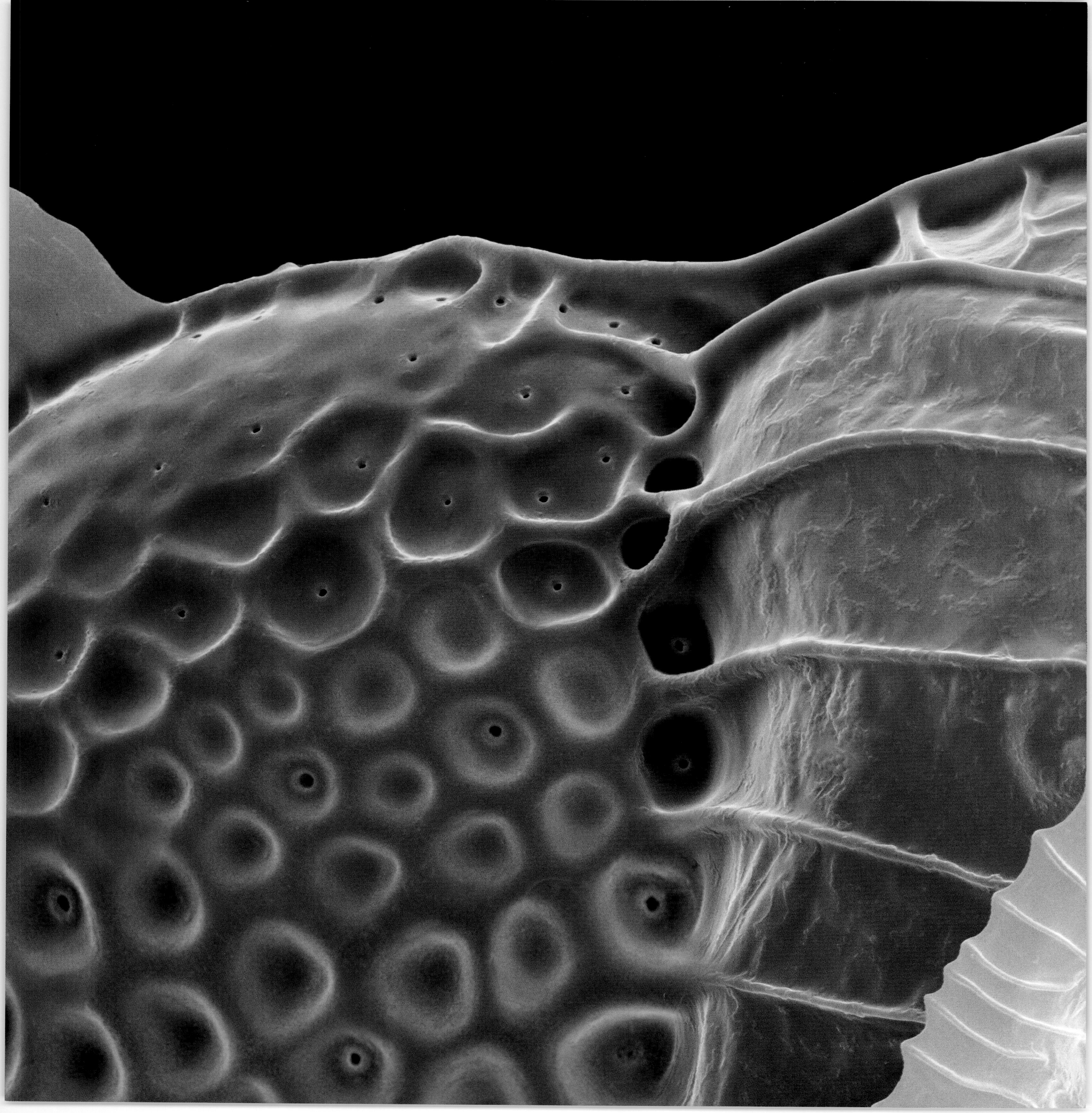

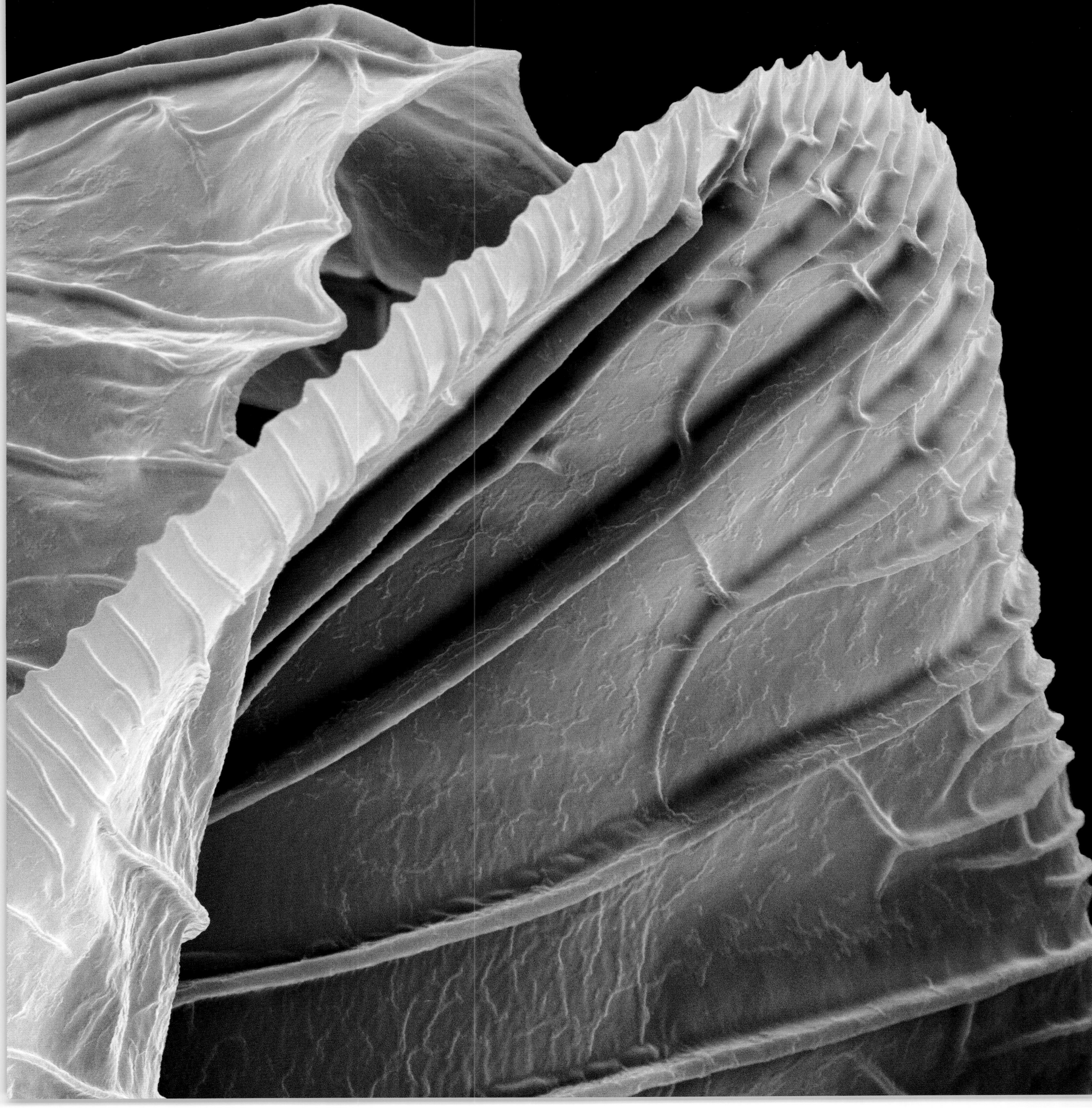

196

**Facing Page**

*Dinophysis caudata* reproducing asexually by binary fission.
A flagellum can be seen—the coiling thread to the upper right. Manatee Cay, Belize.
2500 x. About 80 microns wide, 0.08 millimeter

**Page 198, Page 199, Page 200 →**

*Protoperidinium depressum*, Florida Gulf Stream.
1333 x. About 150 microns across, 0.15 millimeter

**Page 201 →**

Another *Protoperidinium depressum*, Florida Gulf Stream.
1917 x. About 120 microns from top to bottom, 0.12 millimeter

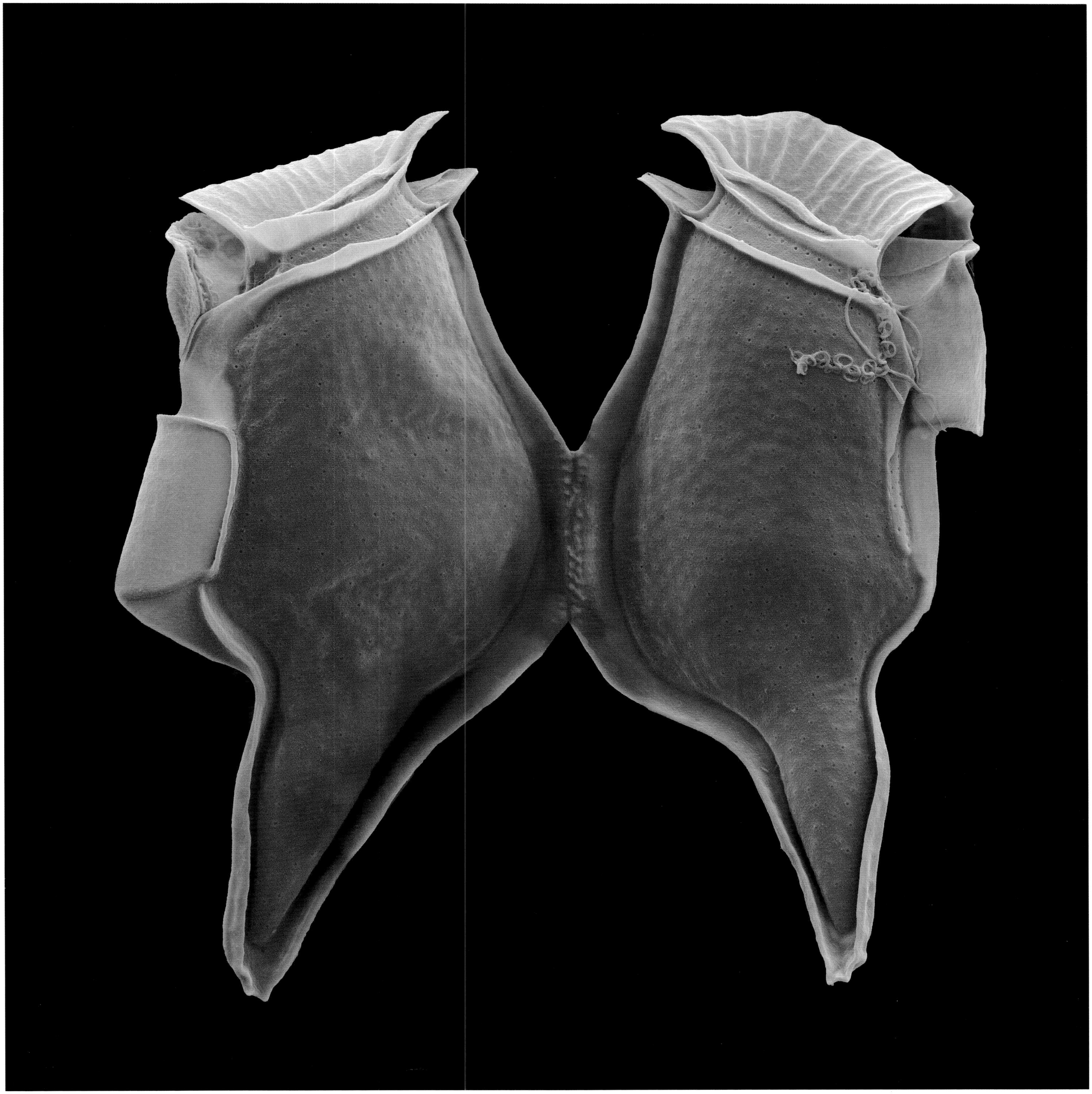

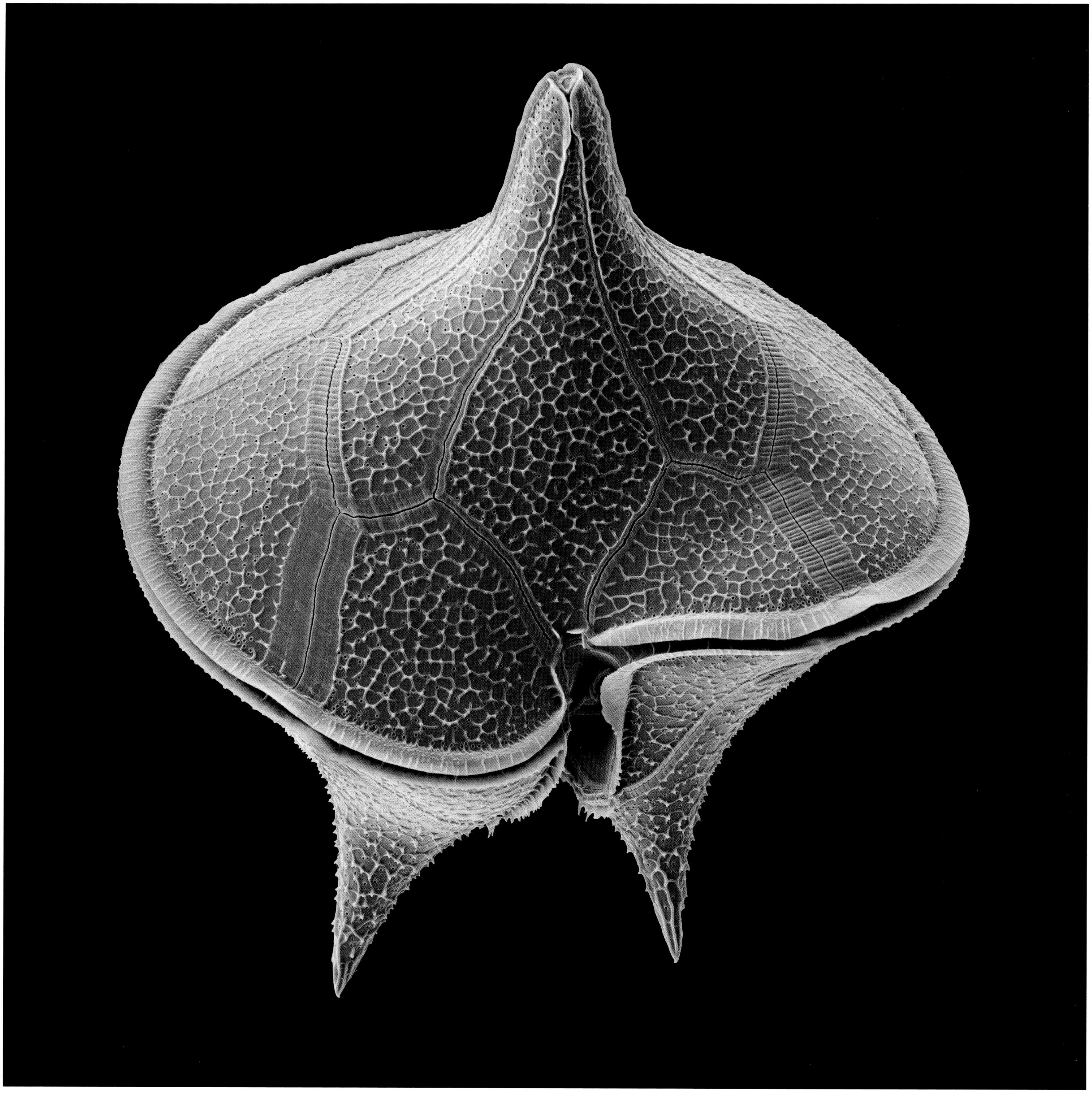

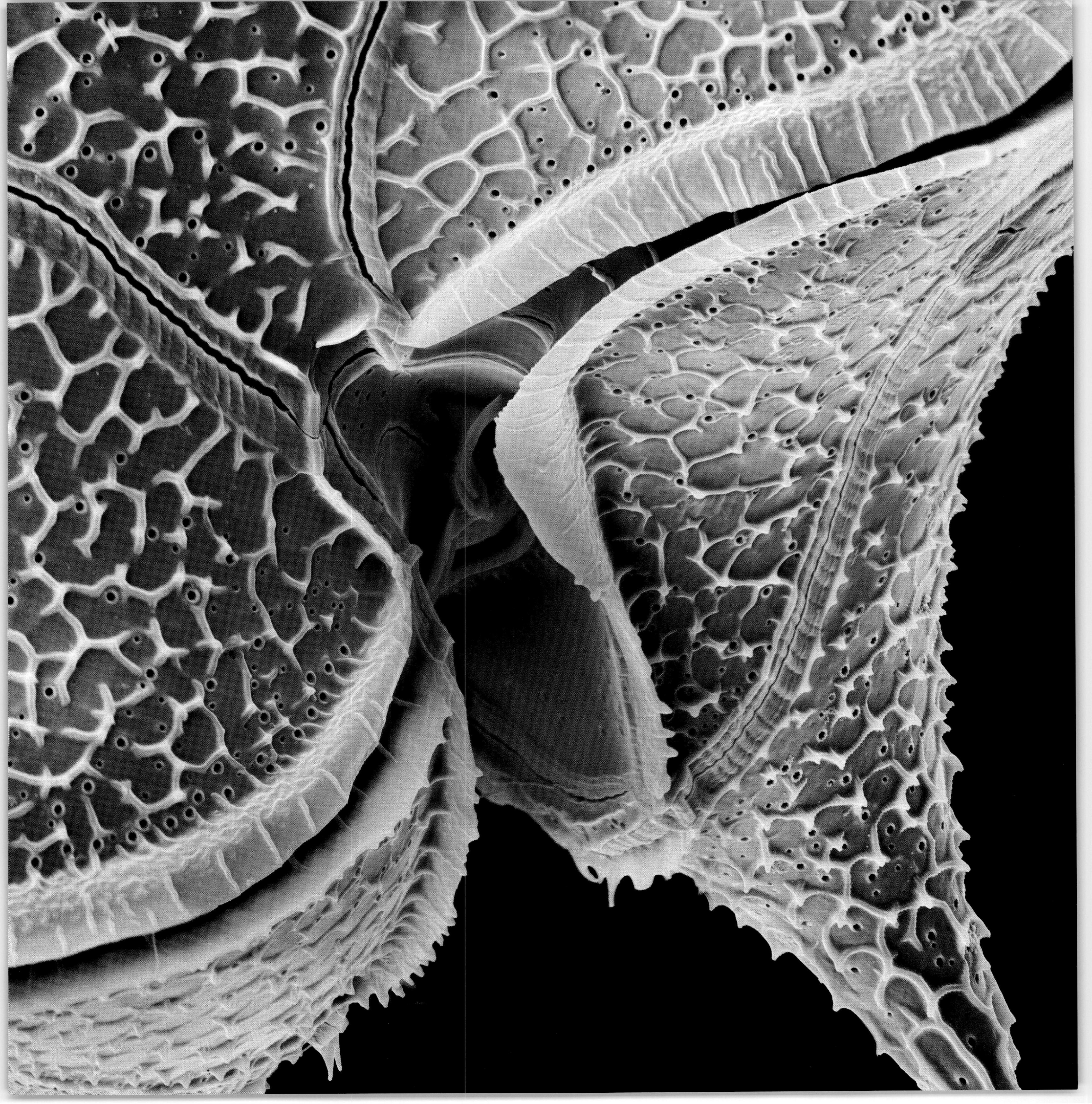

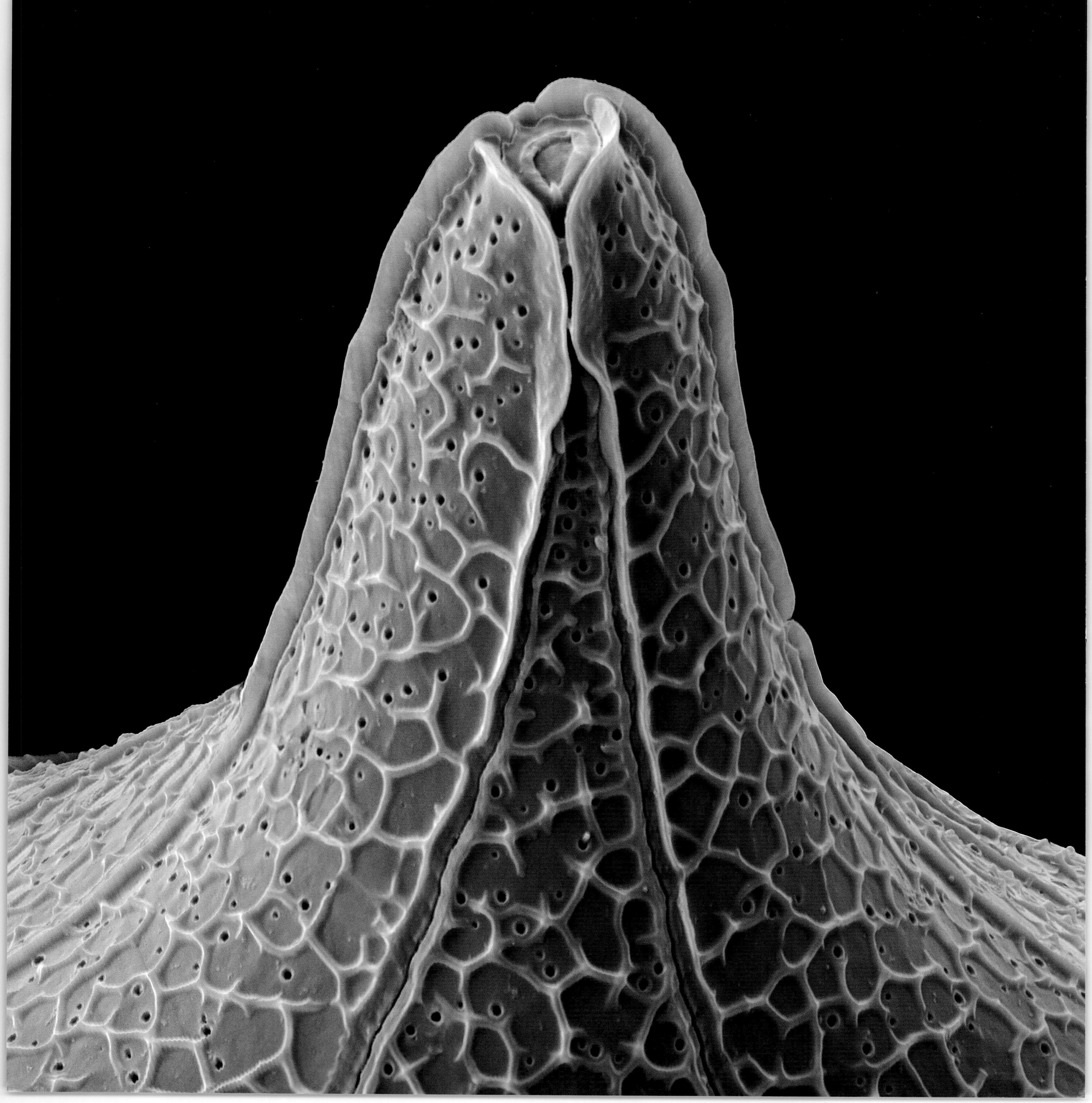

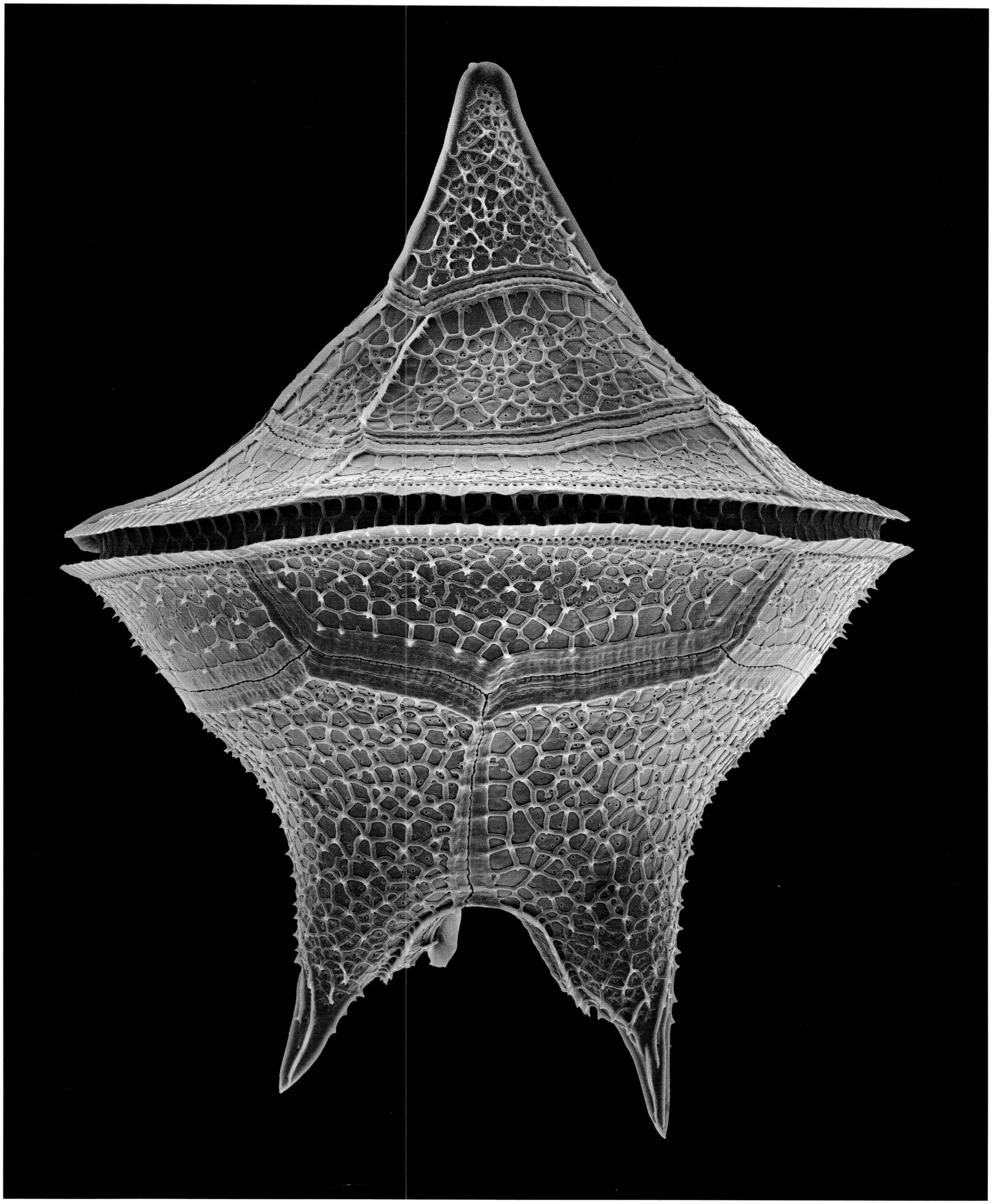

TERRA

# FLORA AND FAUNA

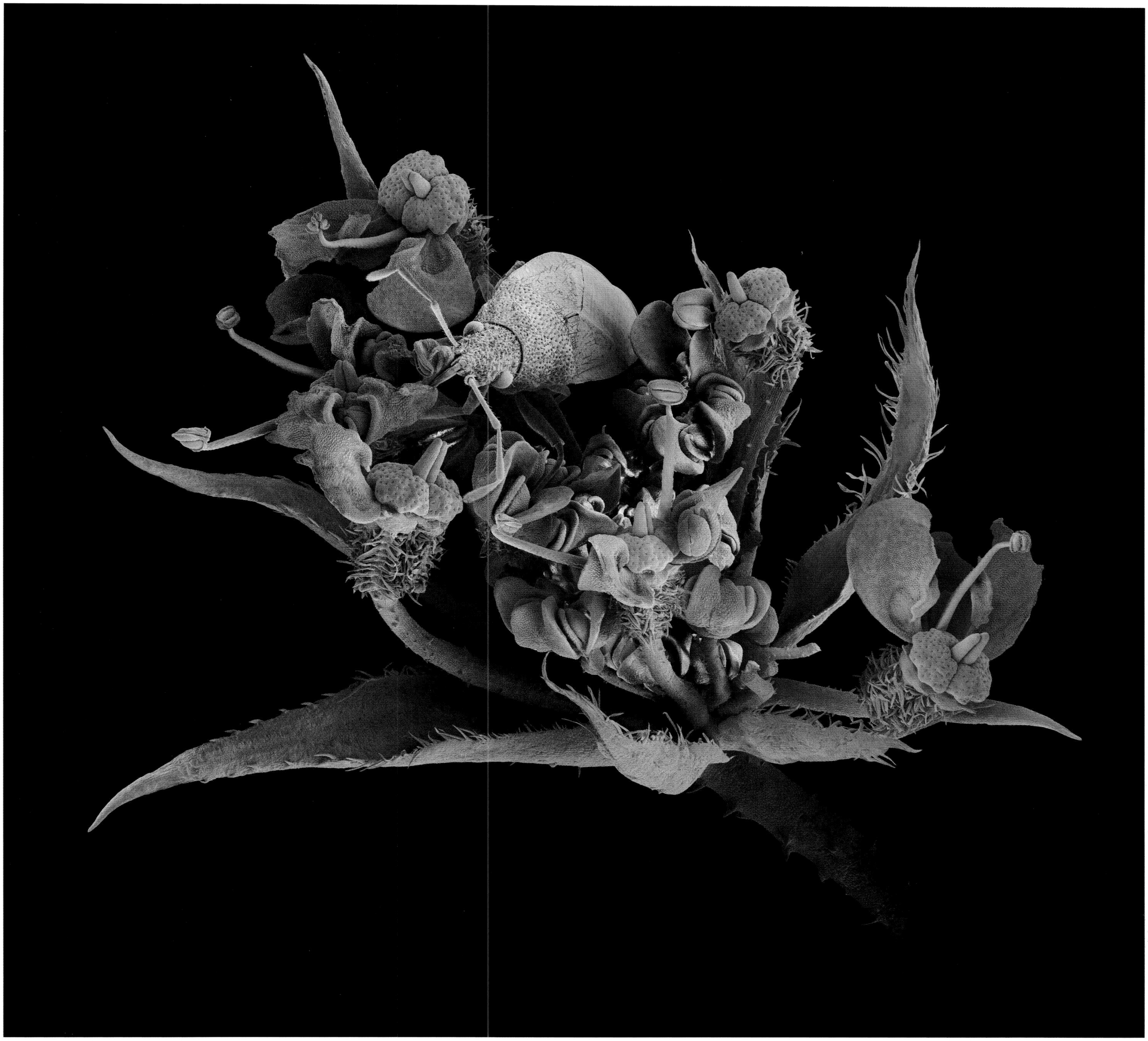

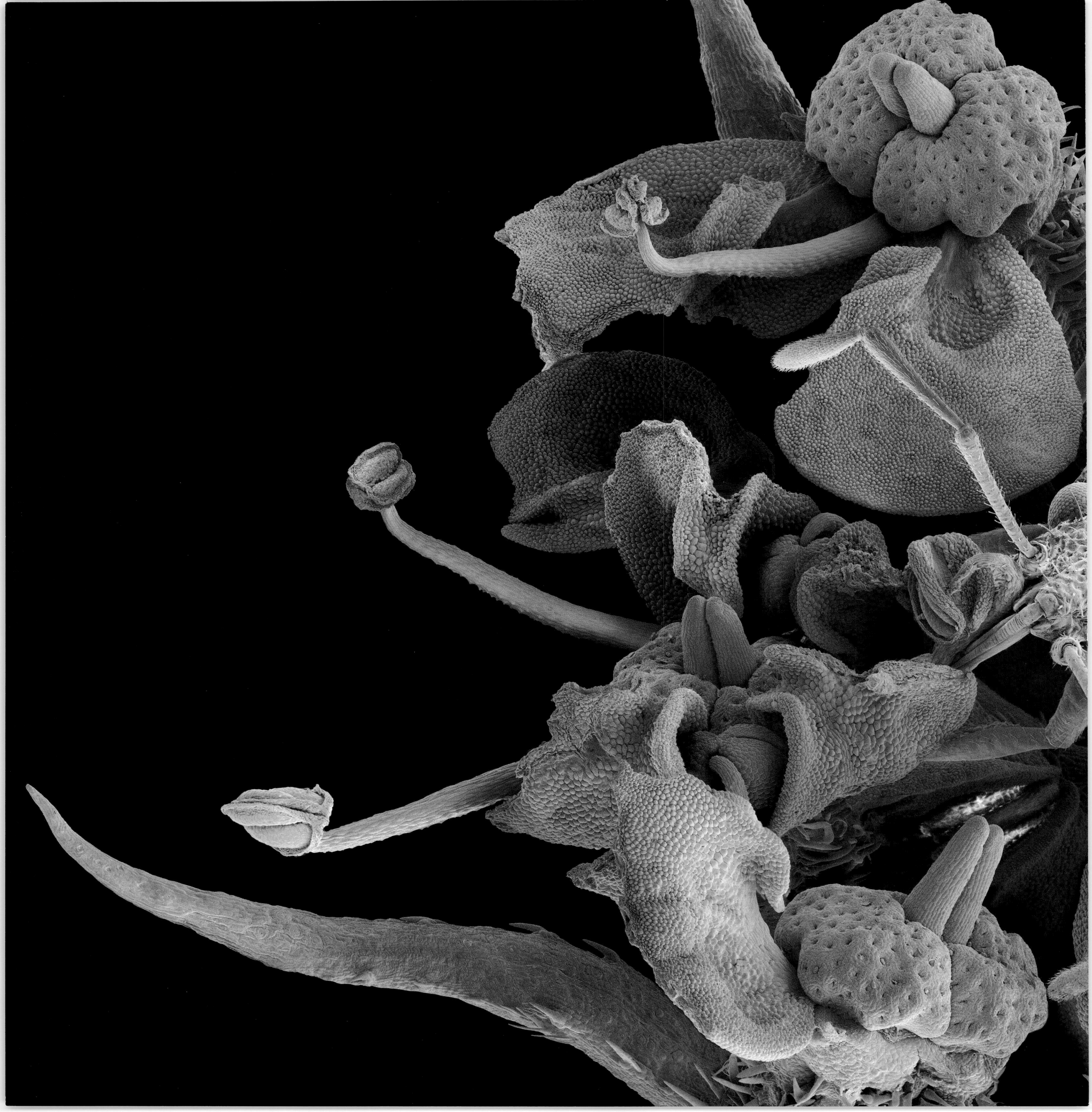

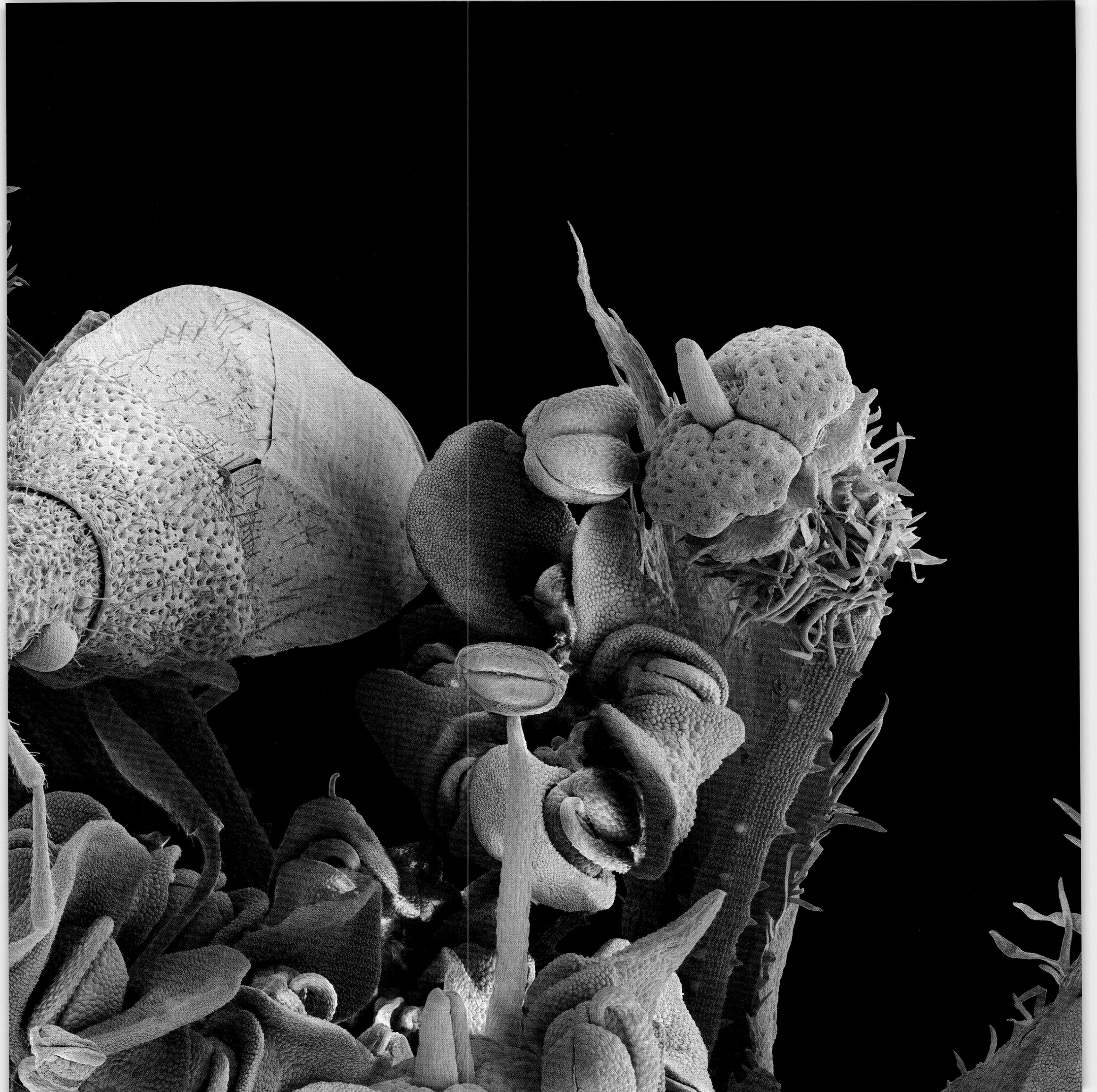

← Page 203 , Pages 204–205

*Polydrusus cervinus* weevil in a flowering plant, Northern Velebit National Park, Croatian Adriatic.
27.5 x. The plant is about 8 millimeters wide

Facing Page

Flowering plant, Northern Velebit National Park, Croatian Adriatic.
26.6 x. About 6 millimeters wide

Page 208 →

Close view of hypericum flower, probably *Hypericum mutilum*, with *Myzus persicae* aphid, Ottawa, Ontario.
193 x. The visible area is about 1.5 millimeters wide

Page 209 →

Wider view of hypericum flower with *Myzus persicae* aphid, Ottawa, Ontario.
47.5 x. The flower is about 4 millimeters wide

Page 210 →

Oxalis flowering plant, probably *Oxalis stricta*, Ottawa, Ontario.
60 x. Just over 2 millimeters wide

Page 211 →

Closer view of central stigma and anthers of oxalis flowering plant.
193 x. The visible area is about 1.5 millimeter wide

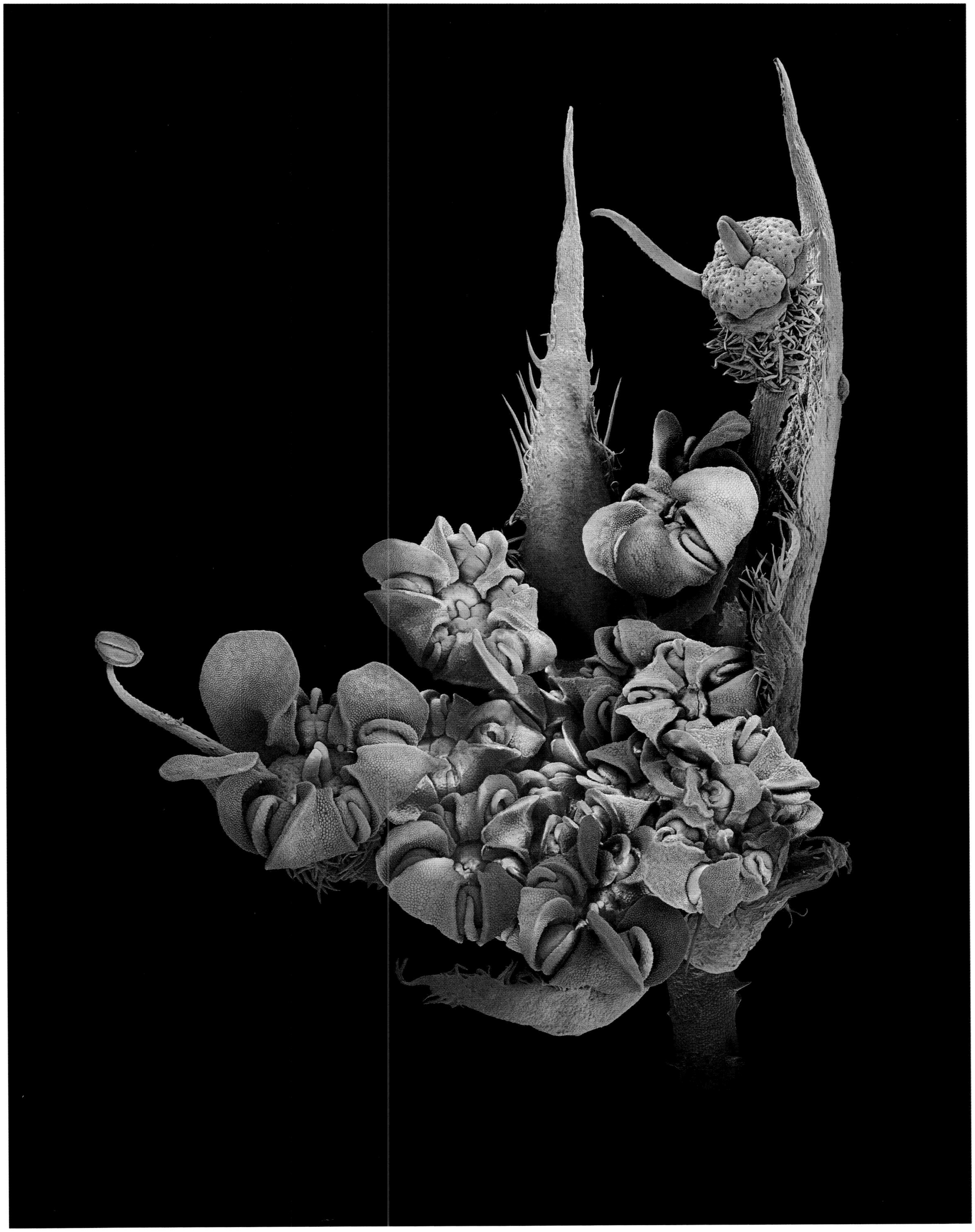

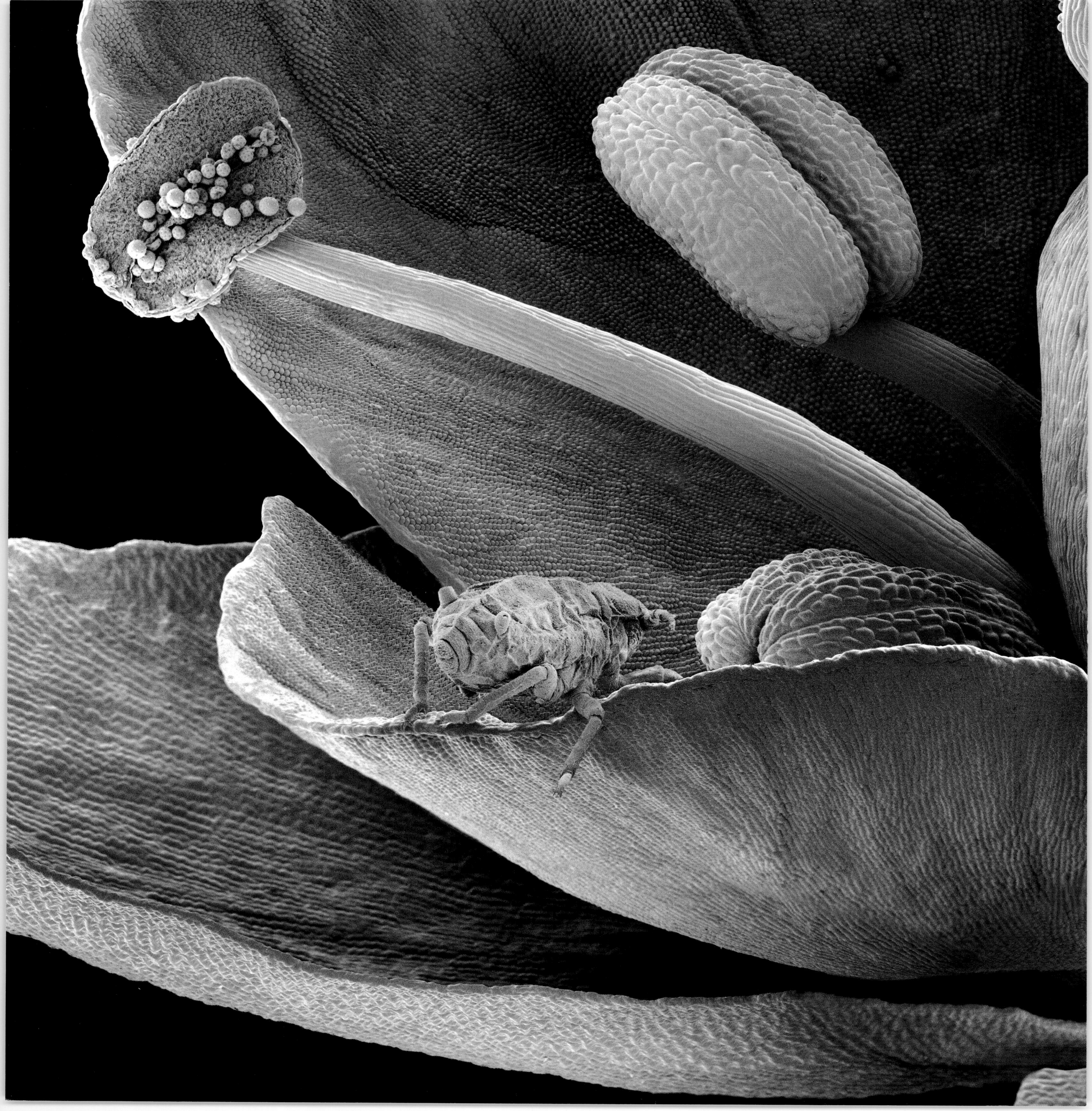

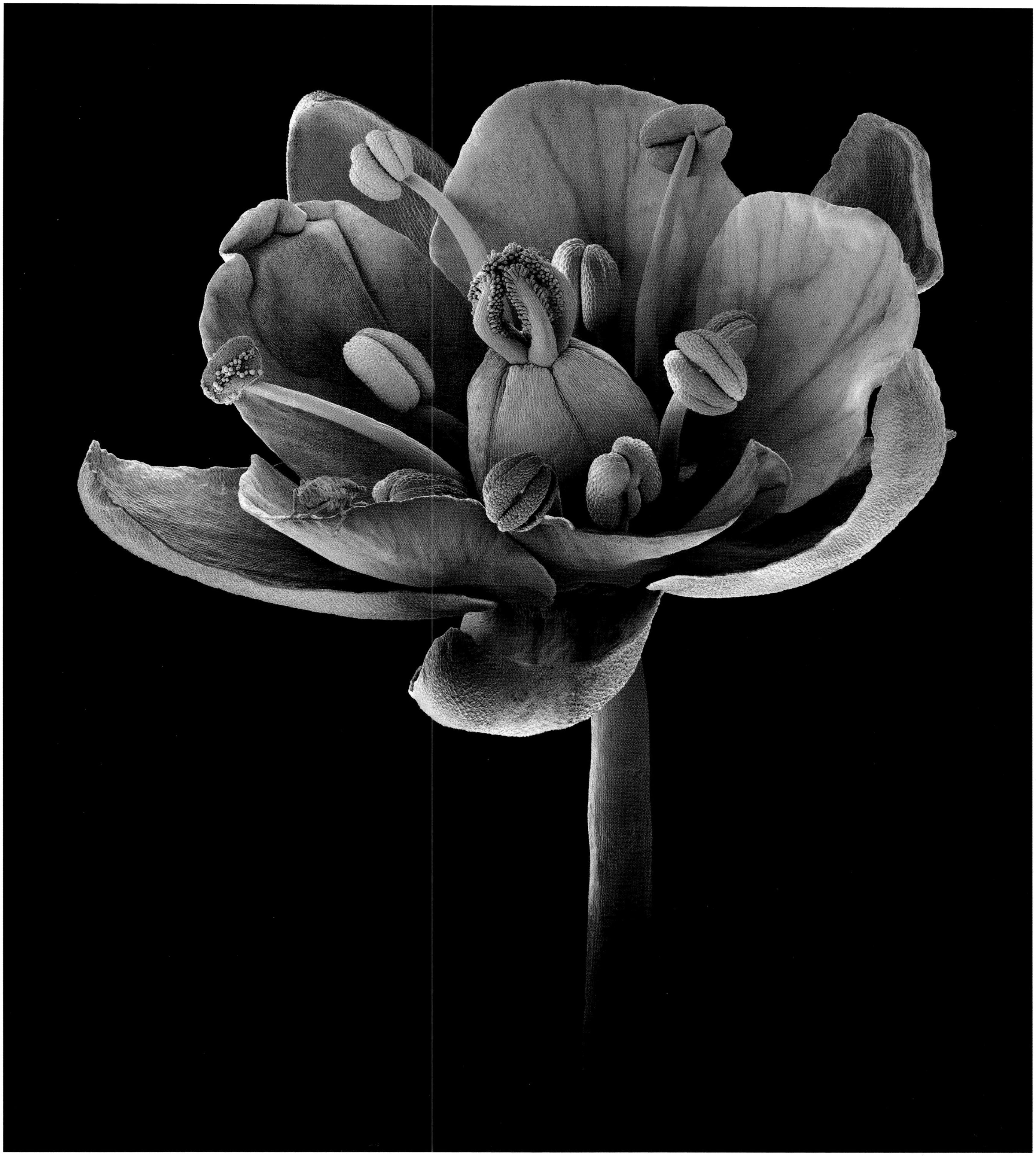

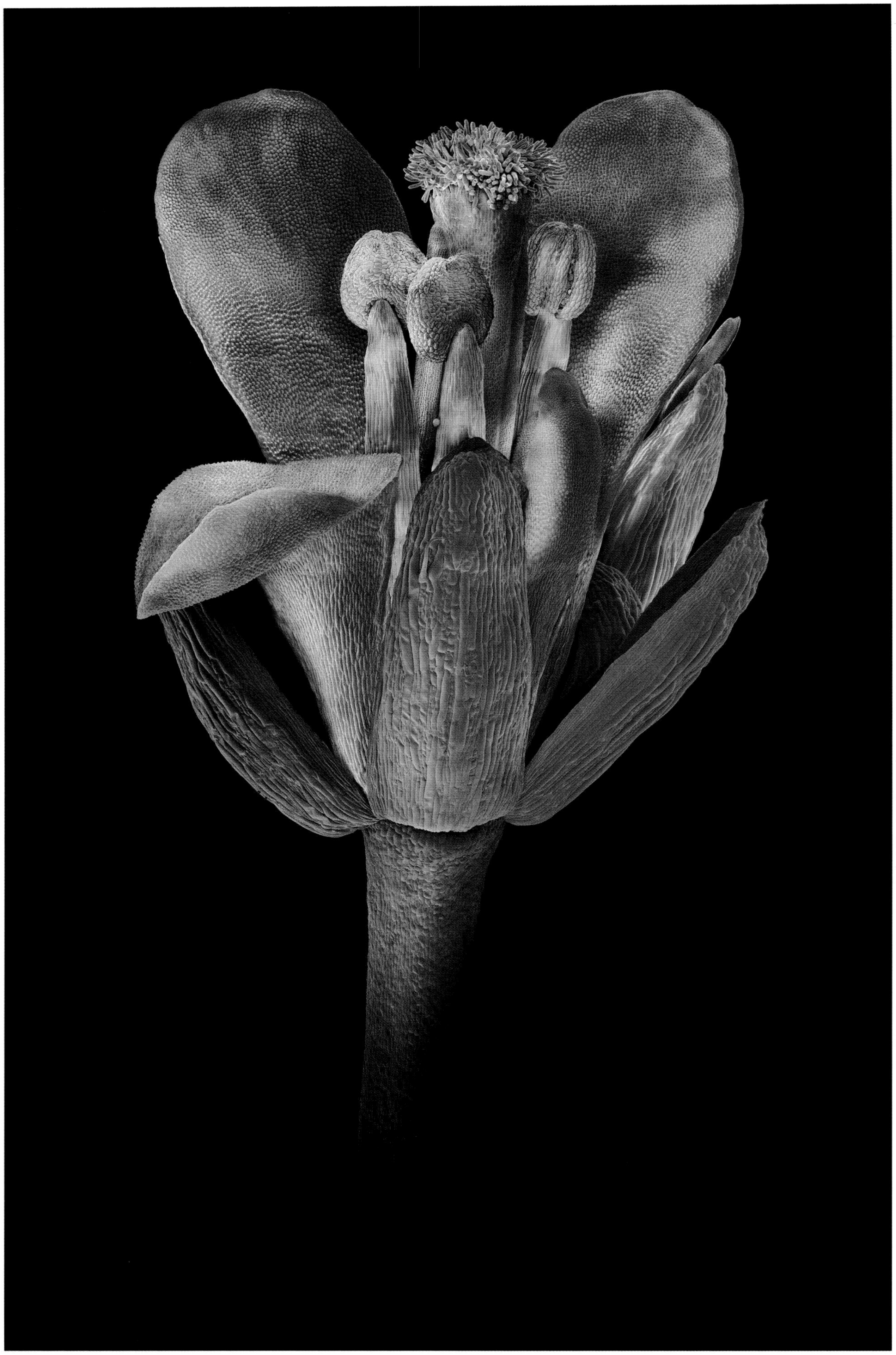

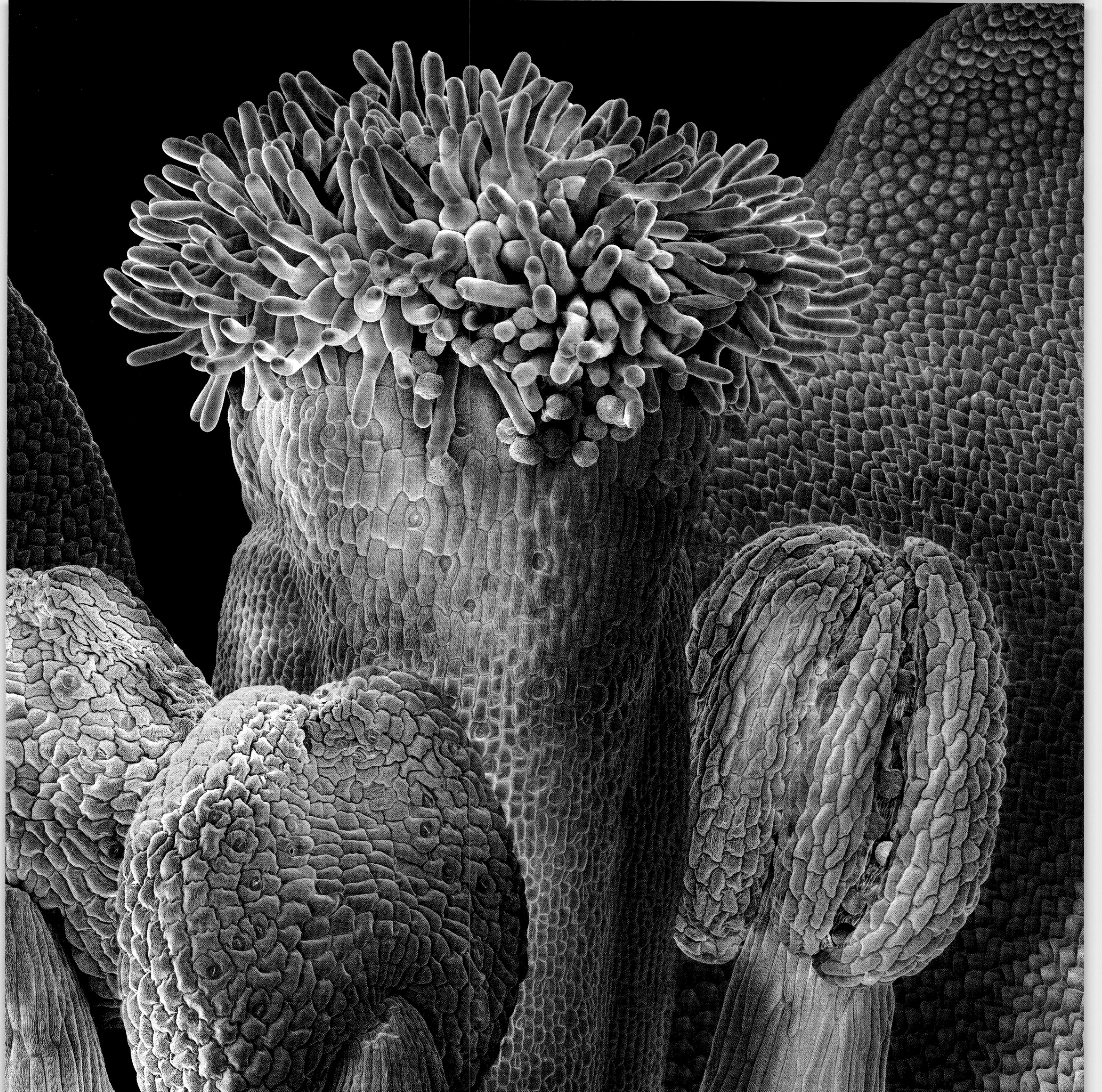

212

Facing Page, Pages 214–215 →

*Euphorbia corollata*, flowering splurge, Ottawa, Ontario.
32 x. About six millimeters wide

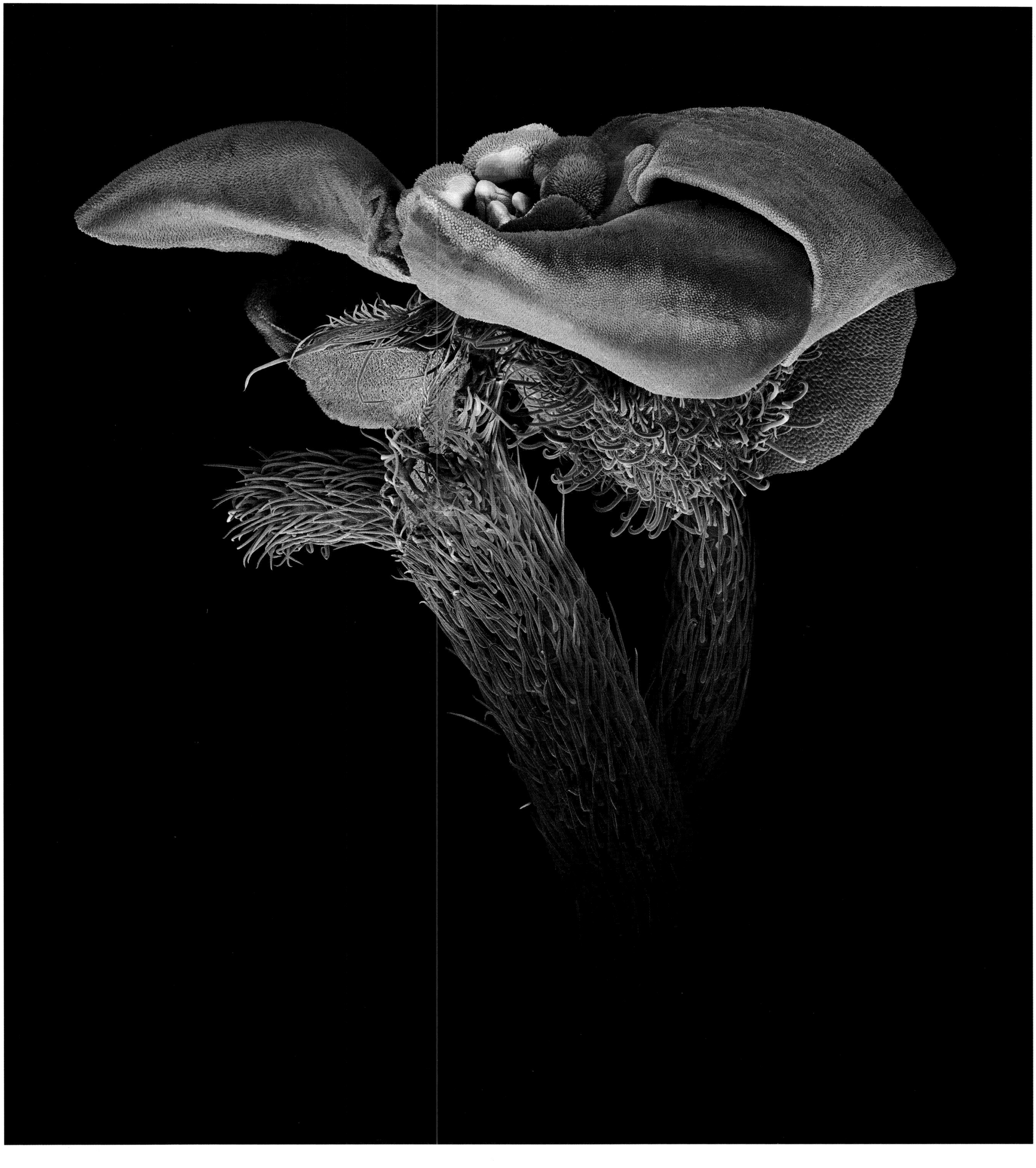

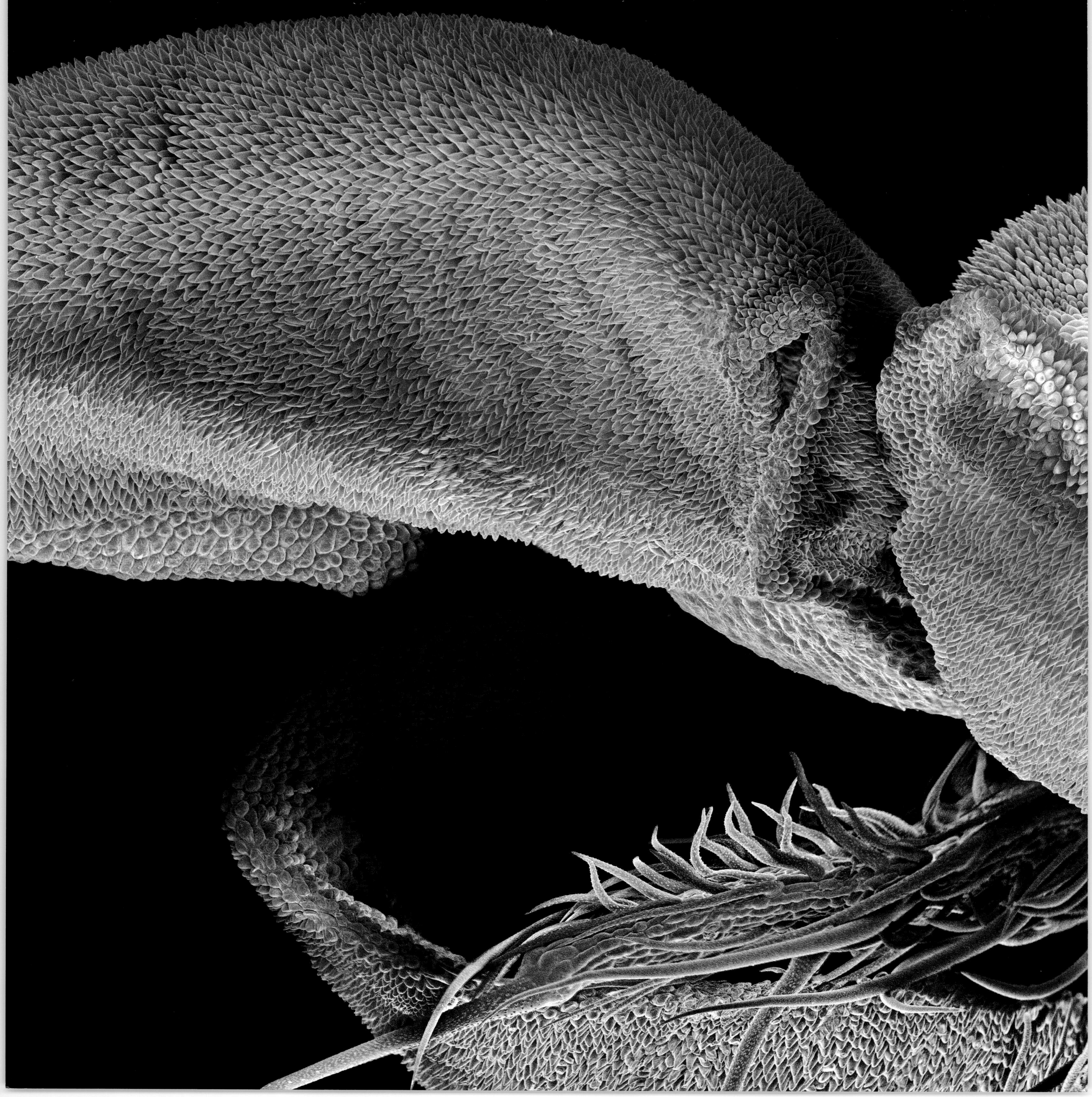

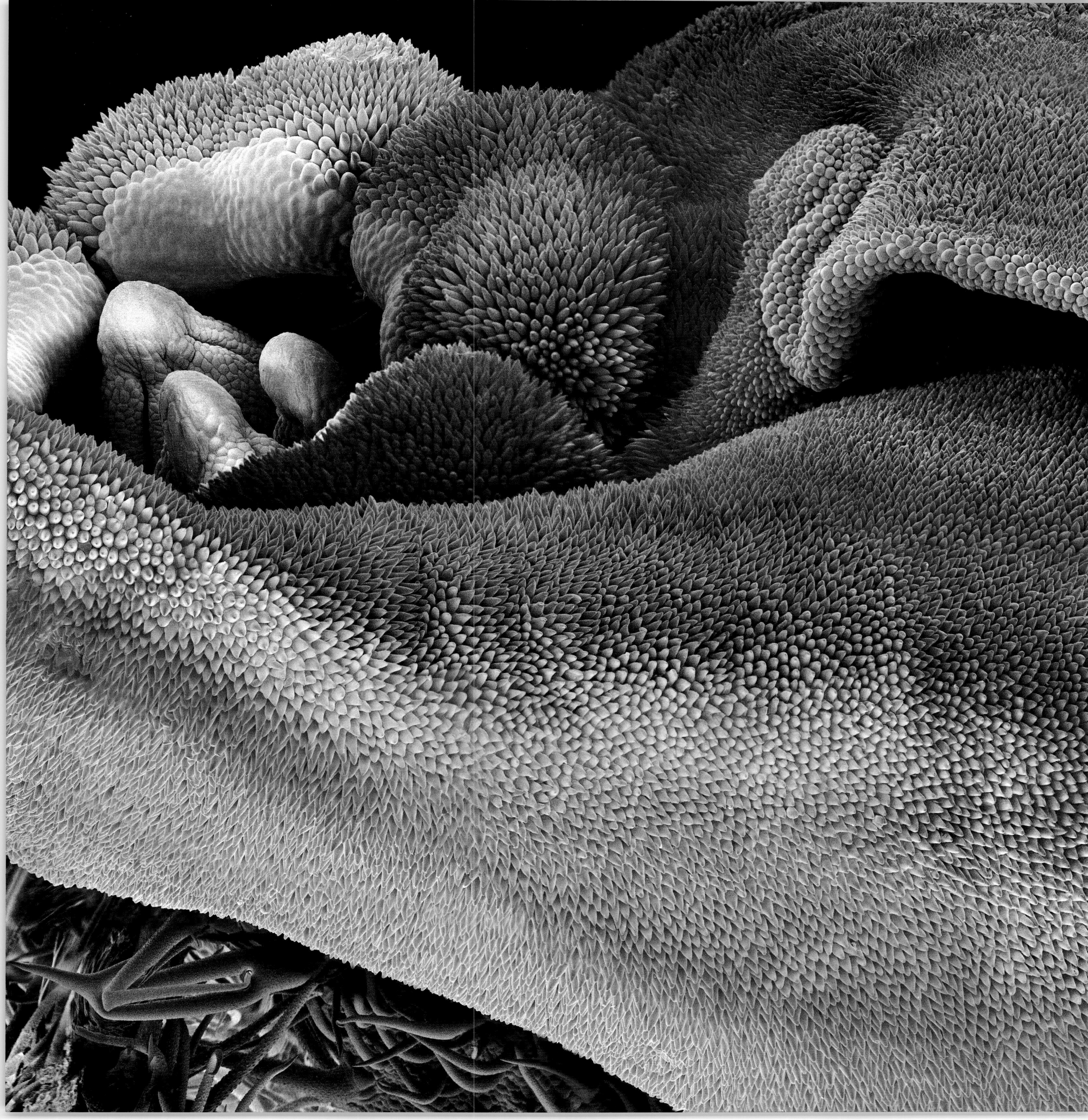

**Facing Page, Pages 218–219 →**

Flowering plant with *Aulacorthum solani*, foxglove aphid, Ottawa, Ontario.
22 x. The aphid is about 2 millimeters long, the plant about one centimeter wide

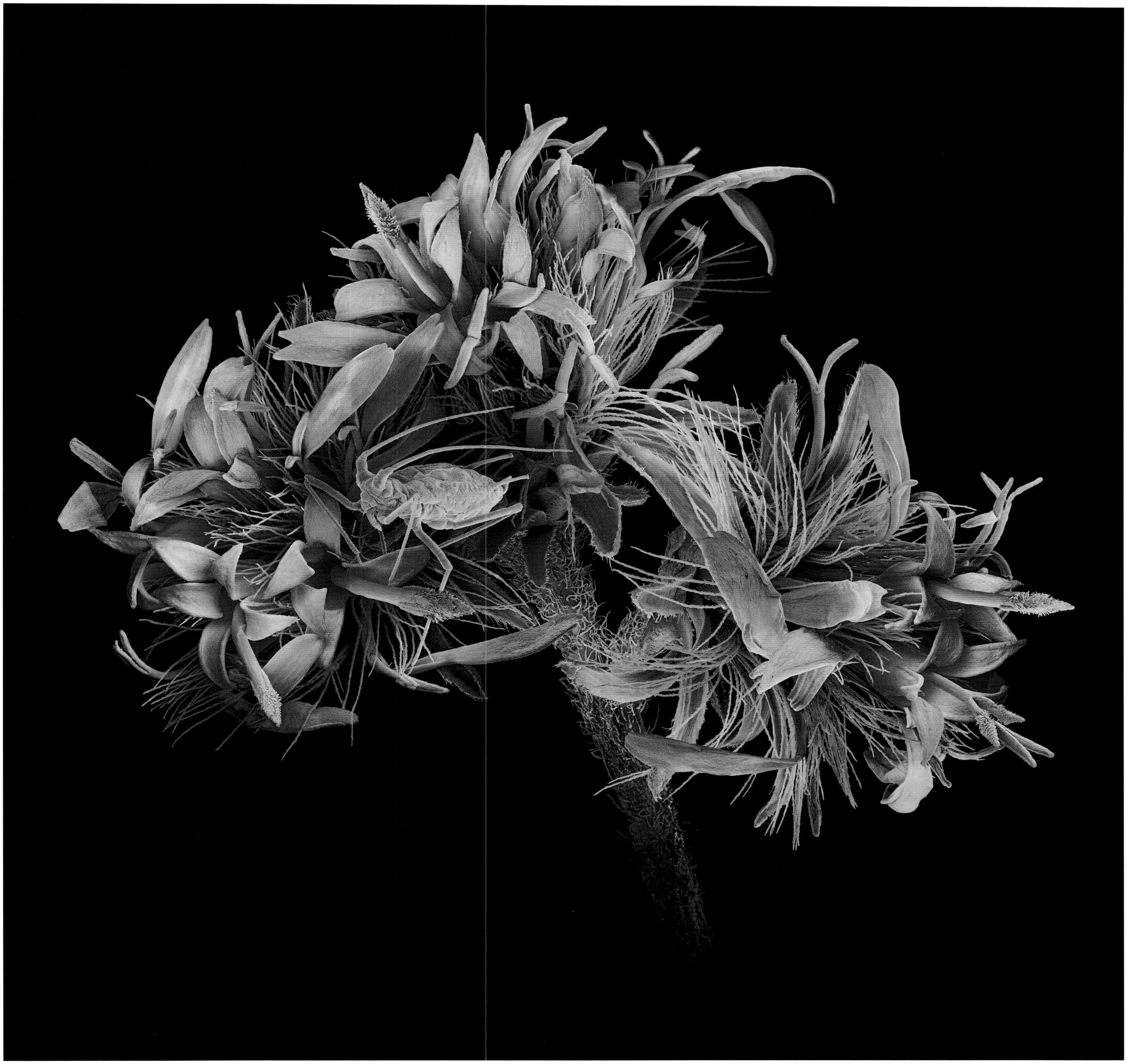

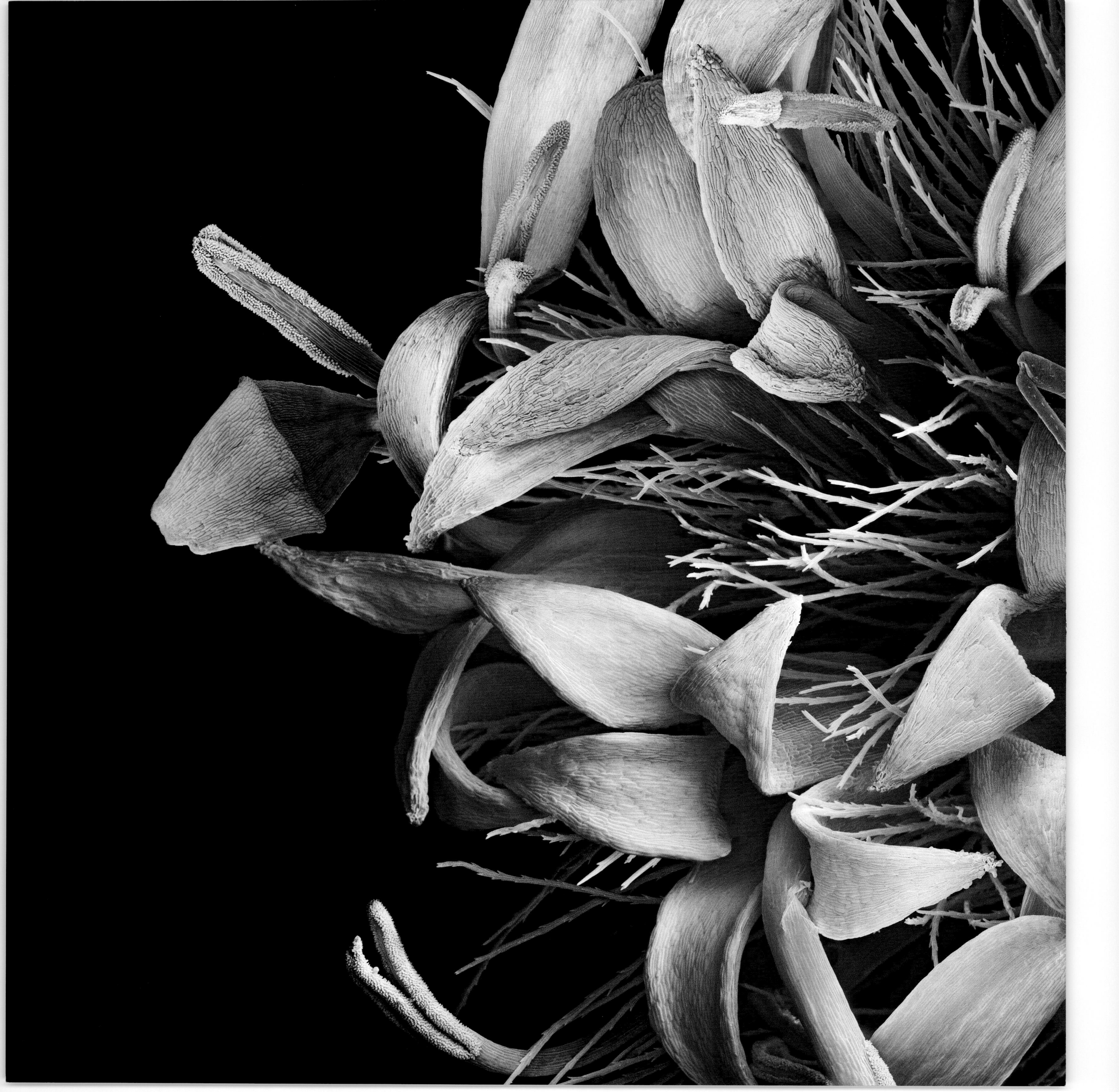

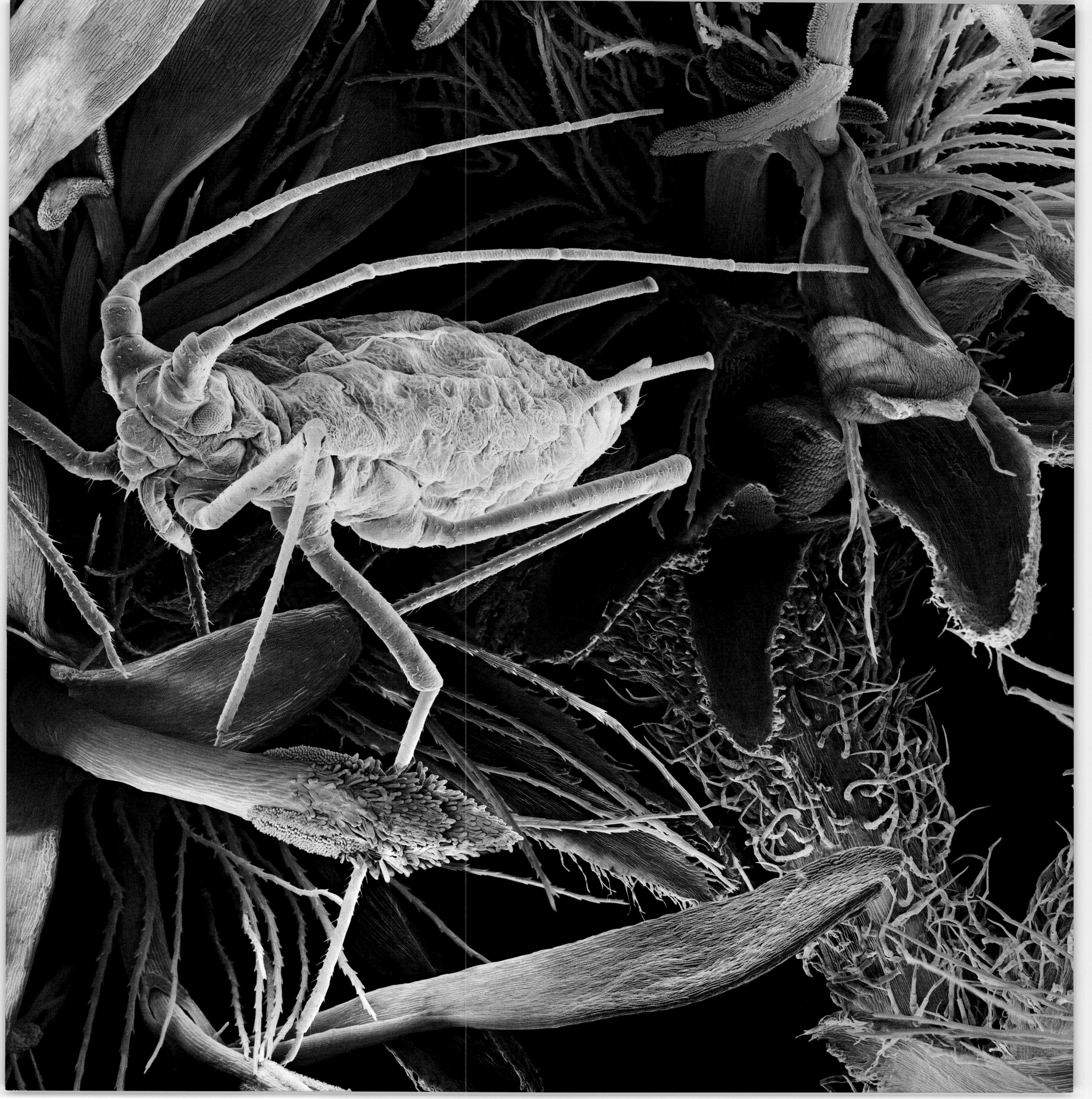

**Facing Page, Pages 222–223, pages 224–225 →**

Flowering plant with *Aphis gossypii* aphids, Northern Velebit National Park, Croatian Adriatic. 33 x. The aphids are about 700 microns long, or 0.7 millimeter, the plant about five millimeters wide

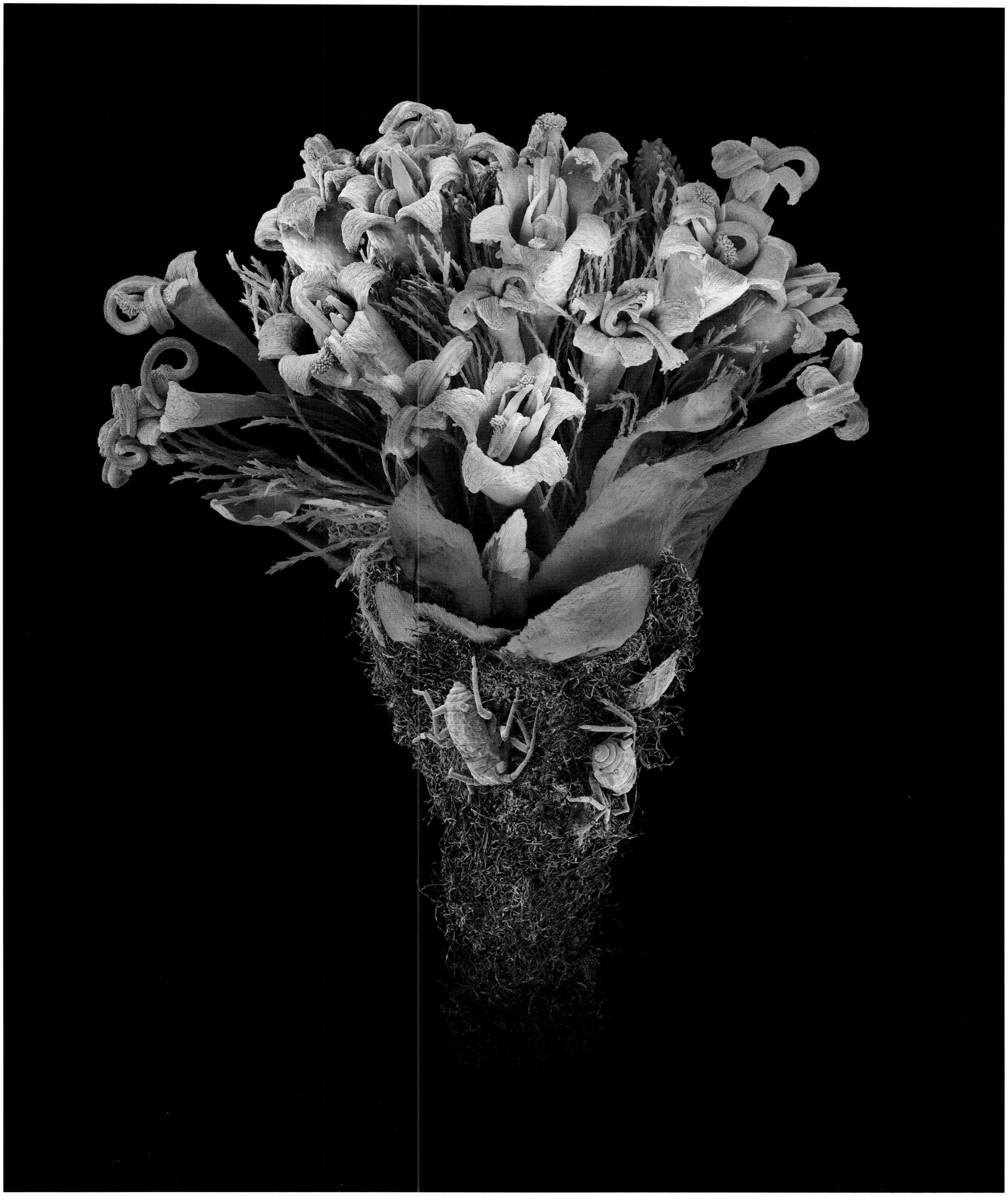

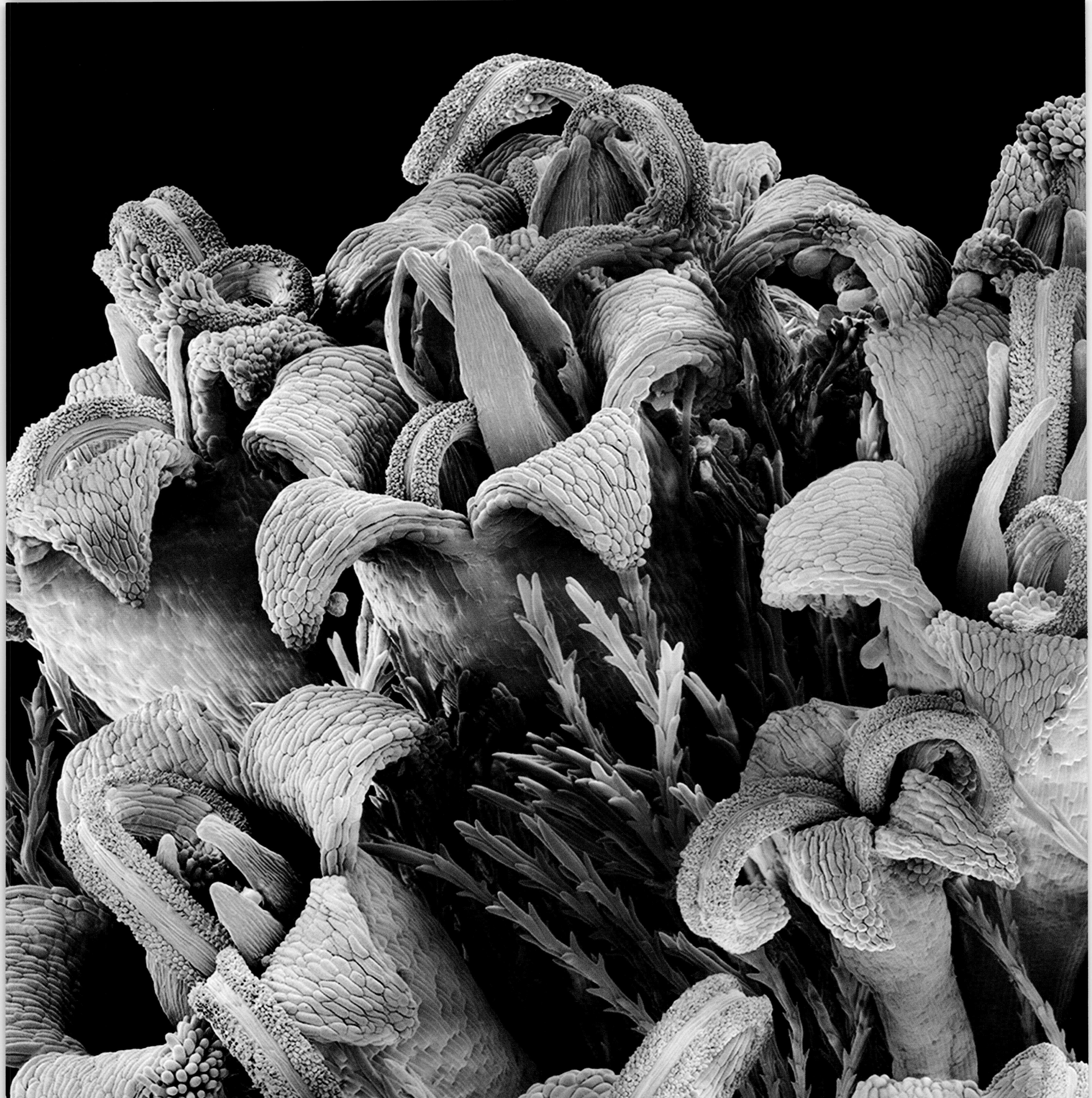

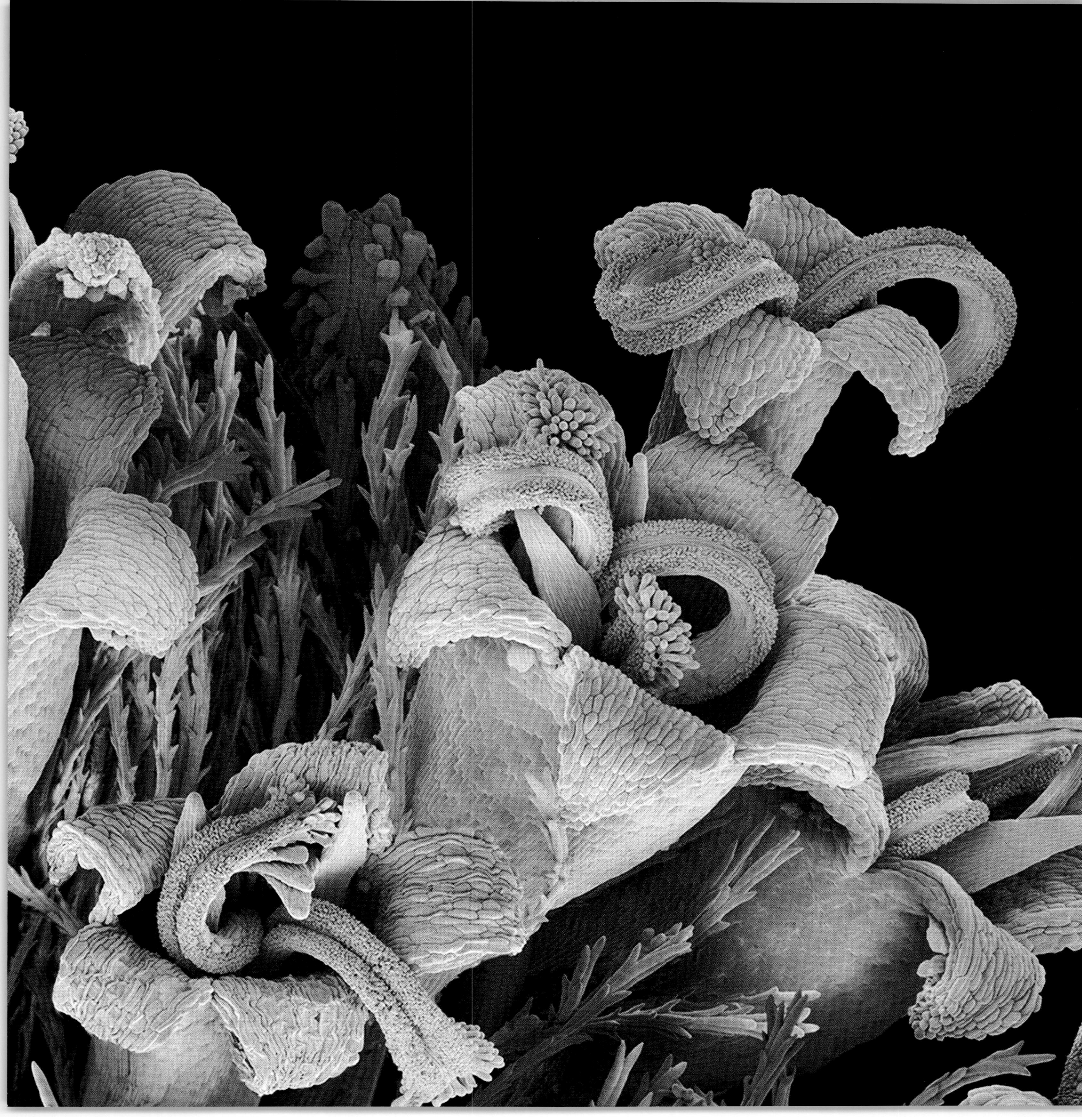

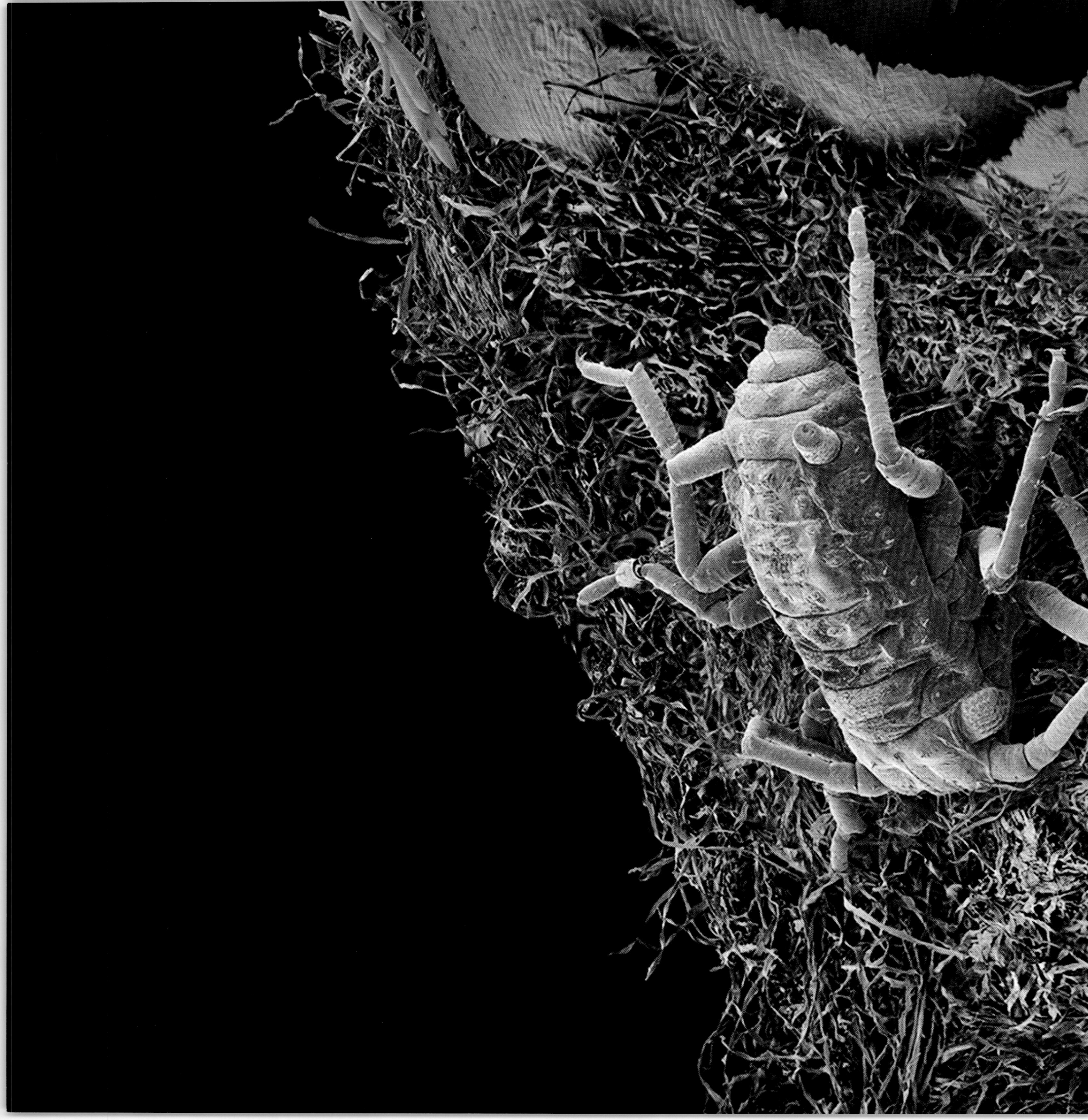

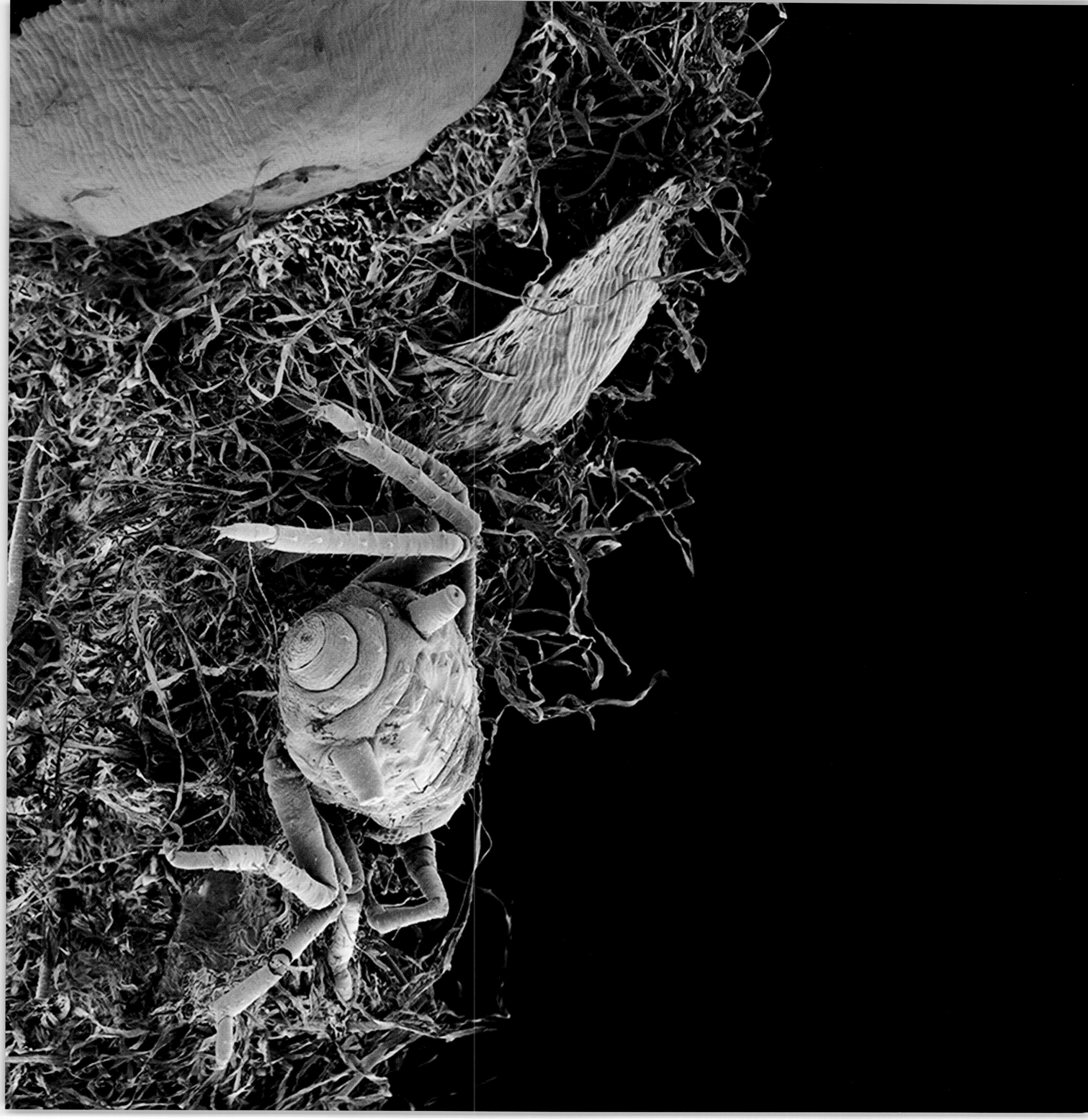

Facing Page, Pages 228–229 →

*Mentha arvensis*, wild mint, with ant of the genus *Leptothorax*, Ottawa, Ontario.
20 x. The ant is about one millimeter wide, the plant about a centimeter tall

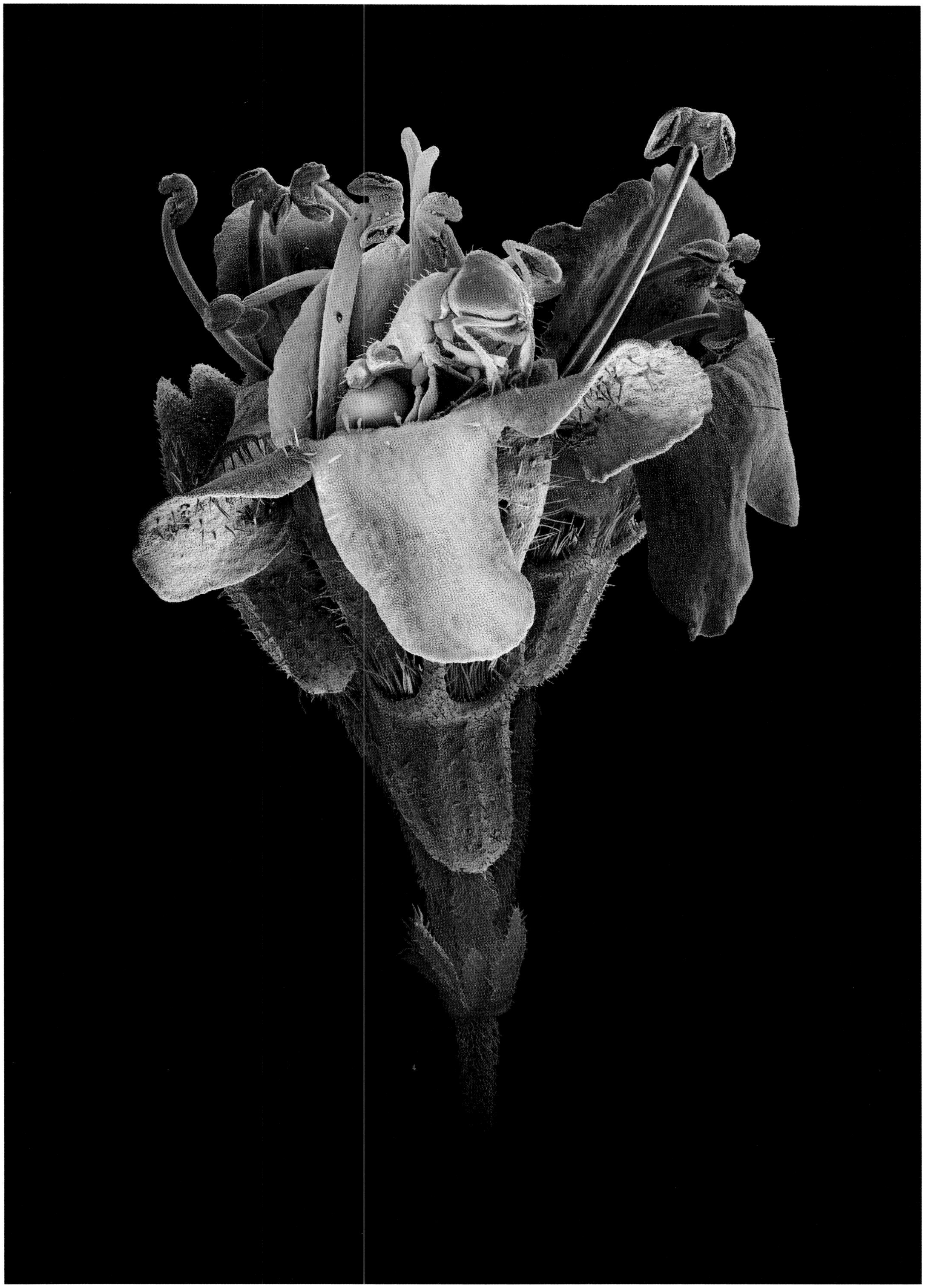

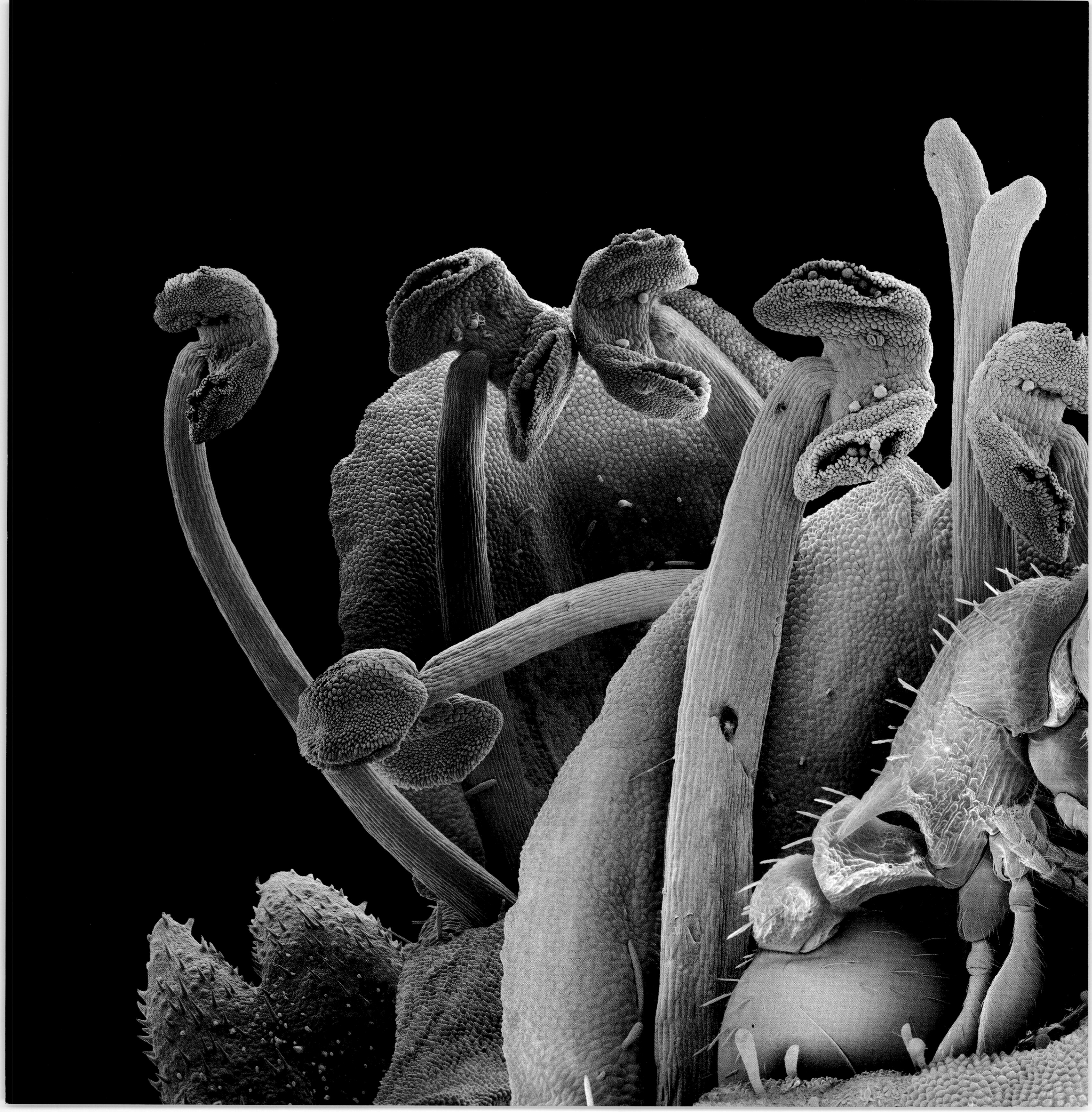

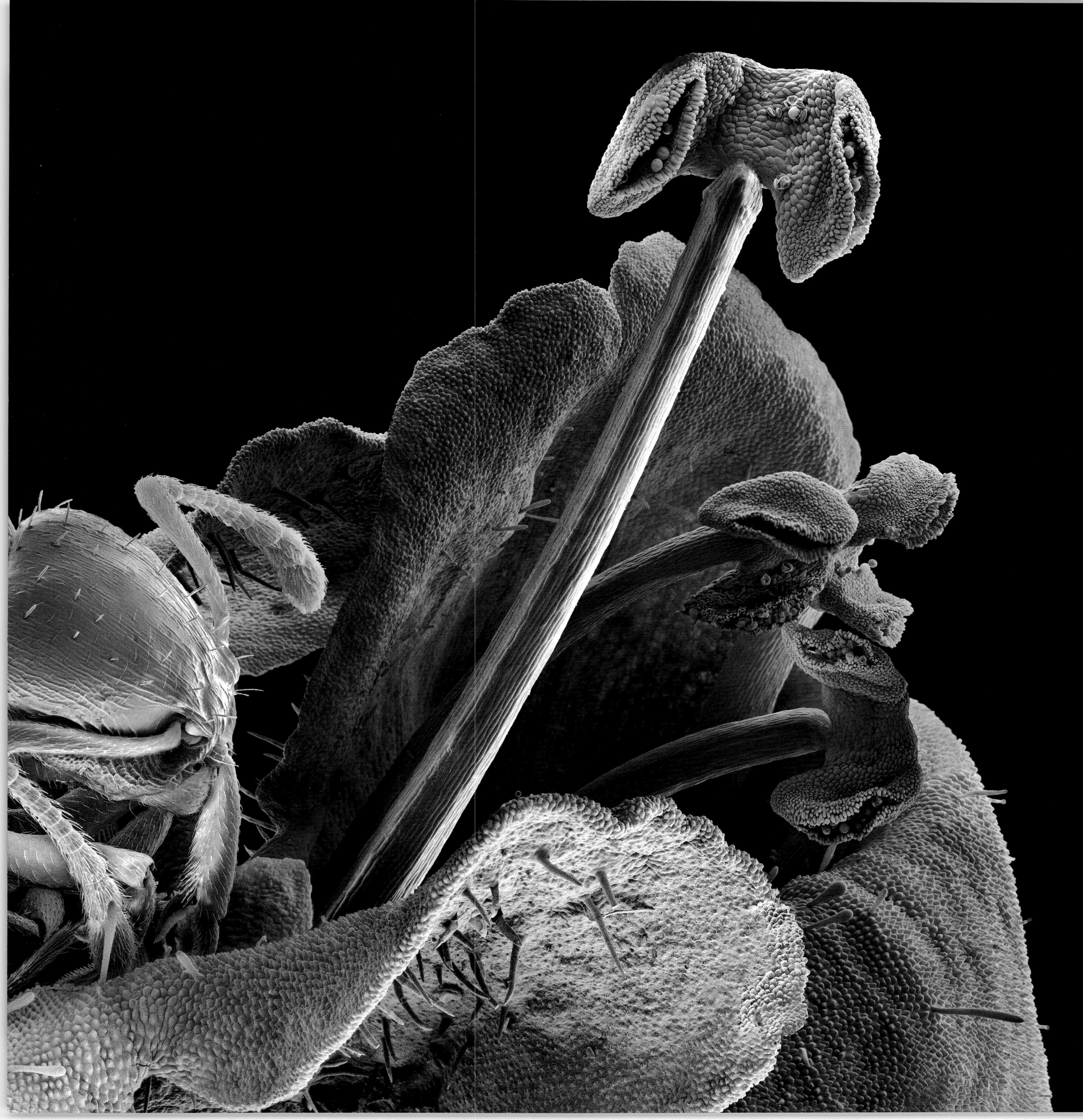

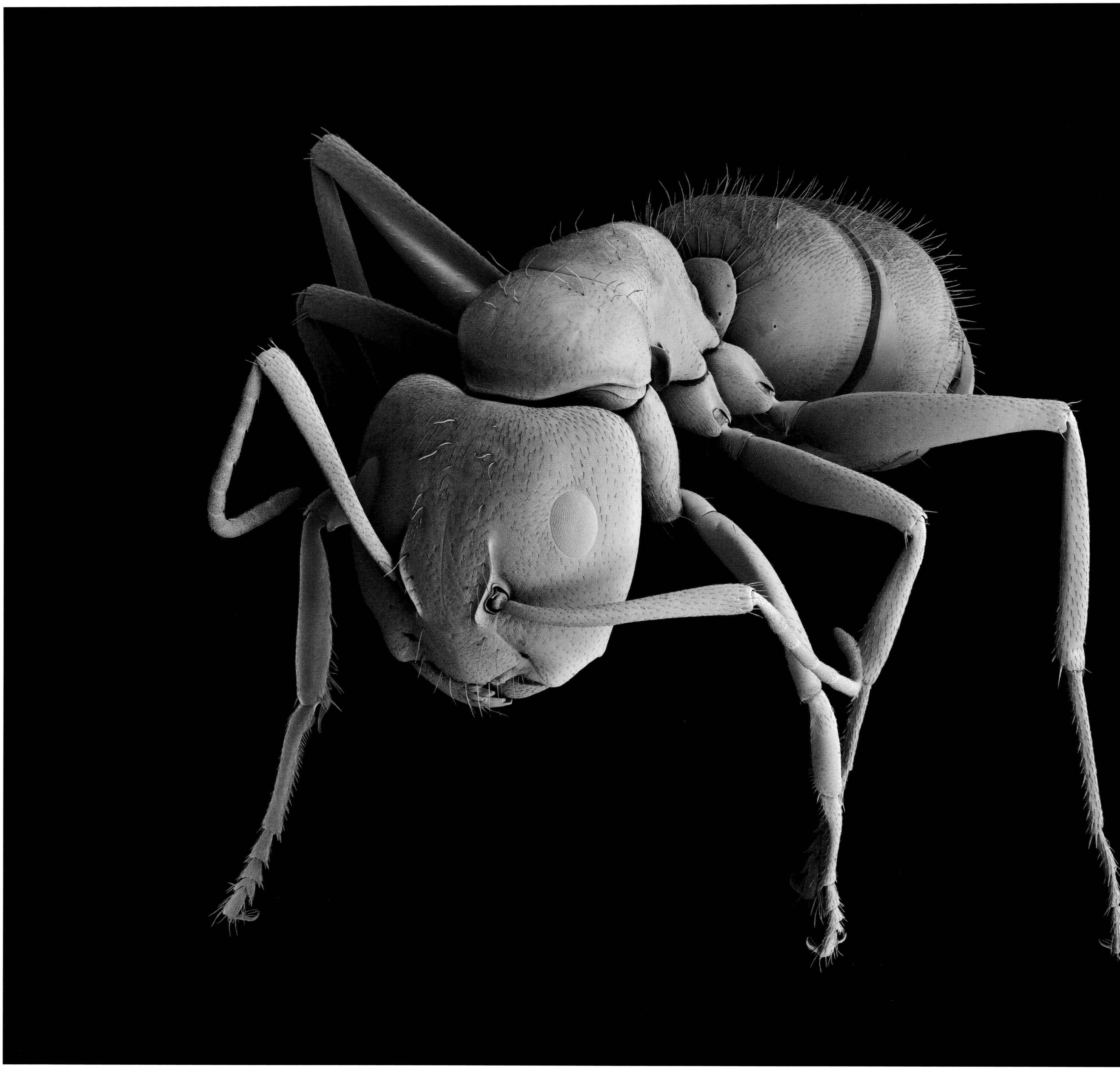

**Facing Page**

*Camponotus pennsylvanicus*, black carpenter ant, Ottawa, Ontario.
31.4 x. Seven millimeters wide

**Page 232 →**

*Lechriops oculatus* weevil, Leon County, Florida.
100 x. About 2 millimeters wide

**Page 233 →**

*Curculio monticola*, acorn weevil, Cochise County, Arizona.
60 x. The visible part is about 2 millimeters wide

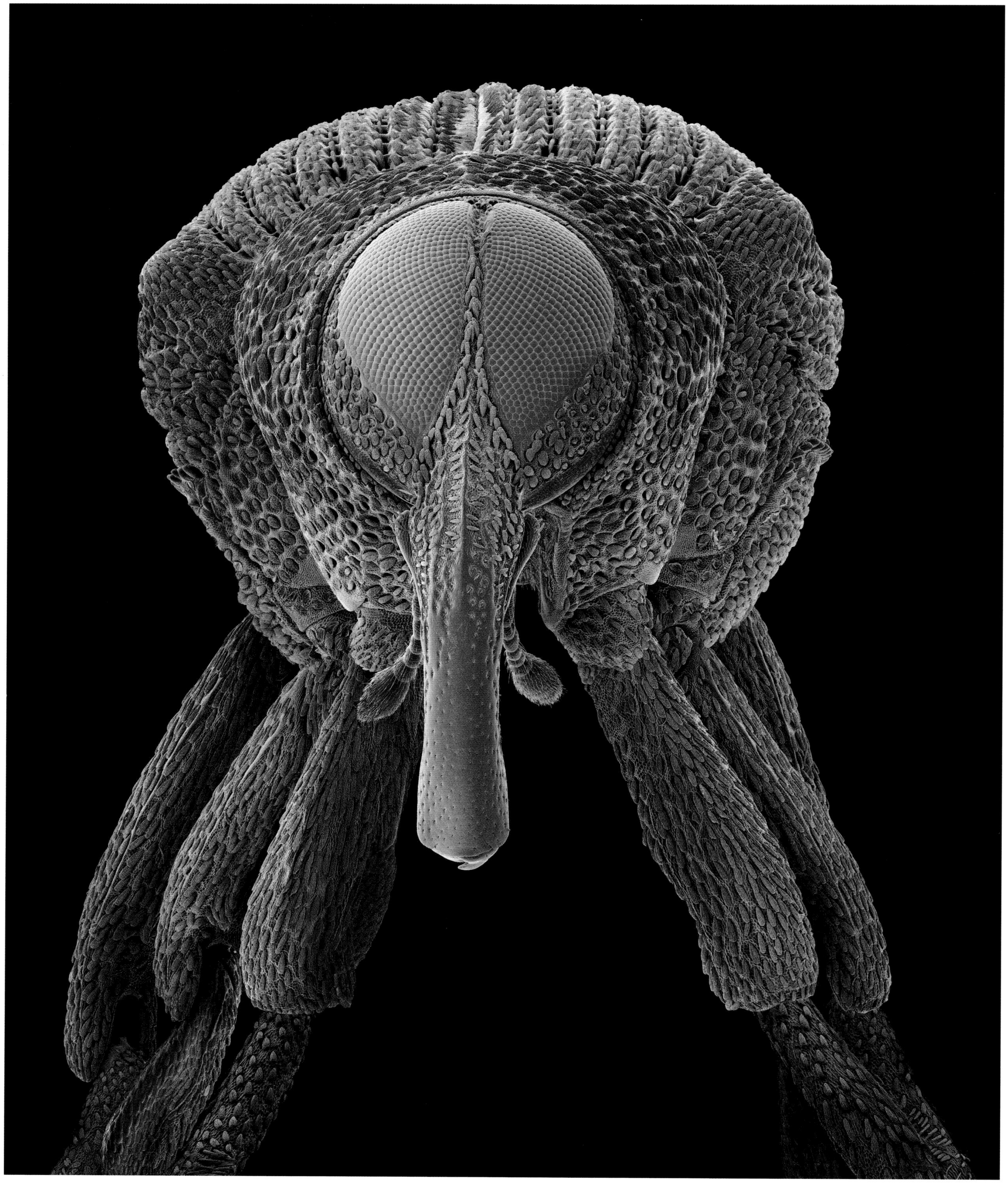

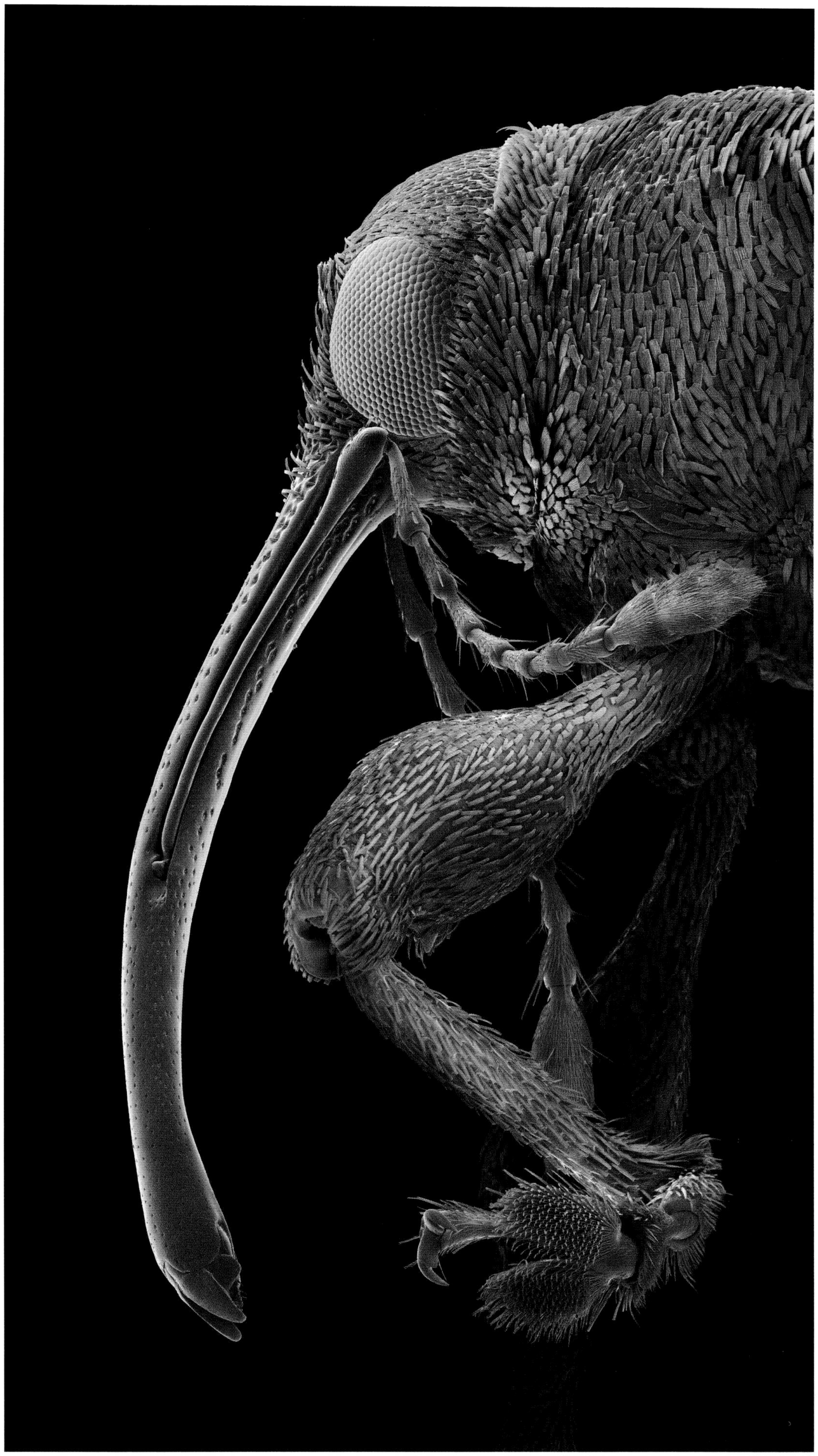

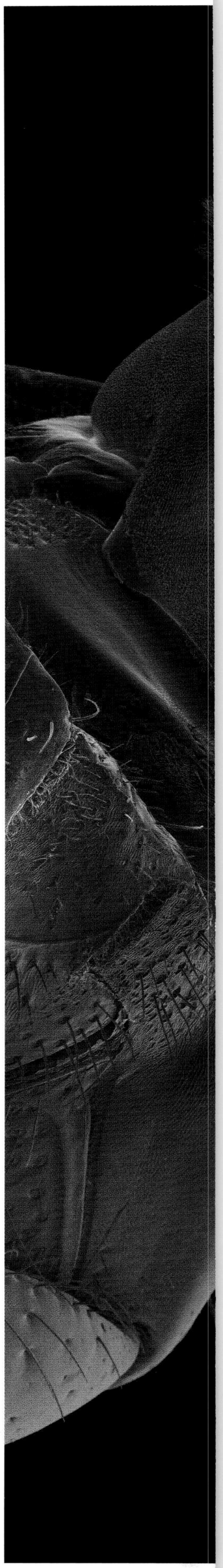

**Facing Page, Pages 236–237 →**

*Anomiopus genieri* dung beetle, Rio Sipao, Bolívar, Venezuela.
41 x. About seven millimeters from tip to left margin

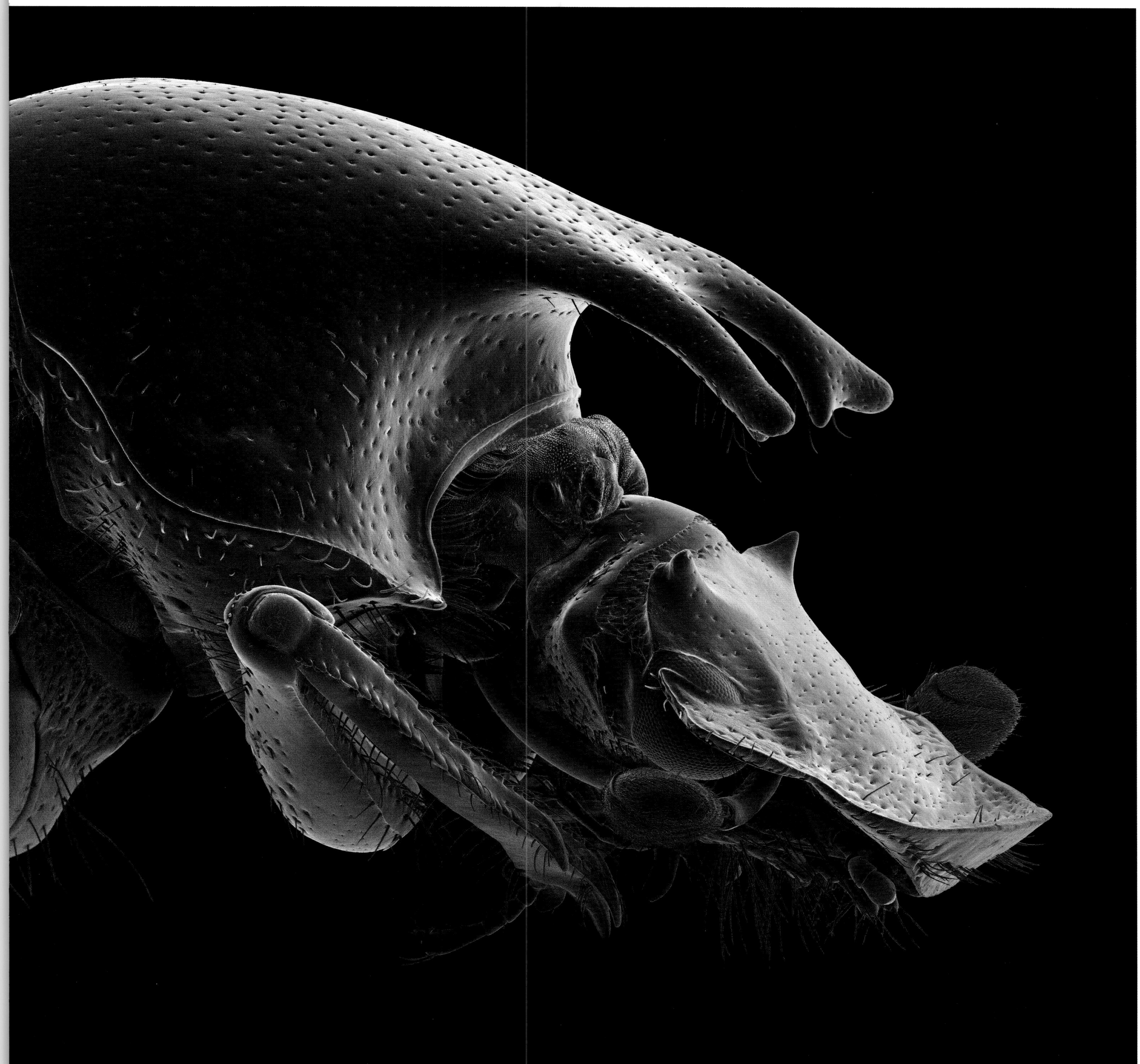

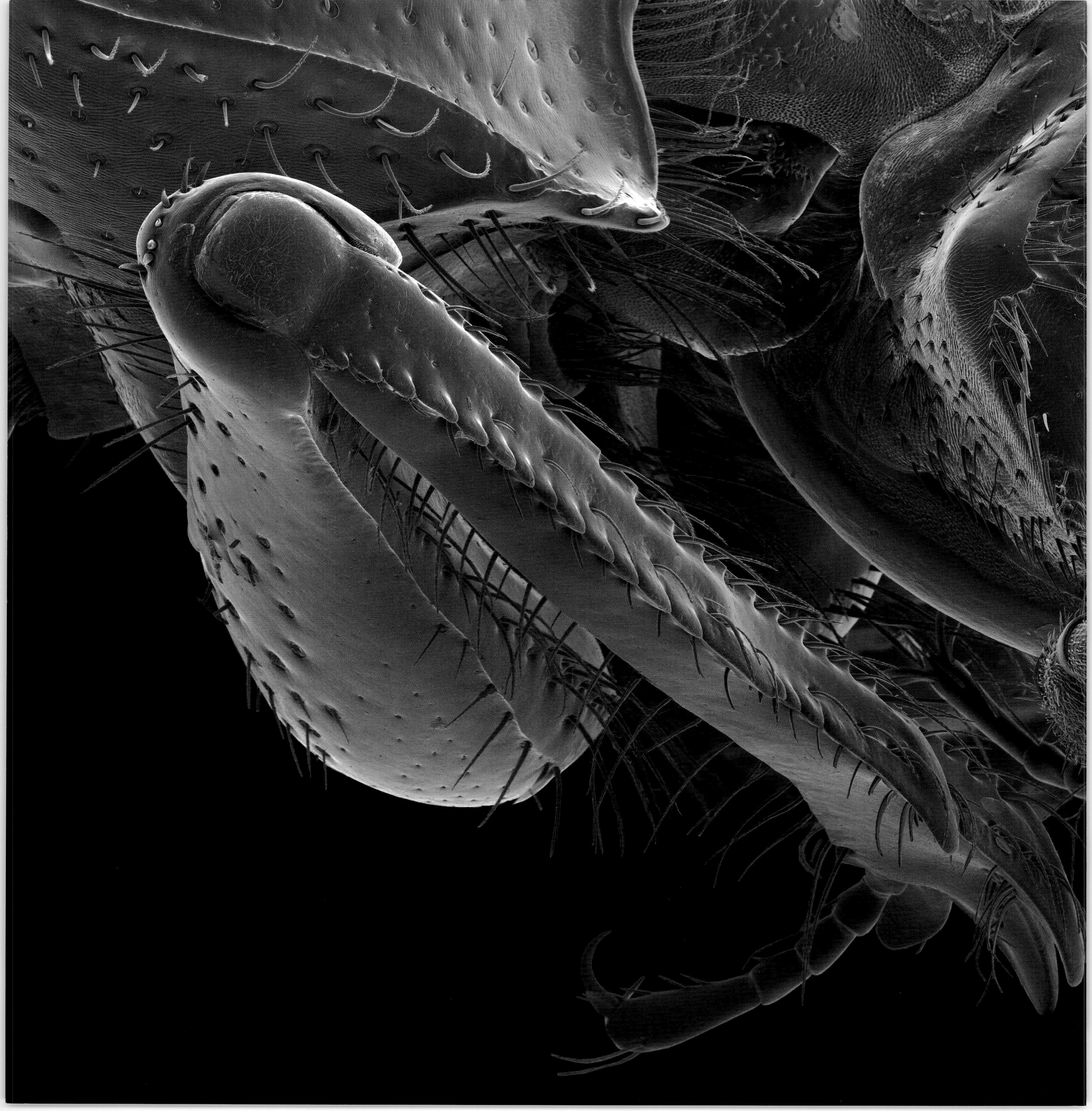

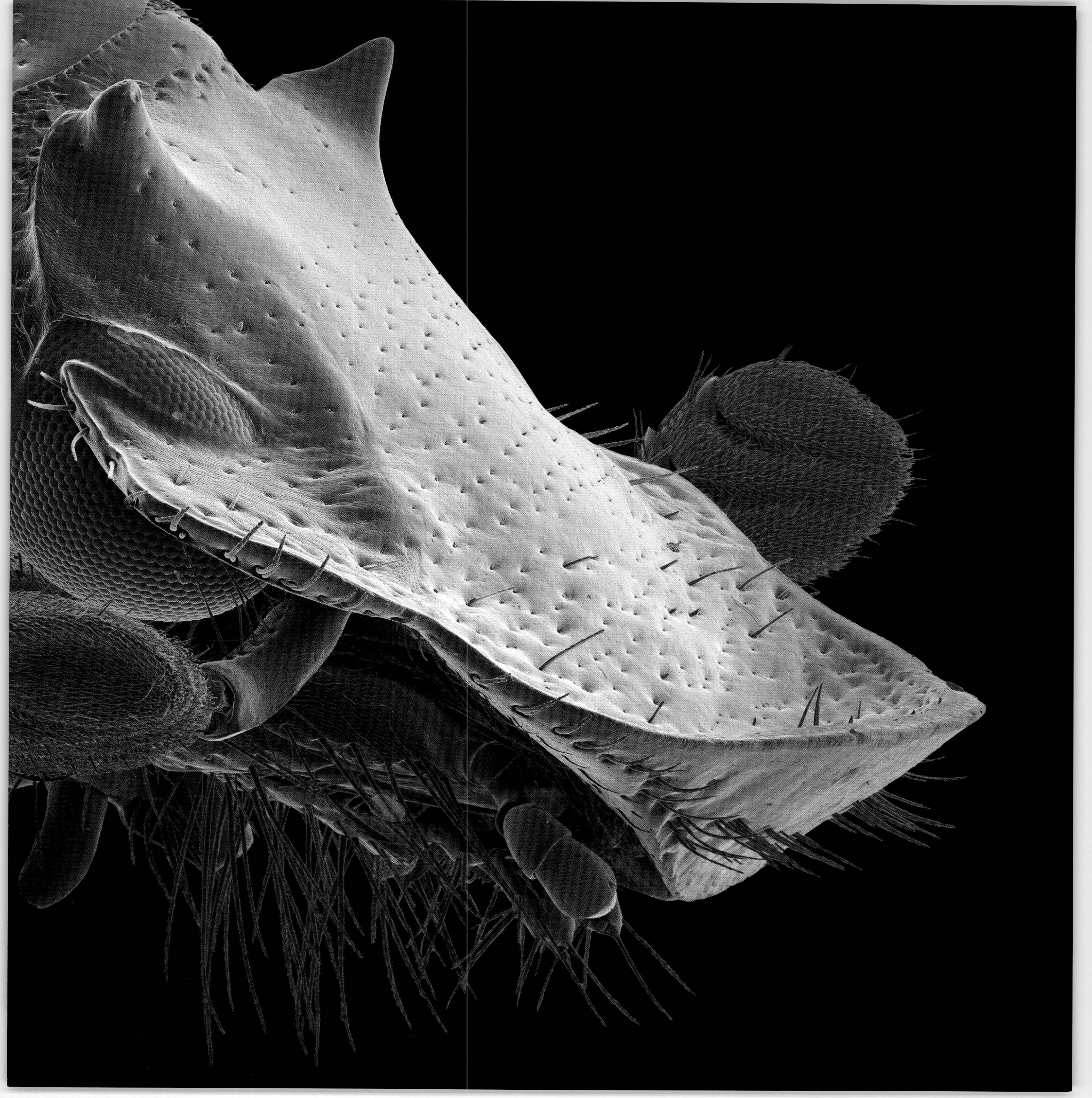

**Facing Page**

*Onthophagus limonensis* dung beetle, Turrialba, Cartago Province, Costa Rica.
93 x. About three millimeters from side to side

**Pages 240–241 →**

Side view of the same *Onthophagus limonensis*.
63 x. About seven millimeters in length

**Page 242, Page 243 →**

Side and front views of *Onthophagus francoisgenieri* dung beetle, Madang Baltabang, Madang, Papua New Guinea.
36.6 x. About 6 millimeters wide

**Page 244, Page 245 →**

Tarsus and side view of *Gymnopleurus humeralis* dung beetle, Pafuri, Kruger National Park, Limpopo Province, South Africa.
Tarsus is 150 x and 1.6 millimeters from picture top to bottom. The area on page 253 is 59 x and 4 millimeters wide

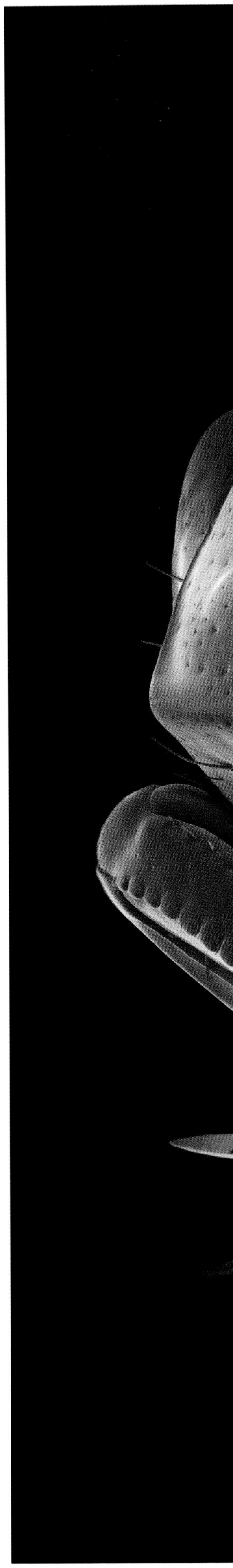

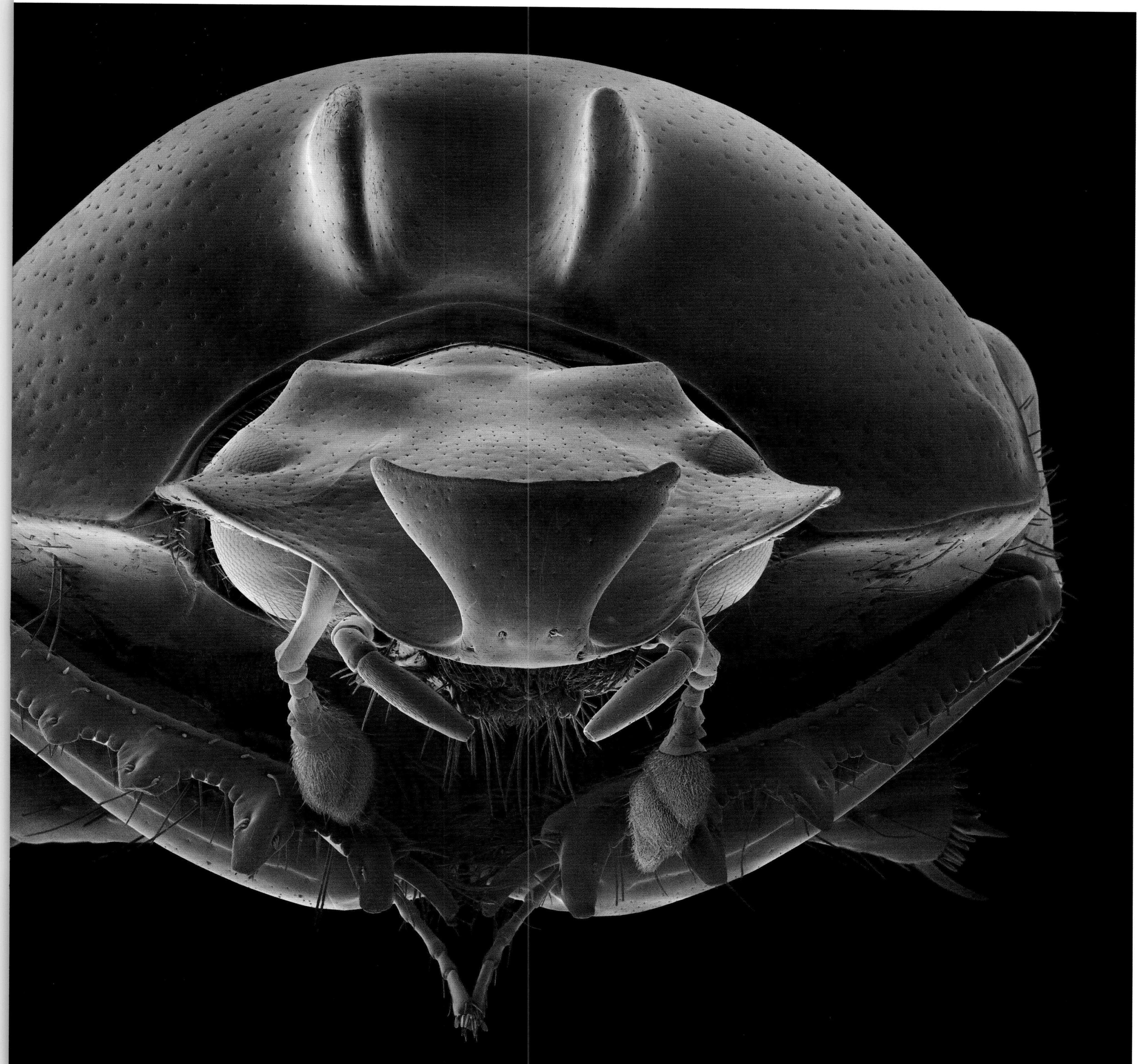

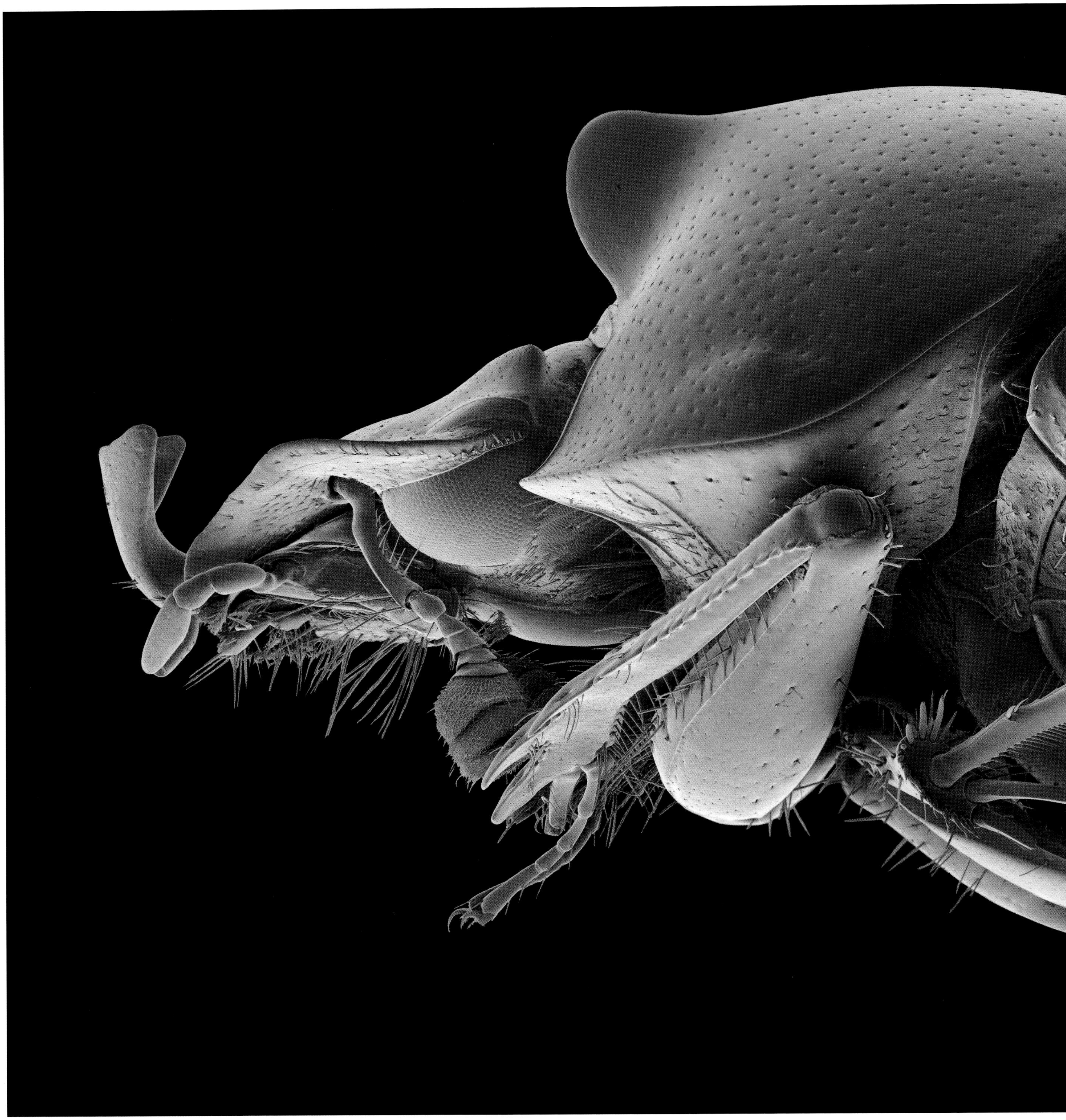

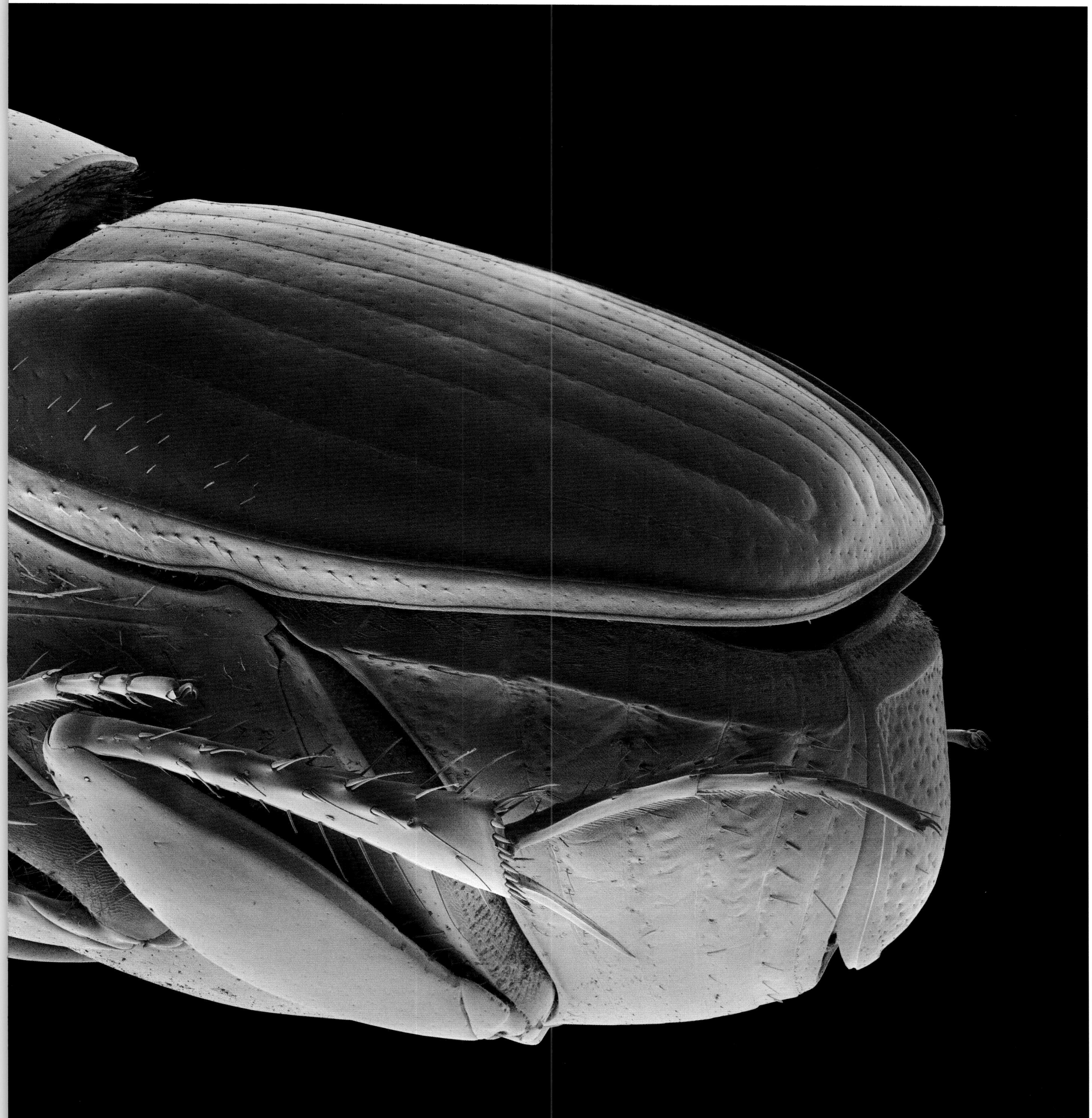

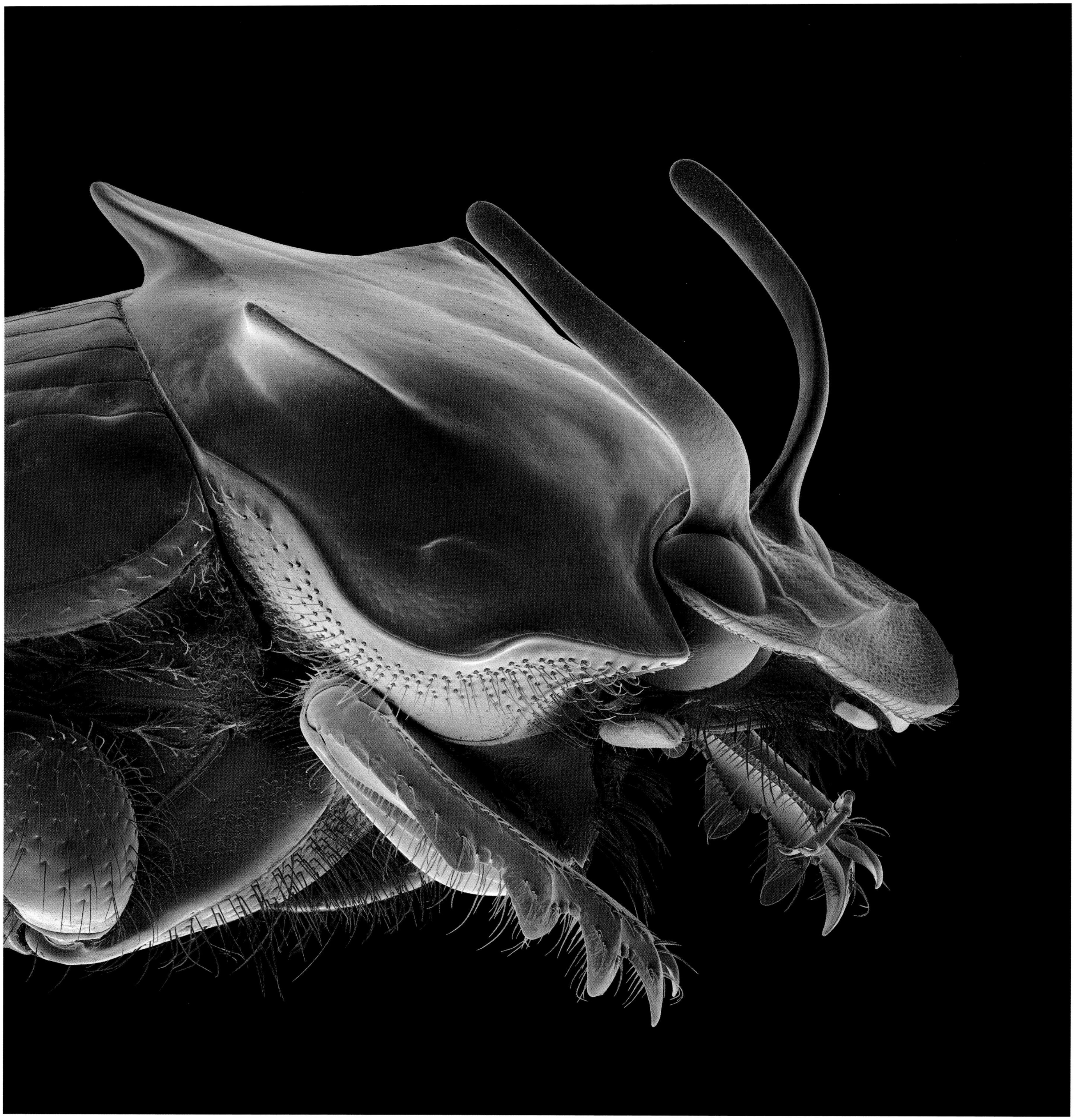

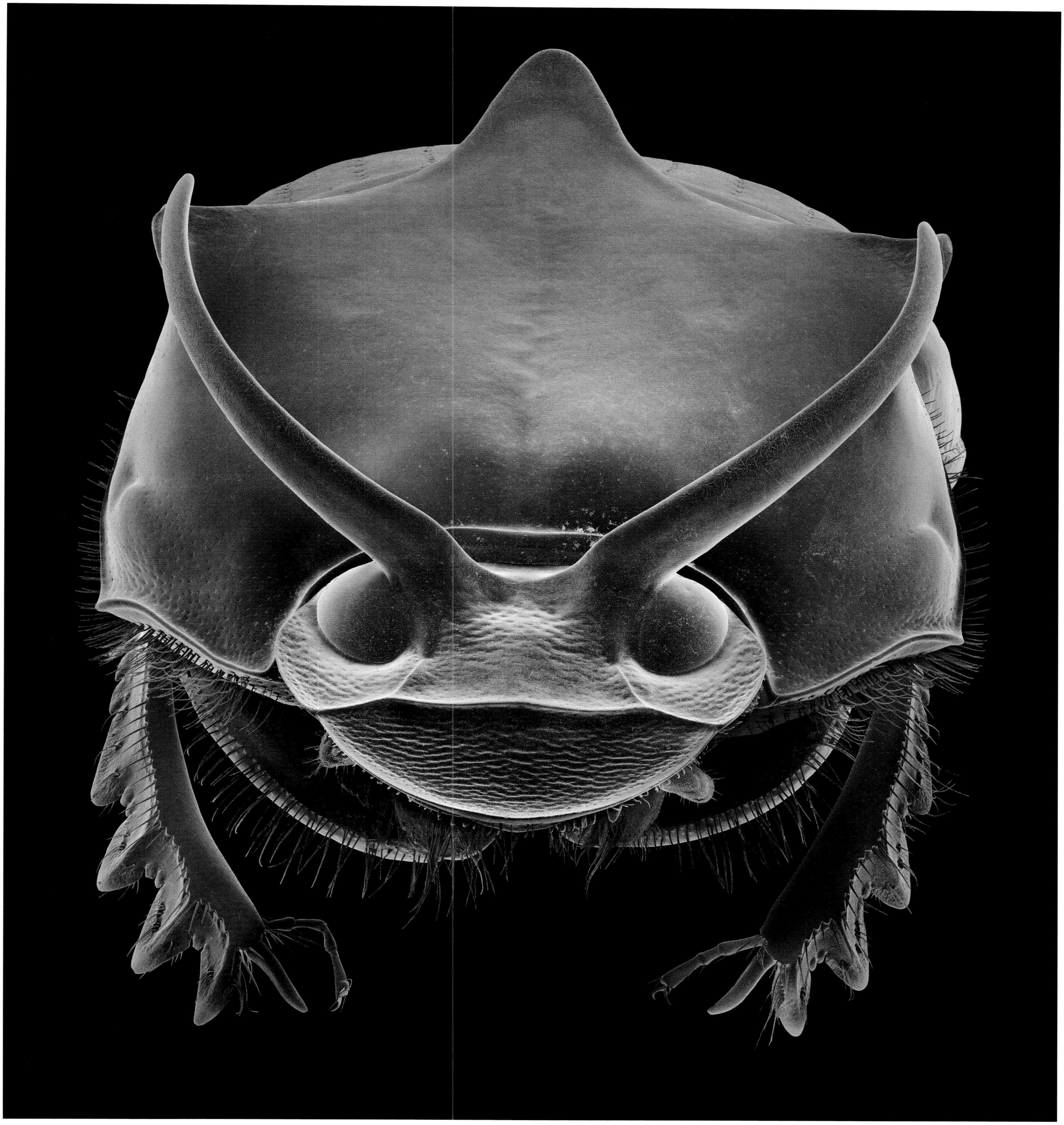

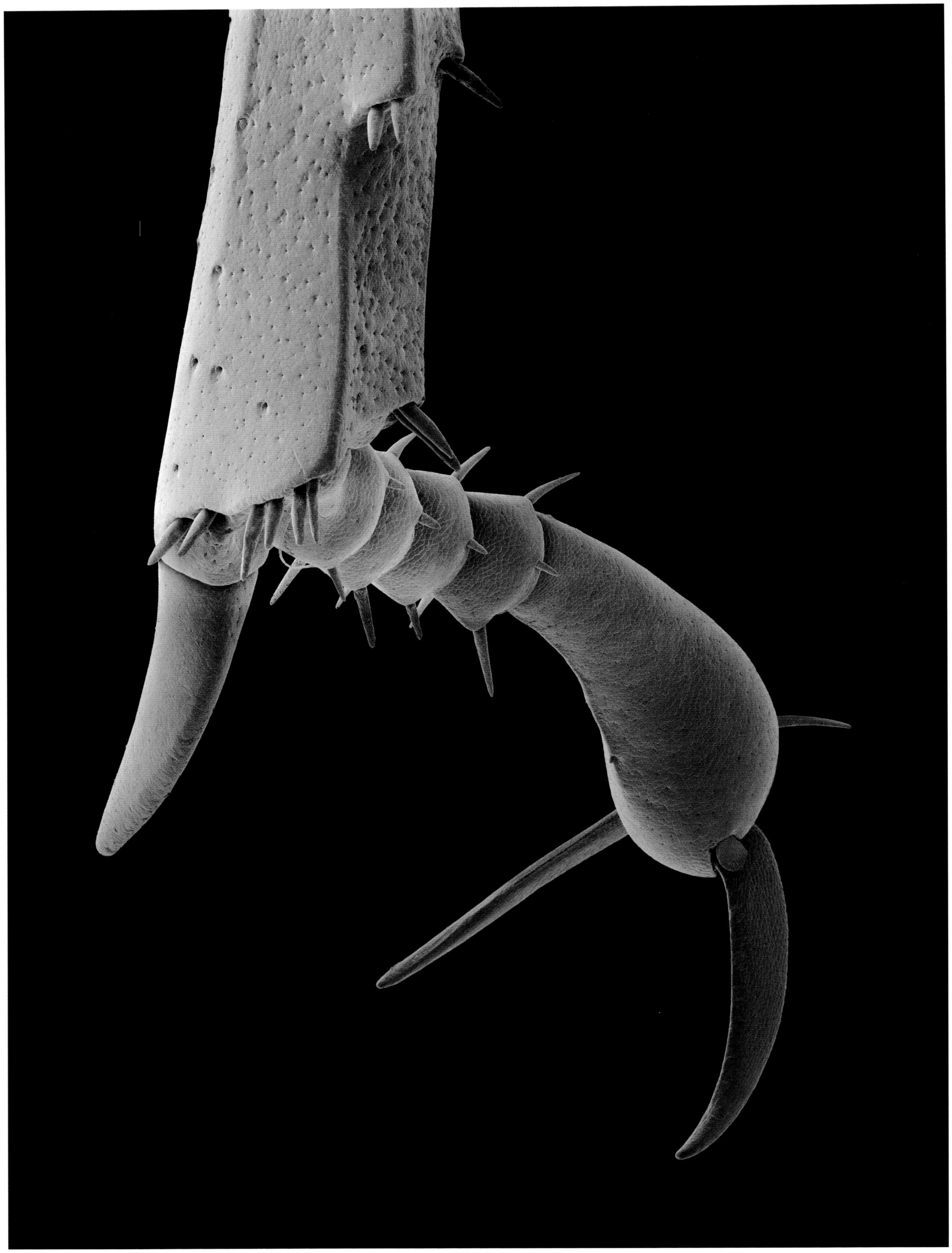

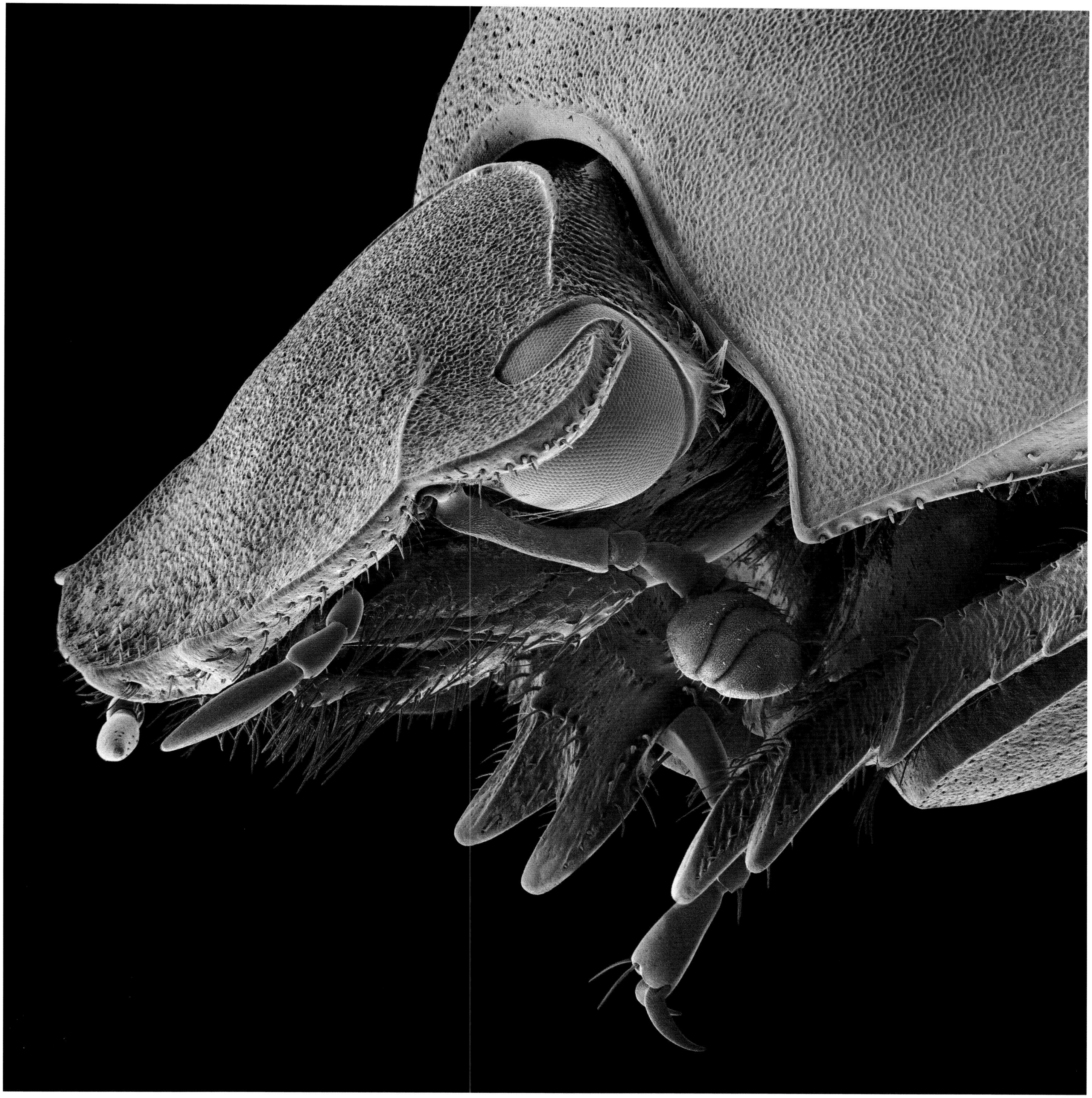

AER

# FLIGHT

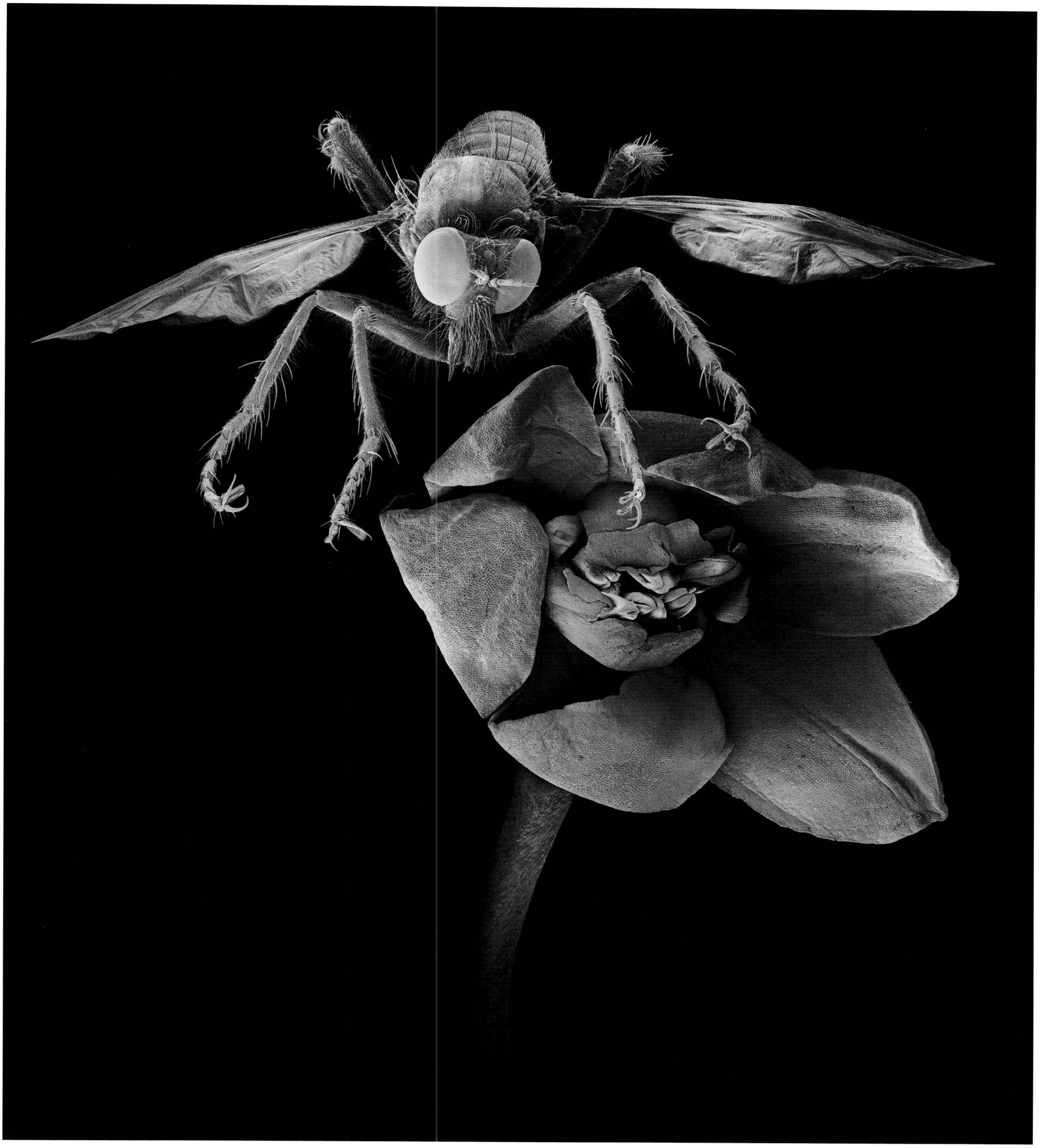

← Page 247

Asilidae family robber fly and flowering plant, Empress Dune Field, Alberta, Canada.
21 x. The visible area is slightly wider than one centimeter

Facing Page, Pages 250–251 →

Two views of a butterfly from the Lycaenidae family, Northern Velebit National Park, Croatian Adriatic.
36 x. The visible parts of the butterfly are about 5 millimeters wide

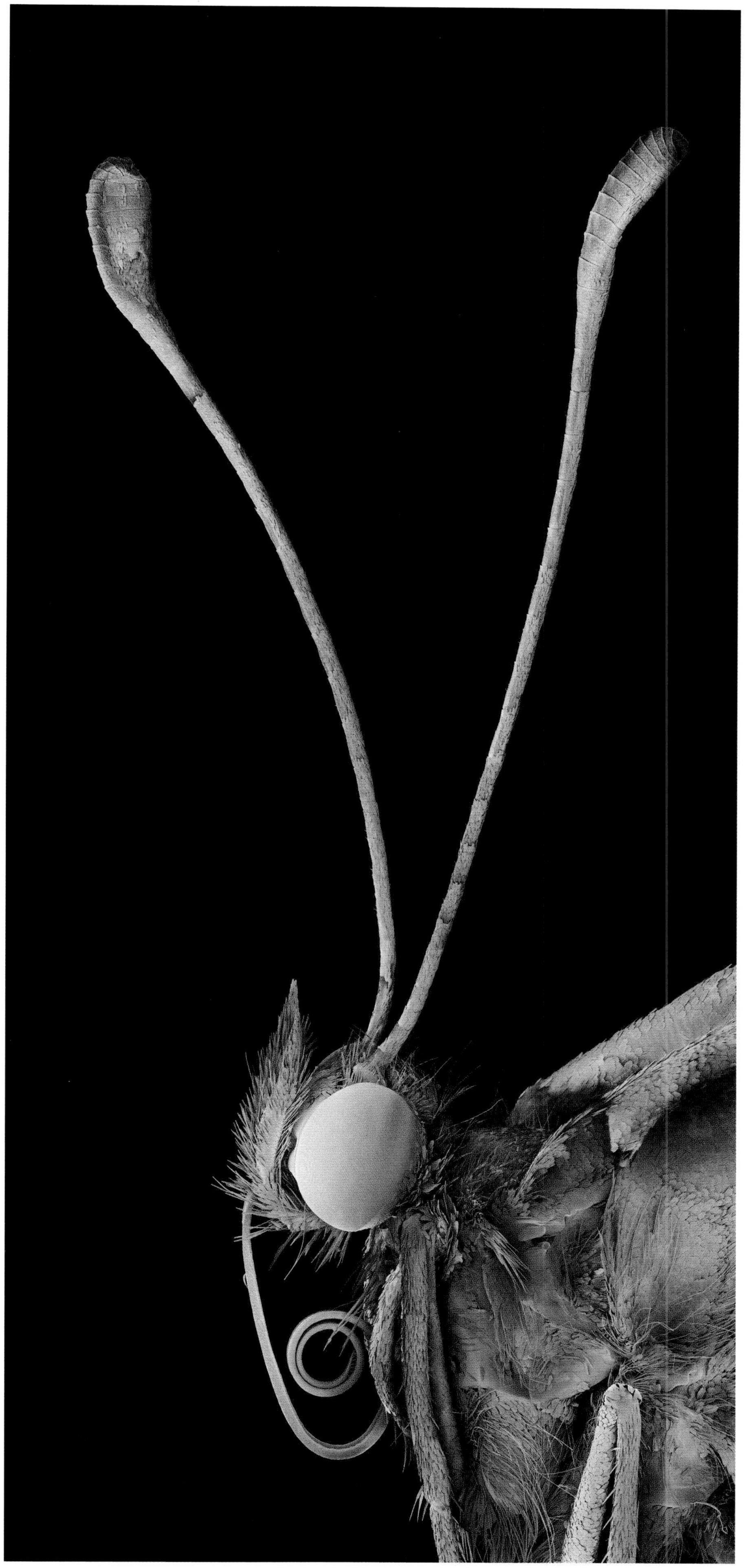

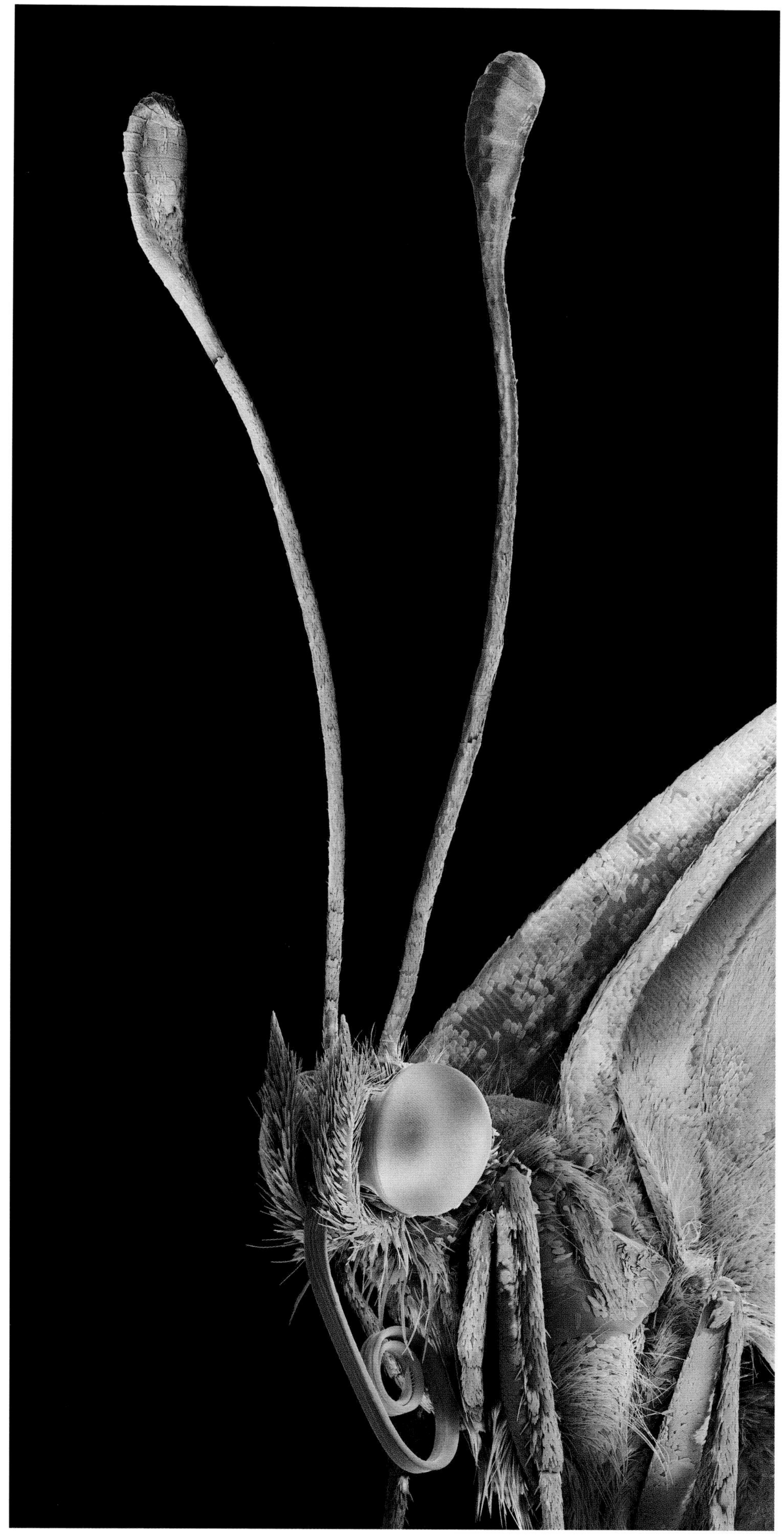

**Facing Page**

Detail view of butterfly from the Lycaenidae family, Northern Velebit National Park, Croatian Adriatic. 60.5 x. About 4.8 millimeters wide

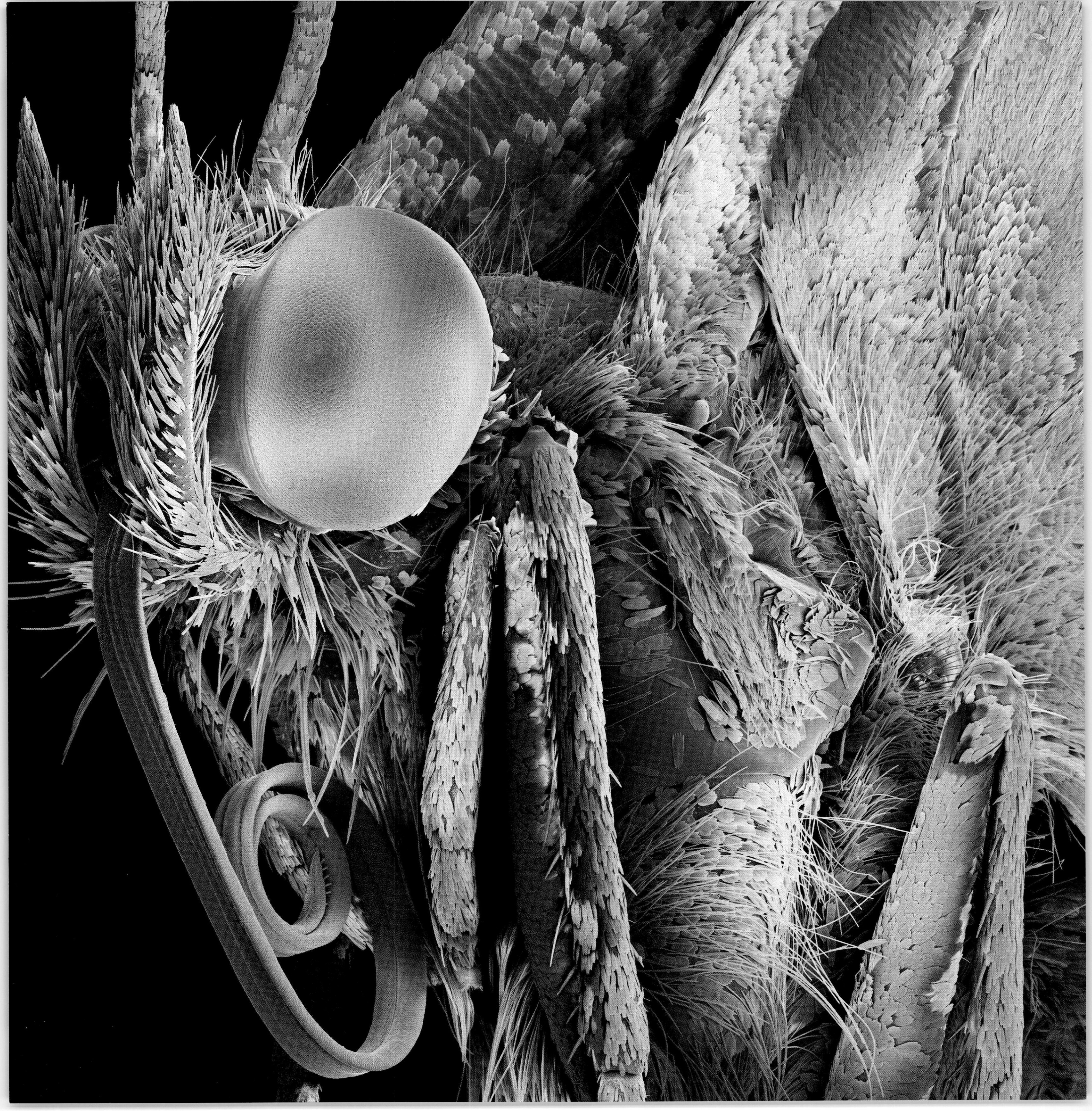

252

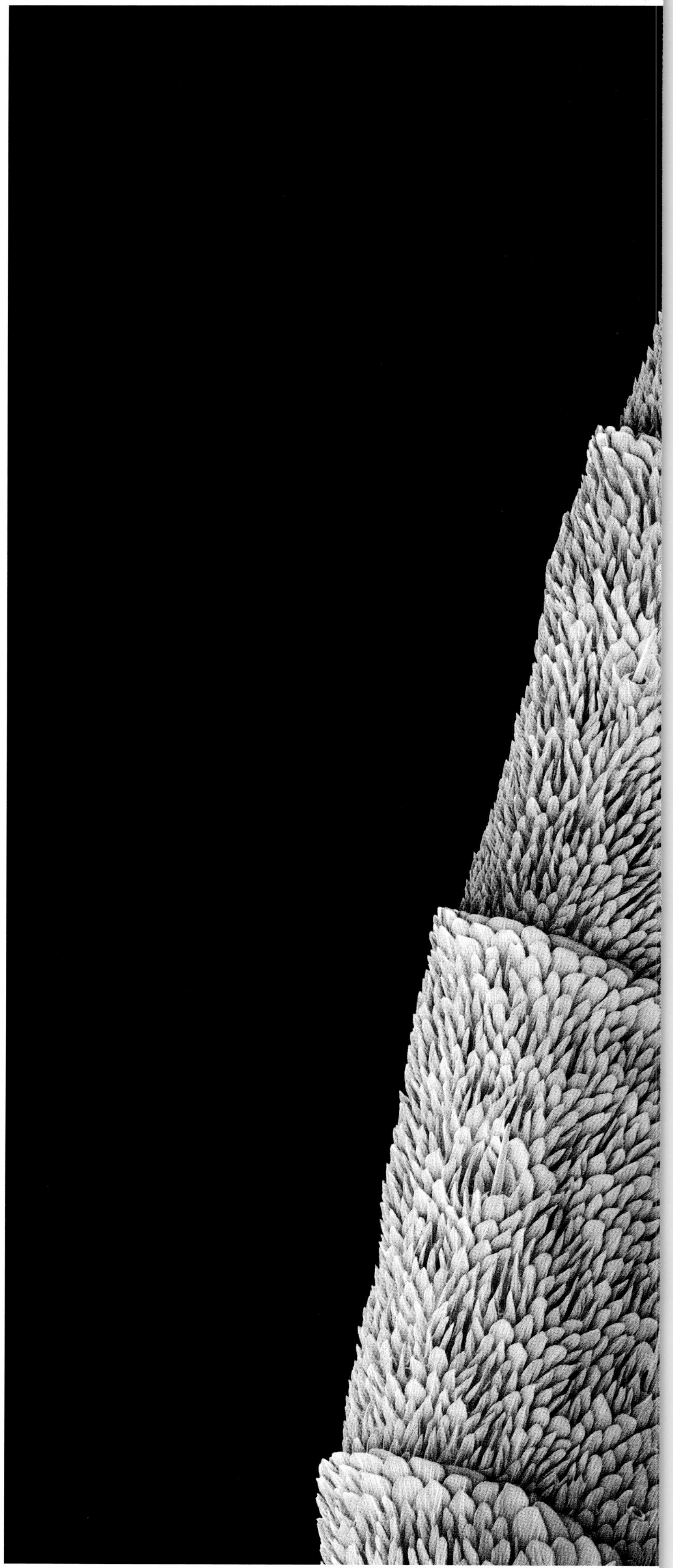

**Facing Page**

Antennae tip of the Lycaenidae family butterfly, Northern Velebit National Park, Croatian Adriatic.
4166 x. The visible area is about 600 microns from top to bottom, 0.6 millimeter

**Page 254, Page 255 →**

Two views of wing scales of the Adriatic butterfly.
On the left: 2083 x and 120 microns wide, 0.1 millimeter.
On the right: 16,666 x and 15 microns wide, 0.015 millimeter

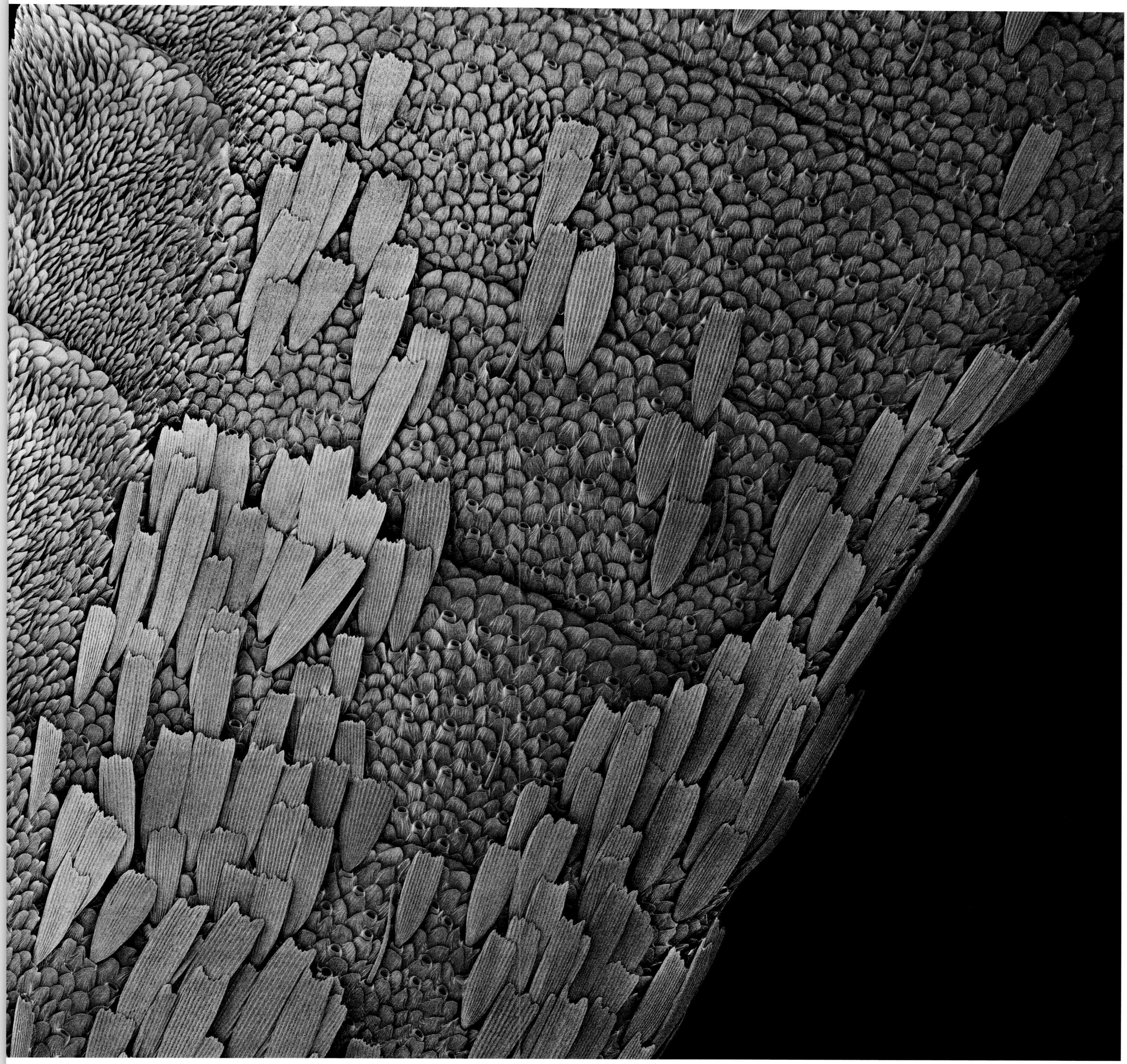

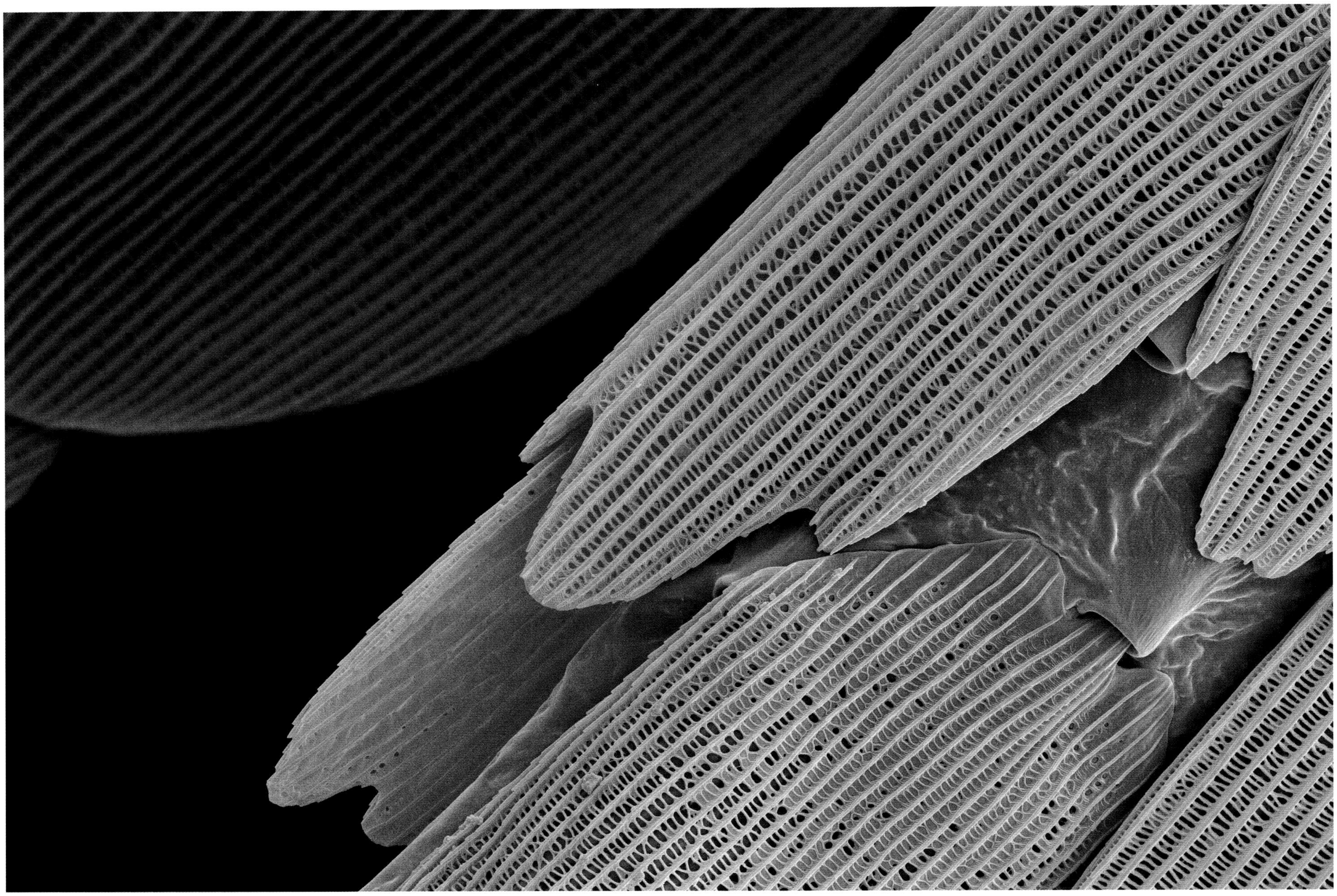

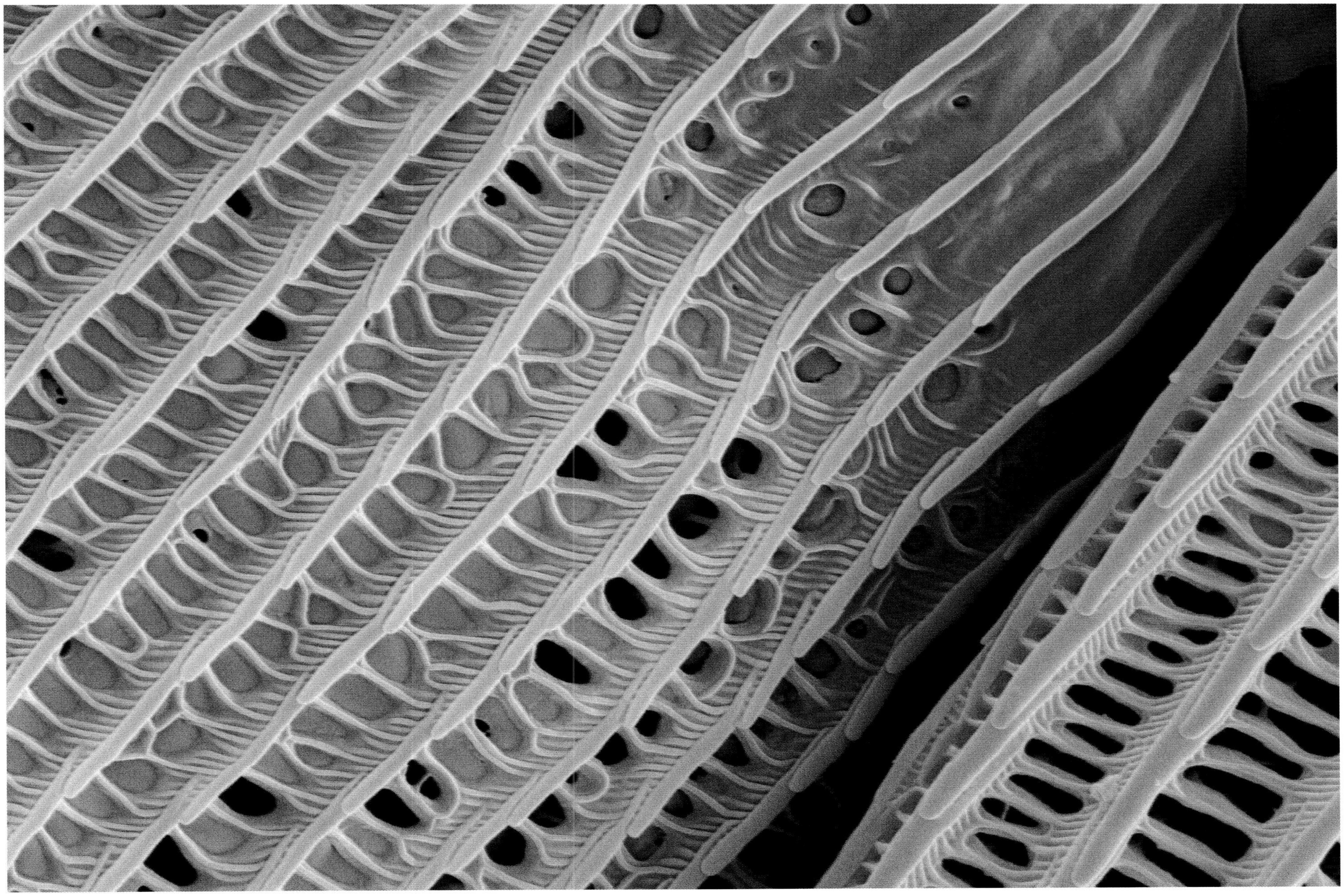

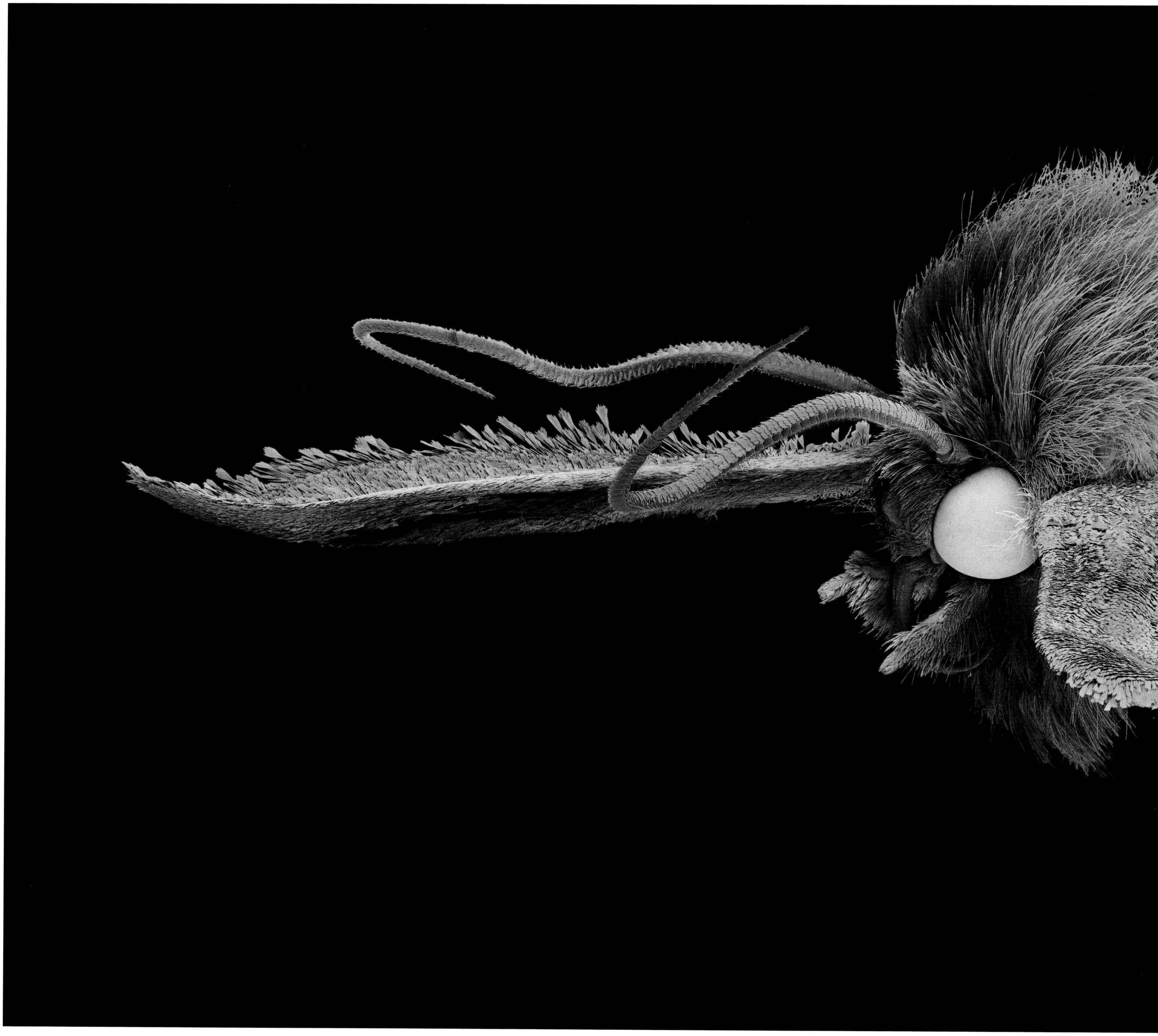

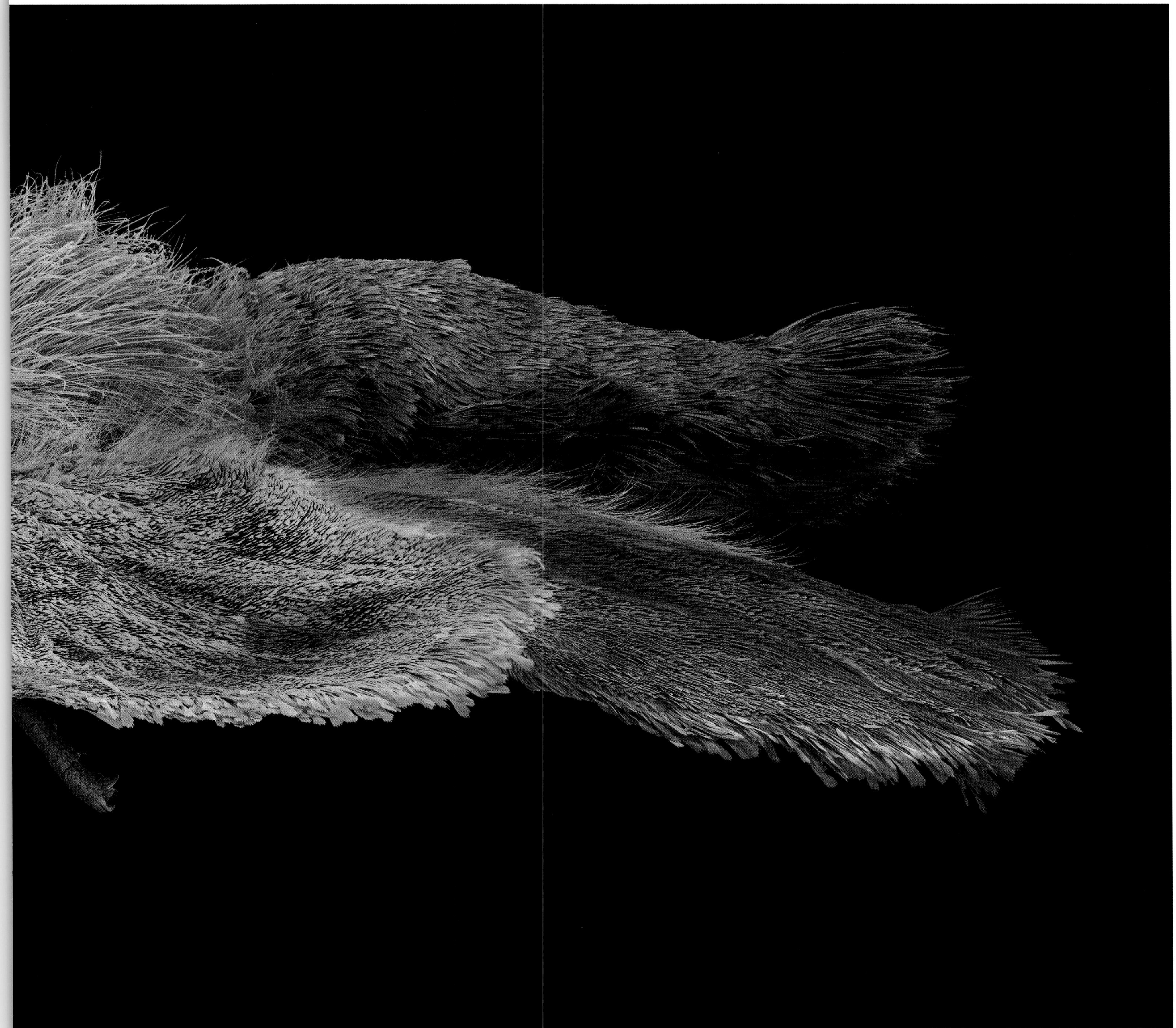

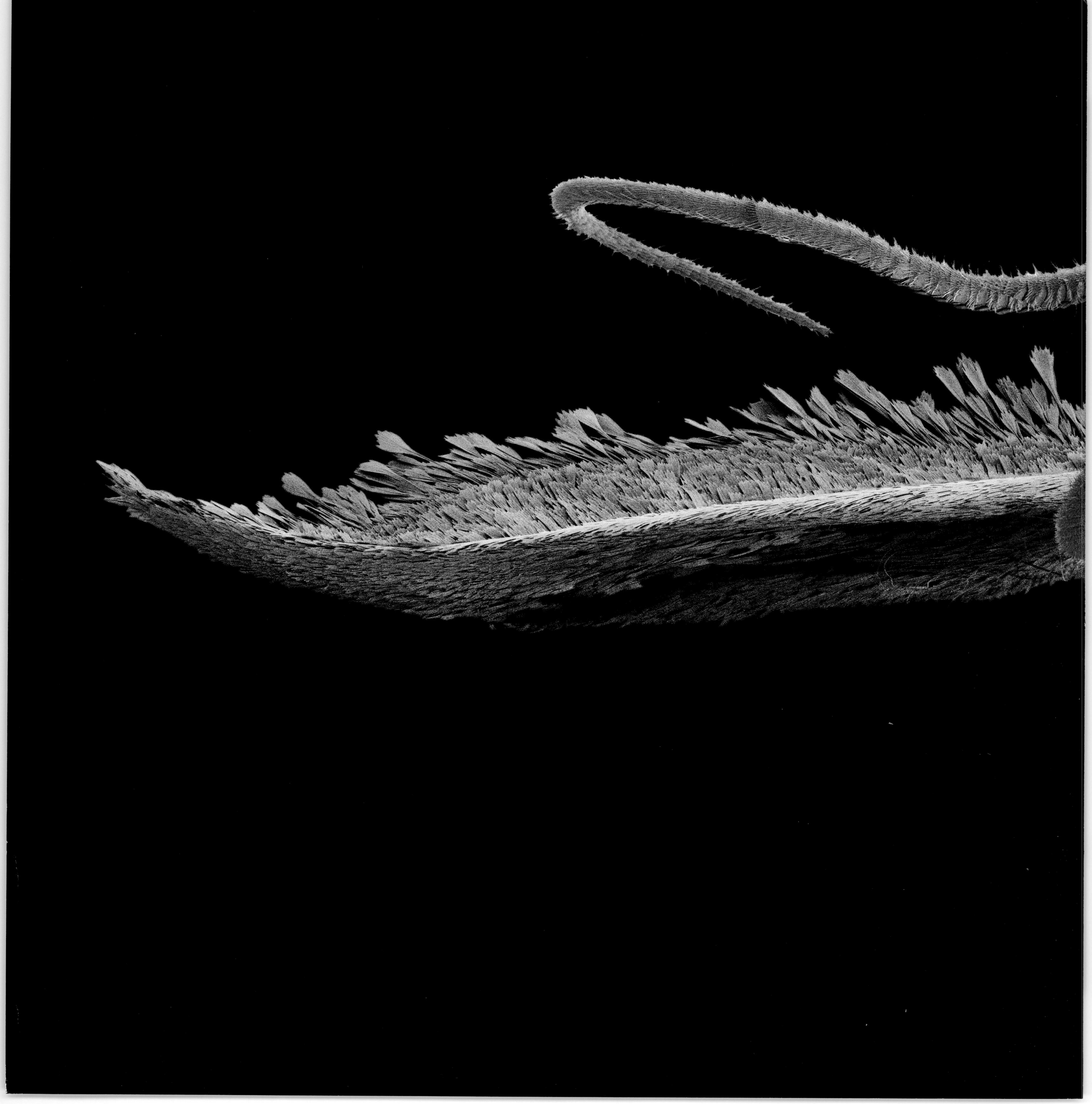

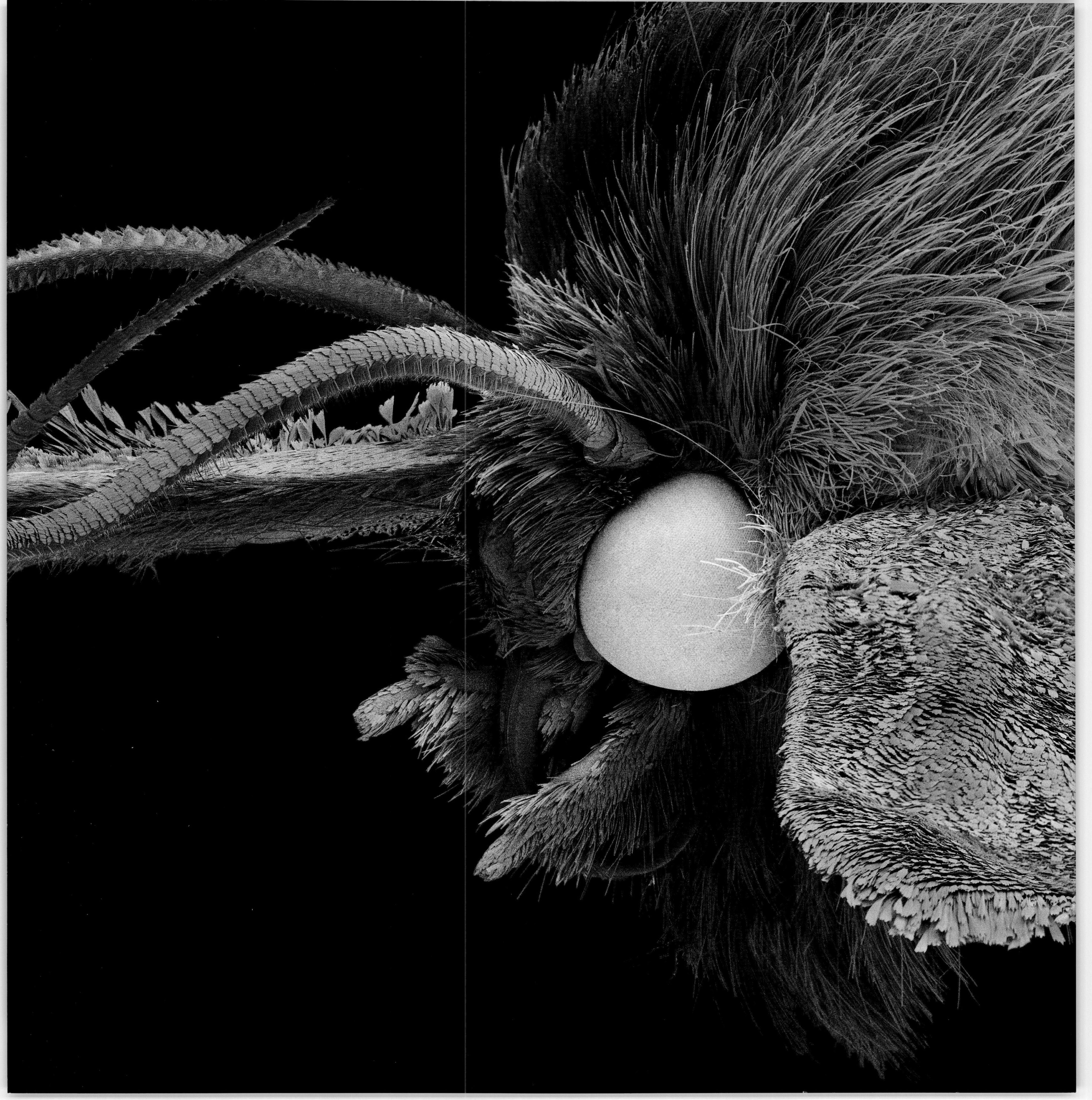

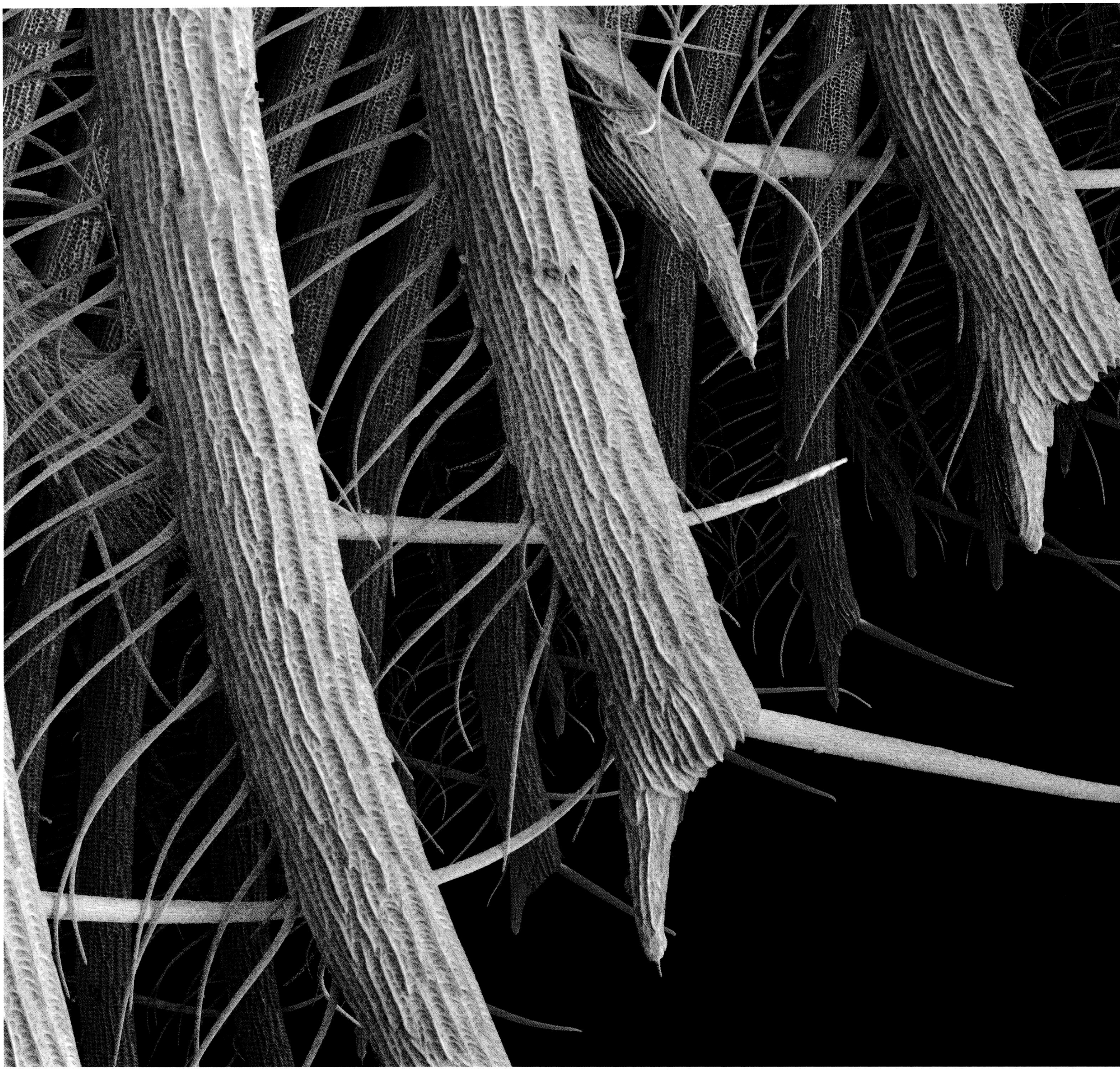

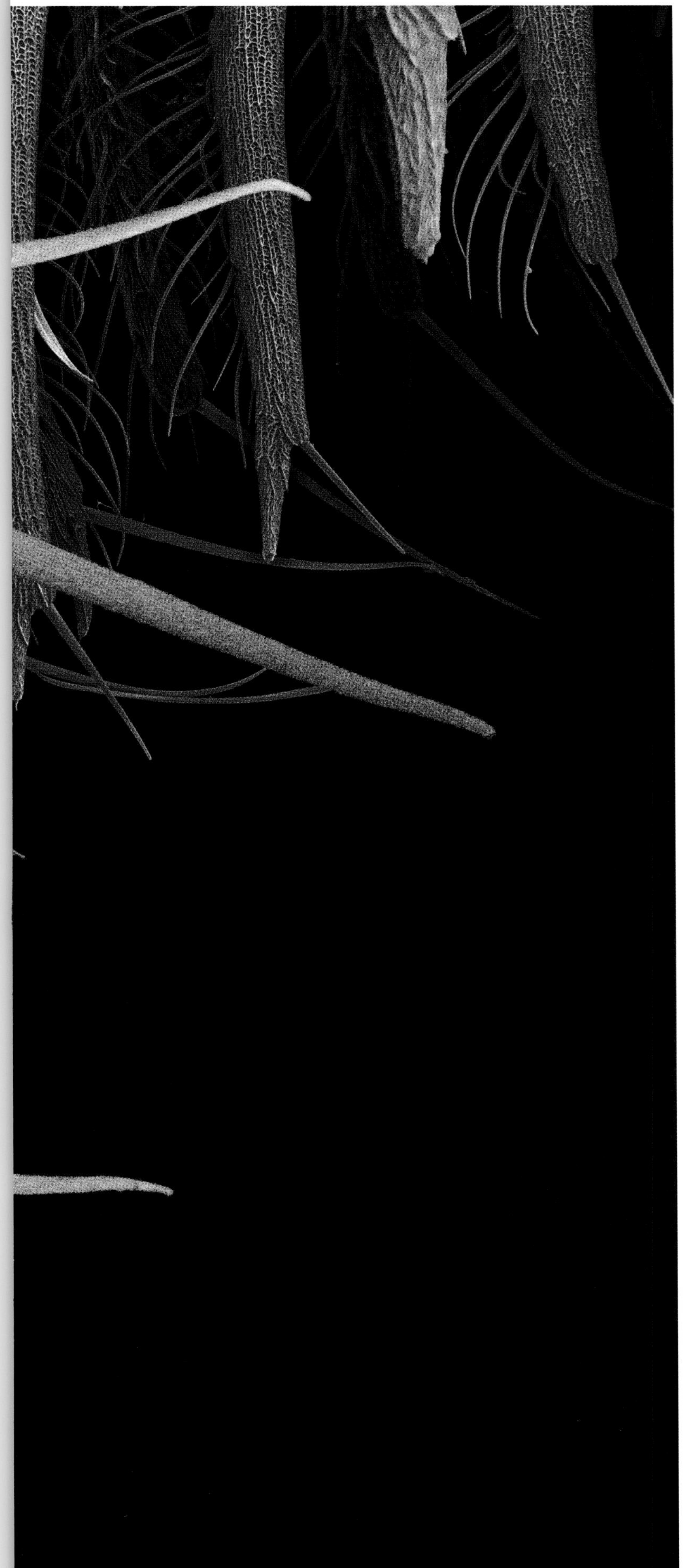



**Facing Page**

Paired antennae of *Lymantria dispar dispar* moth, Ottawa, Ontario.
195 x. About 2 millimeters wide

← Pages 256–257

*Apamea niveivenosa* moth, Richmond, Ontario.
16.7 x. Slightly wider than 3 centimeters from left to right

← Pages 258–259

Closer view of the *Apamea niveivenosa* moth.
48 x. Slightly wider than 1 centimeter from wing tip to right edge

**Facing Page**

*Apis mellifera*, European honeybee, Ottawa, Ontario.
12.5 x. The bee is slightly under 2 centimeters from wingtip to wingtip

**Page 264–265 →**

*Triepeolus pectoralis*, cuckoo bee, Empress Dune Field, Alberta.
107 x. The visible area is about half a centimeter wide

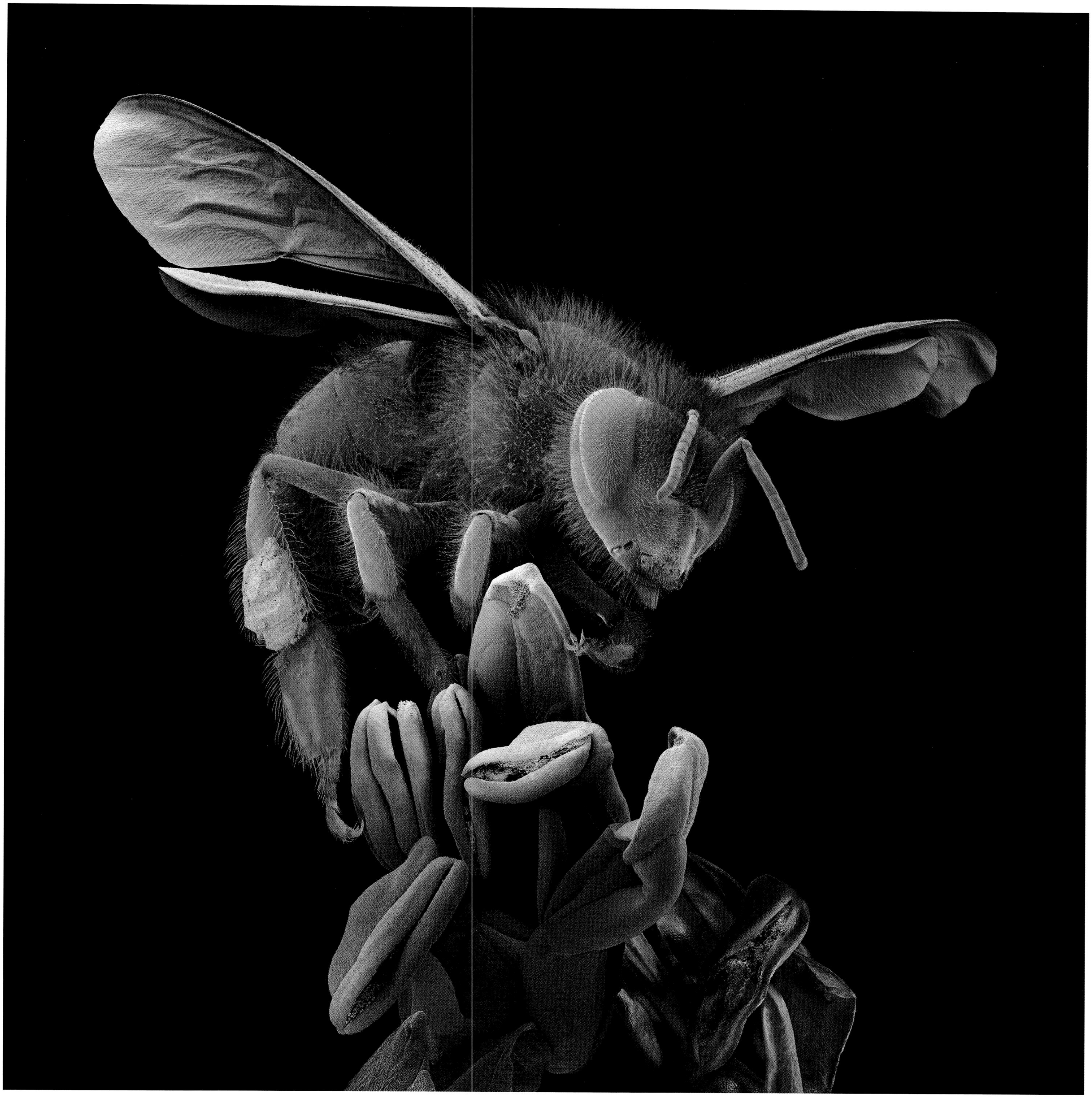

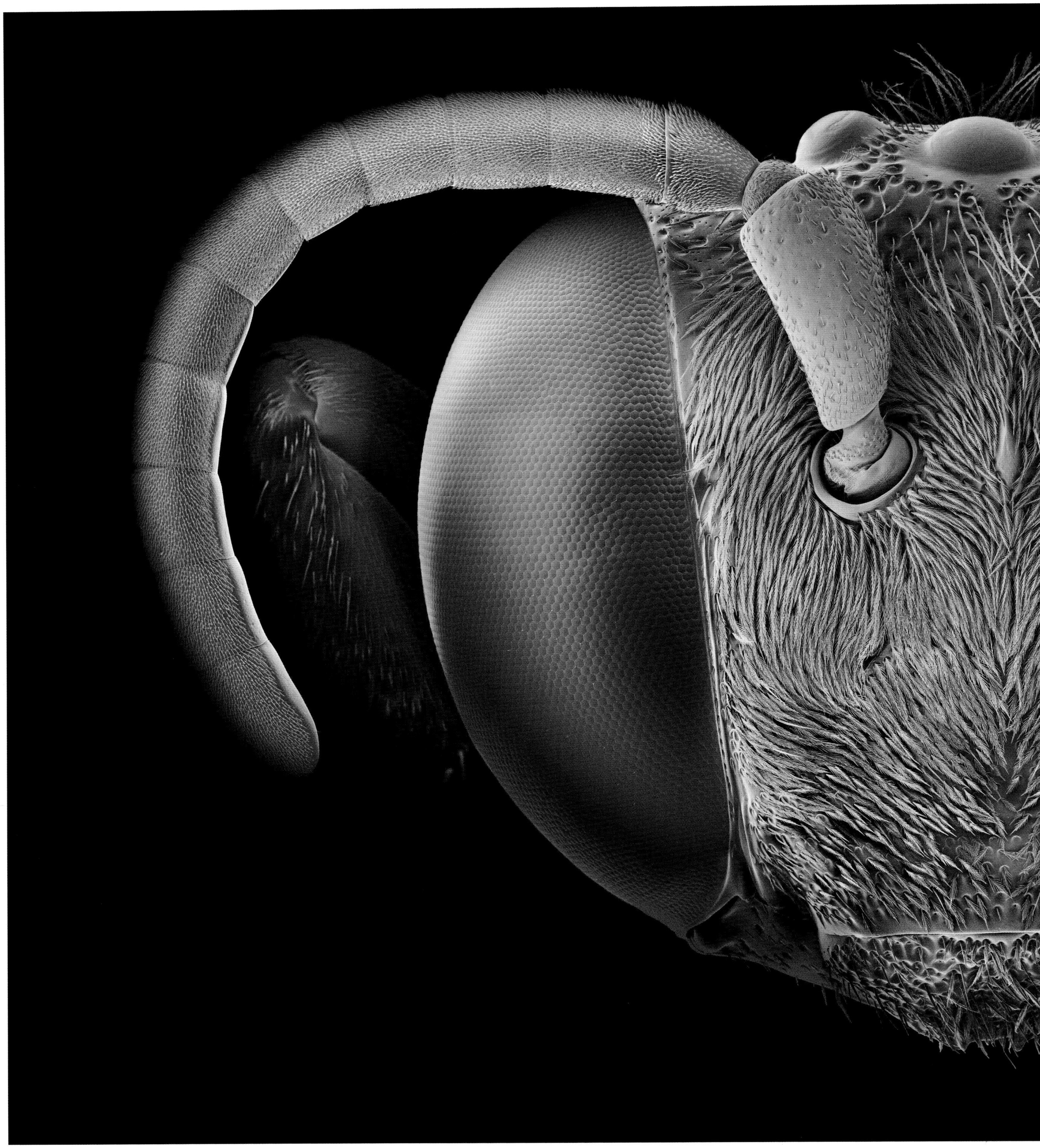

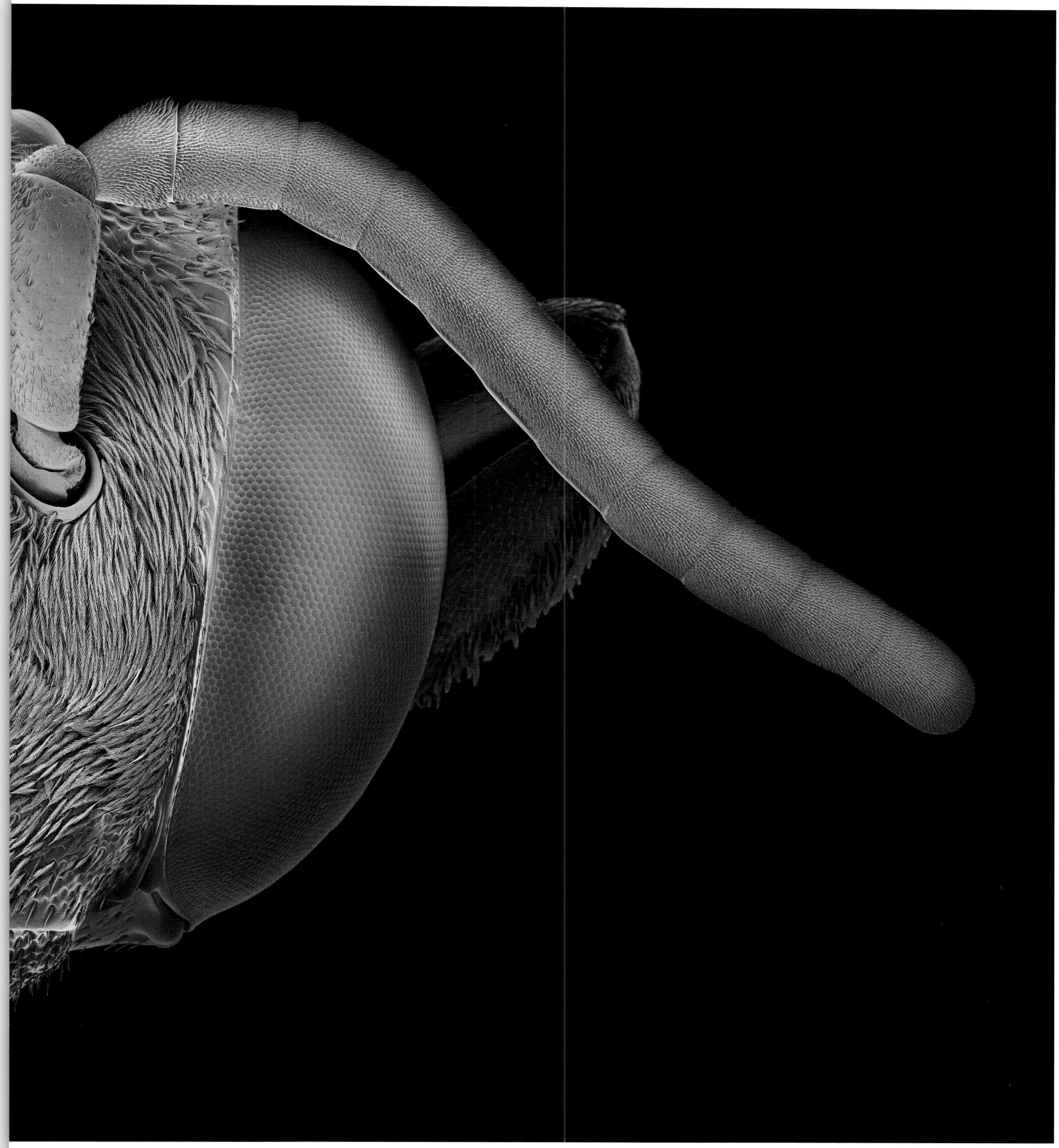

**Facing Page**

*Triepeolus pectoralis*, cuckoo bee, Empress Dune Field, Alberta.
70 x. The visible area is about 4 millimeters wide

**Page 268 →**

Joint between antennal pedicel and flagellum of *Triepeolus pectoralis* bee.
1,155 x. About 225 microns wide, 0.2 millimeter

**Page 269 →**

Scape and antennal socket of *Triepeolus pectoralis* bee.
371 x. About 700 microns wide, 0.7 millimeter

**Page 270 →**

Proboscis, mandibles and maxillae of *Culex pipiens* mosquito, Ottawa, Ontario.
575 x. About 400 microns from tip to top, 0.4 millimeter

**Page 271 →**

Head of *Erythemis simplicicollis*, common pondhawk dragonfly, Gatineau Park, Quebec.
75 x. Visible area is about two millimeters wide

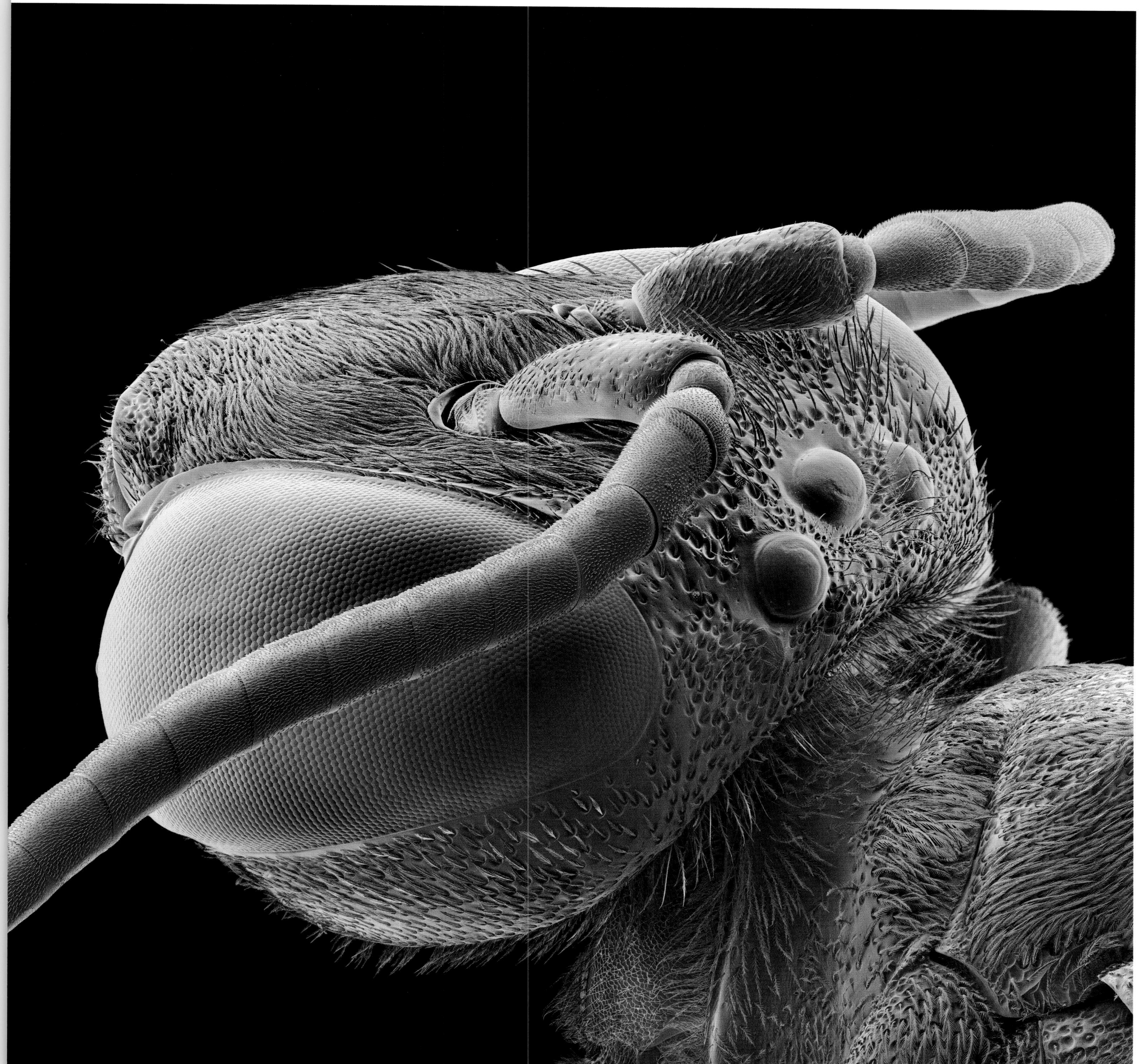

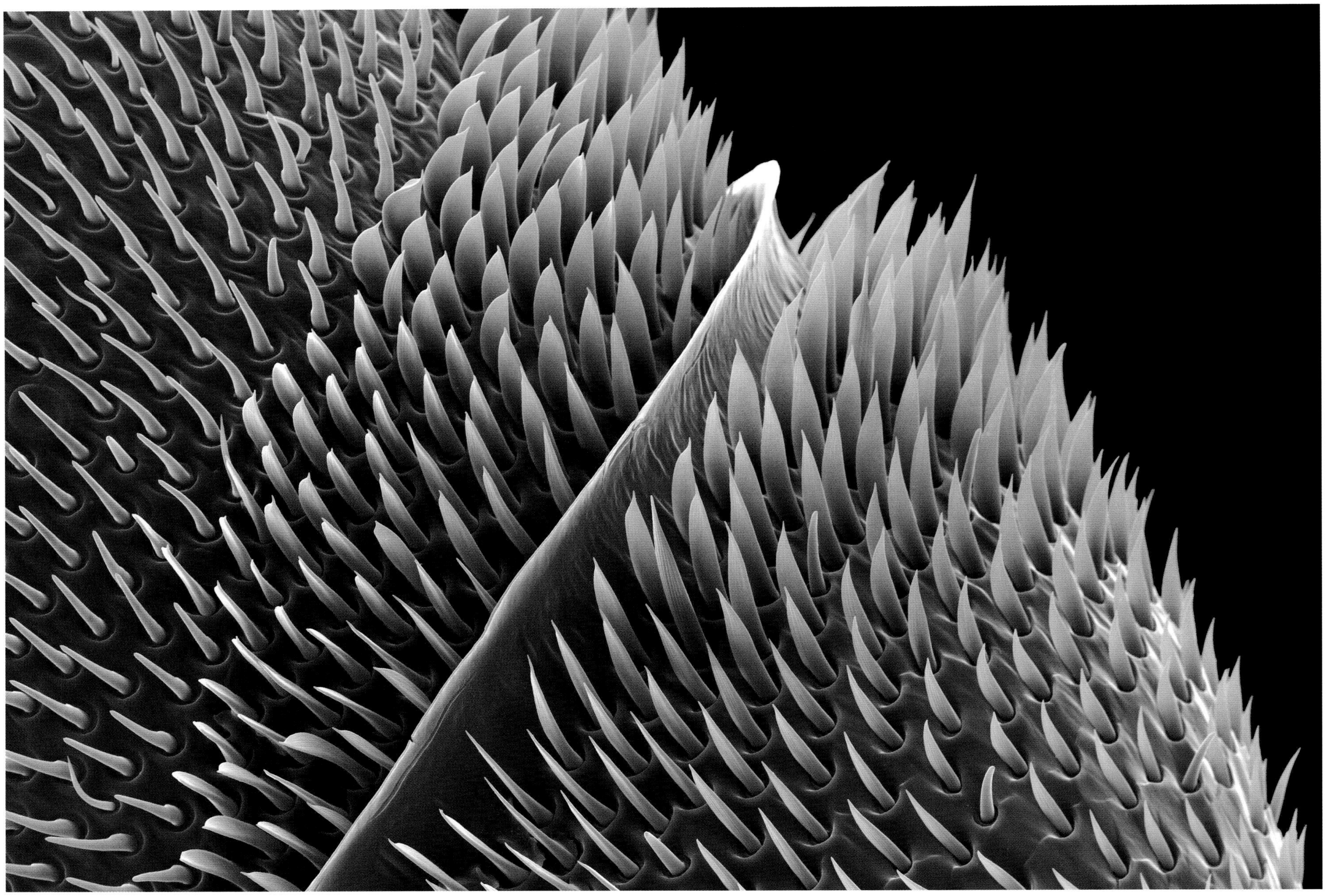

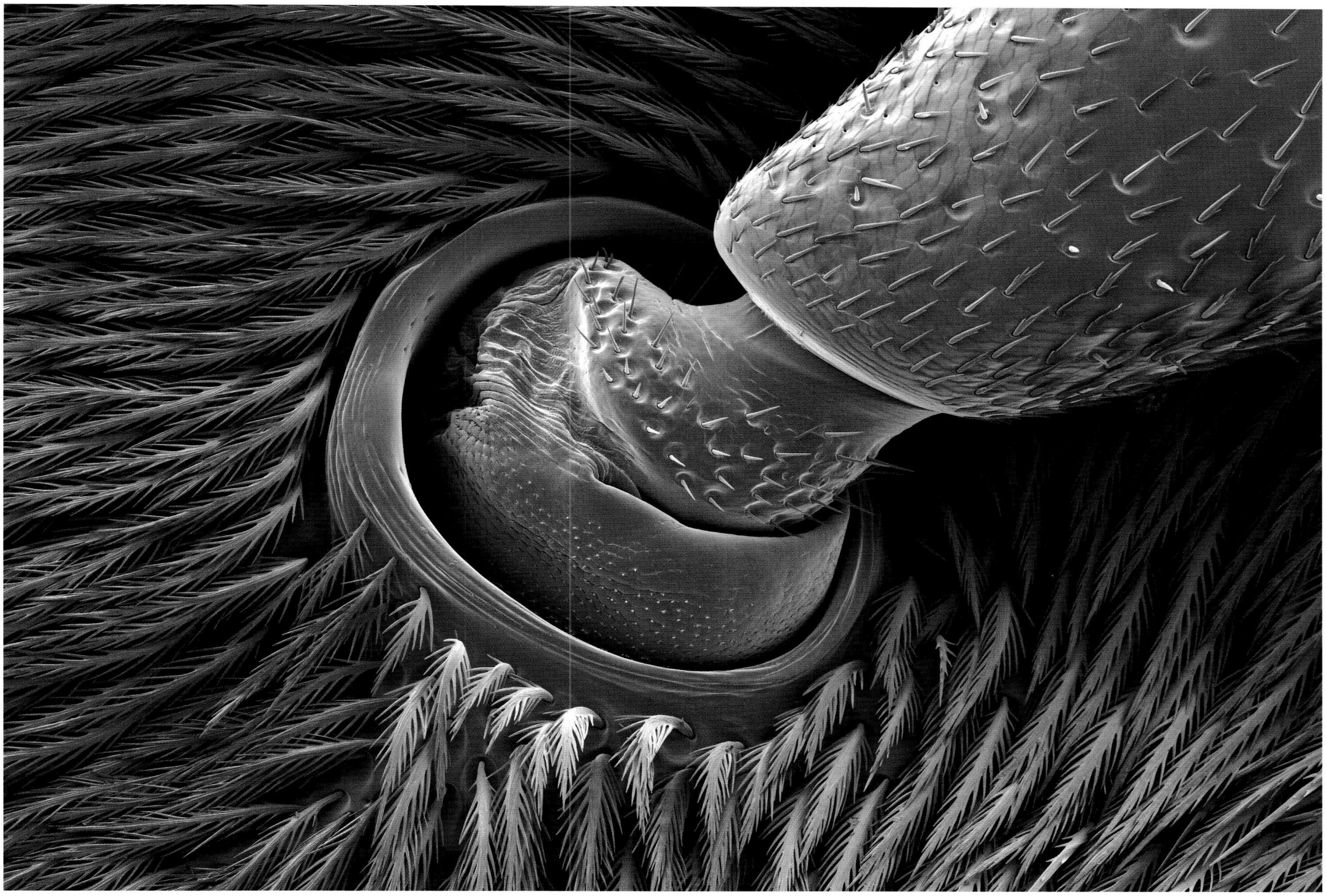

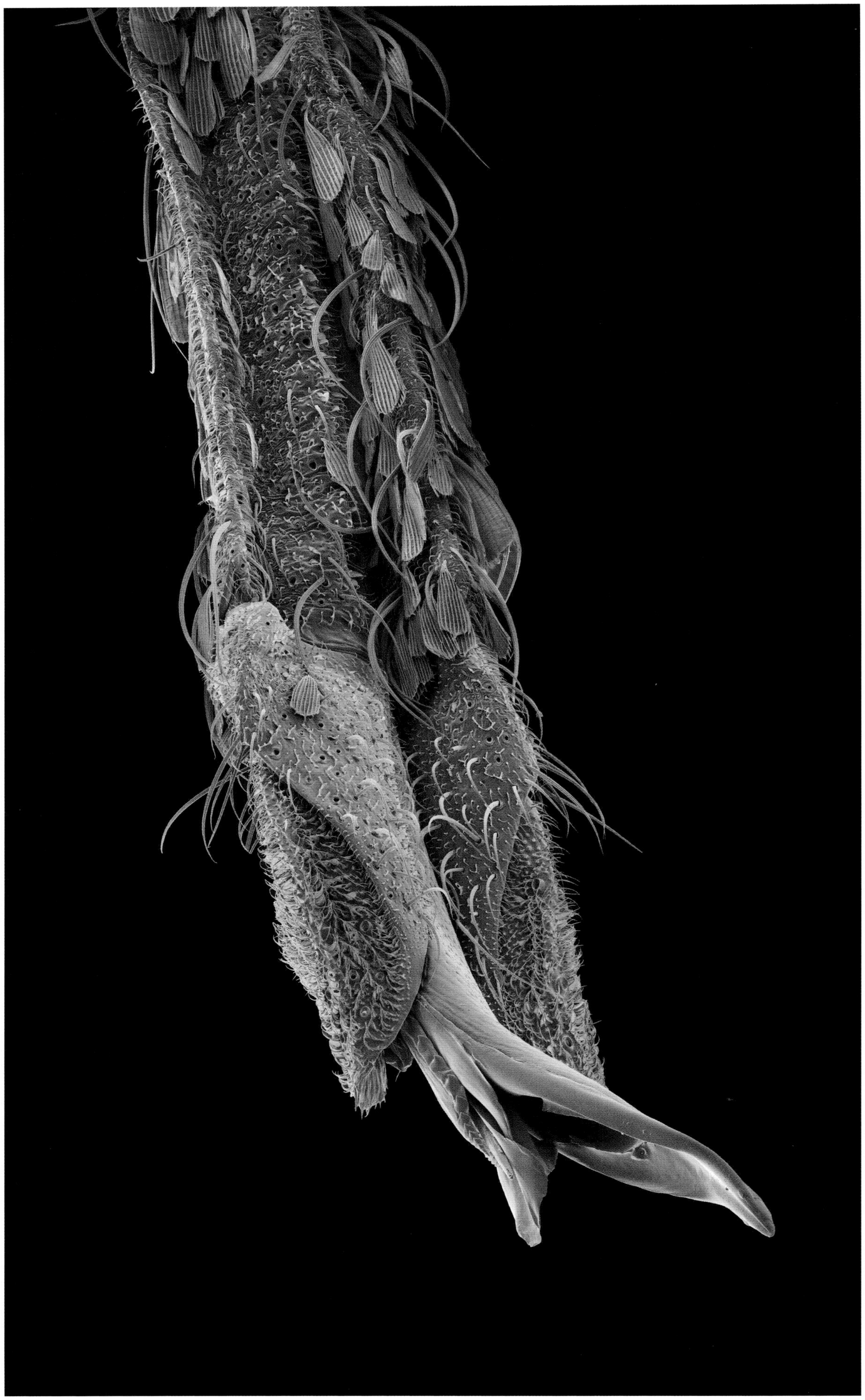

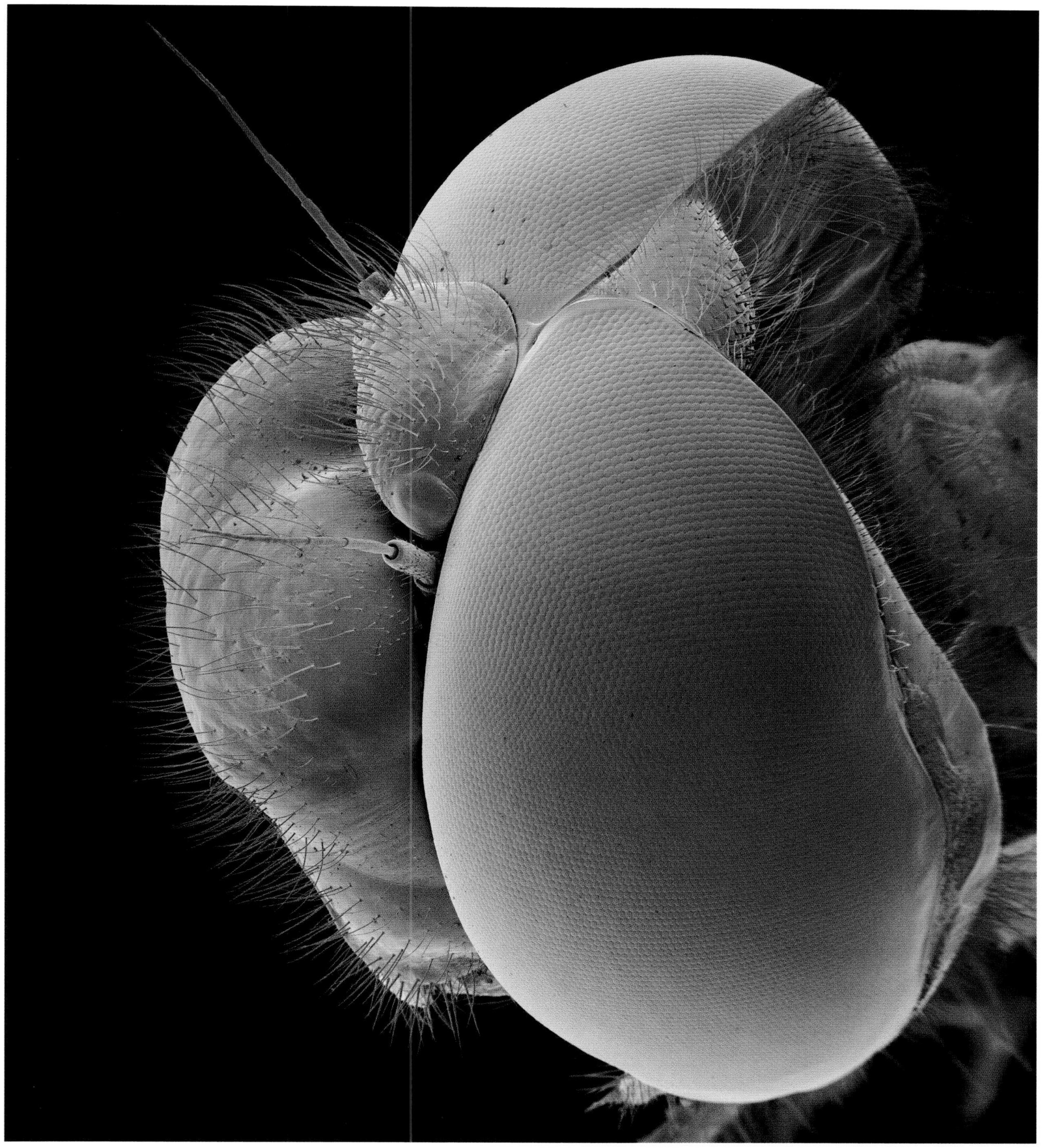

**Facing Page**

Trailing edges of paired wings of *Erythrodiplax bromeliicola* dragonfly, western Cuba. 62 x. Slightly under 4 millimeters from wing tip to margin

**Pages 274–275 →**

Leading wing of *Erythemis simplicicollis*, common pondhawk dragonfly, Gatineau Park, Quebec. 90 x. Visible area is about 6 millimeters wide

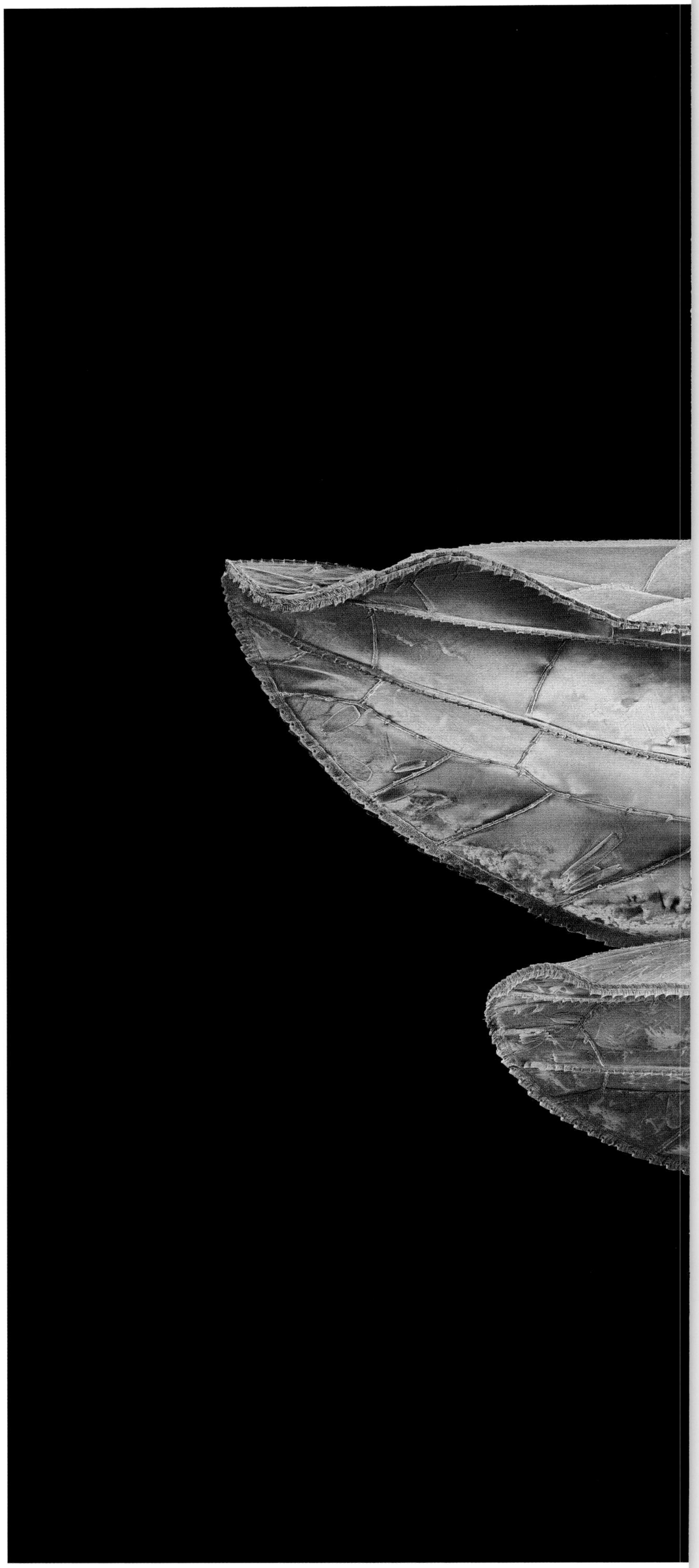

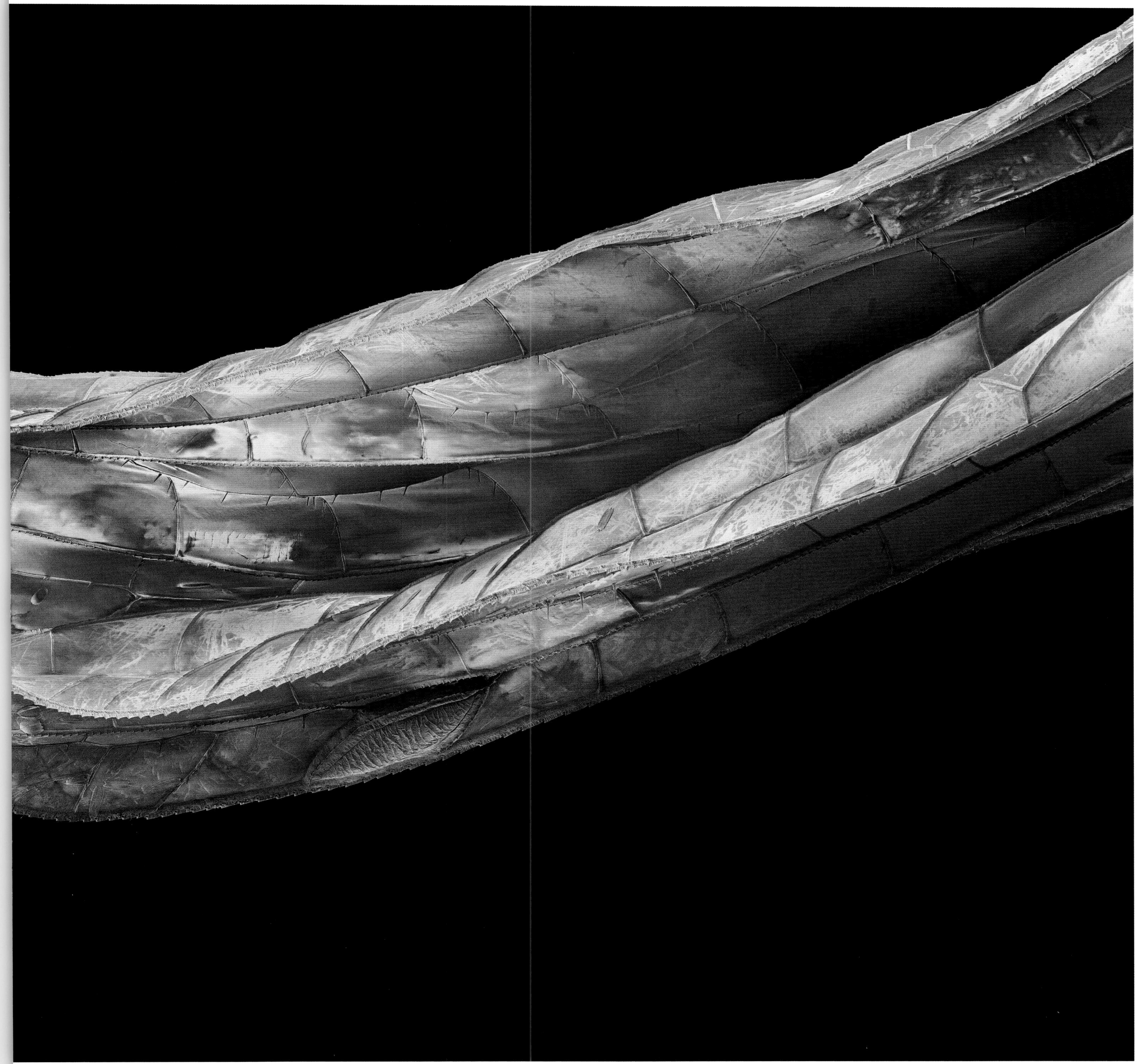

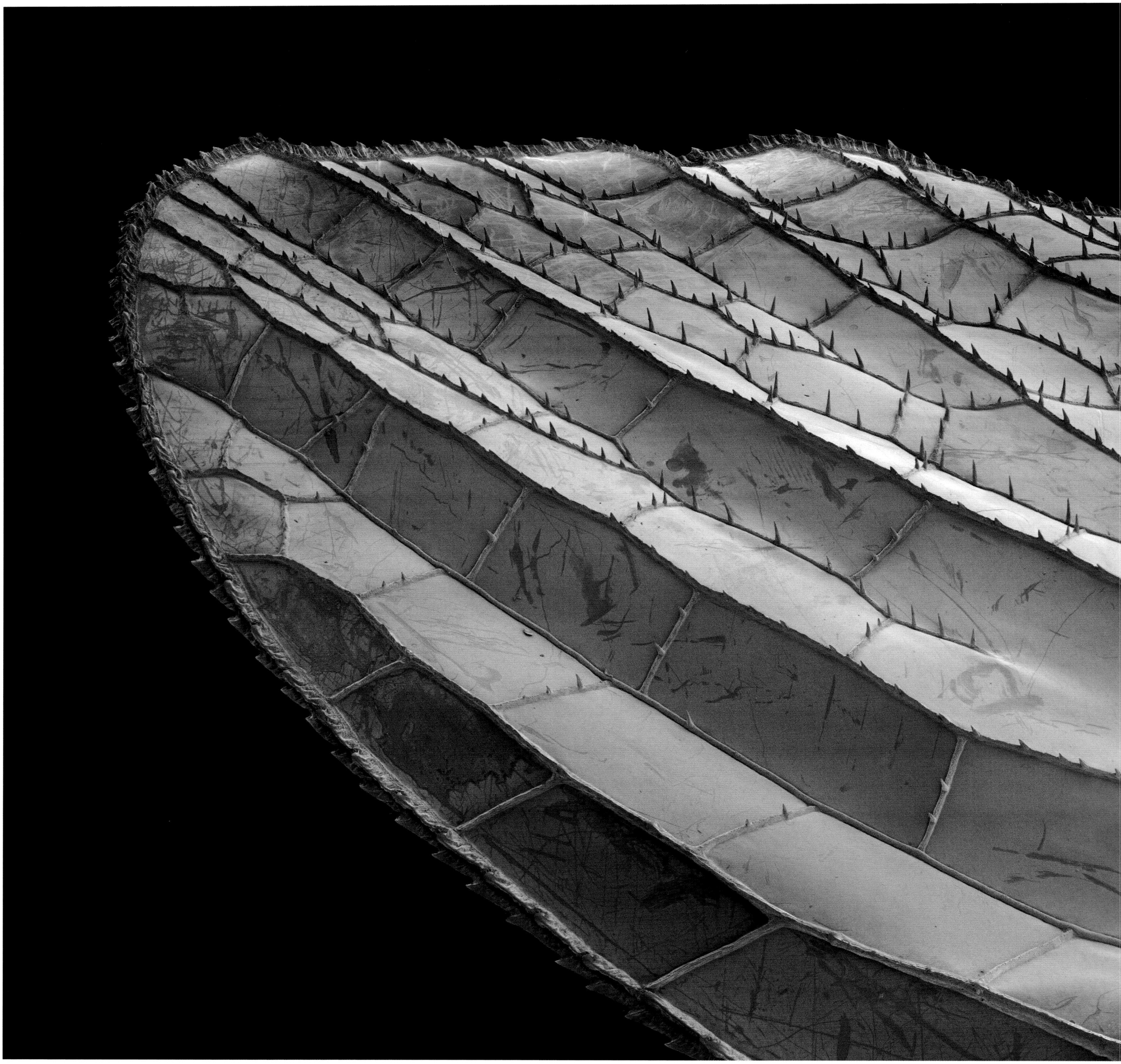

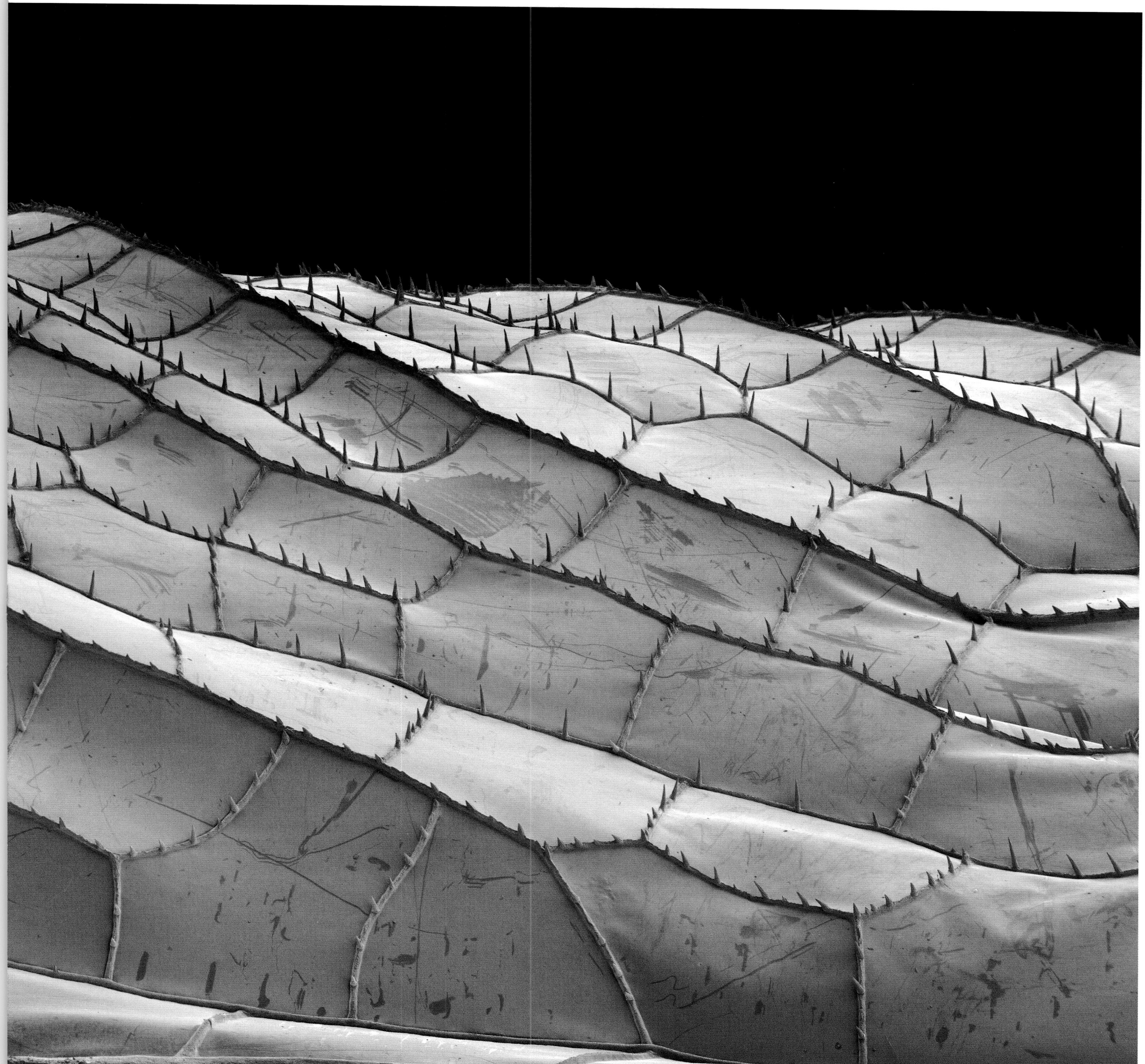

**Facing Page**

View from leading edge across paired wings of *Erythemis simplicicollis* dragonfly, Gatineau Park, Quebec.
53 x. Visible area is about 4.5 millimeters wide

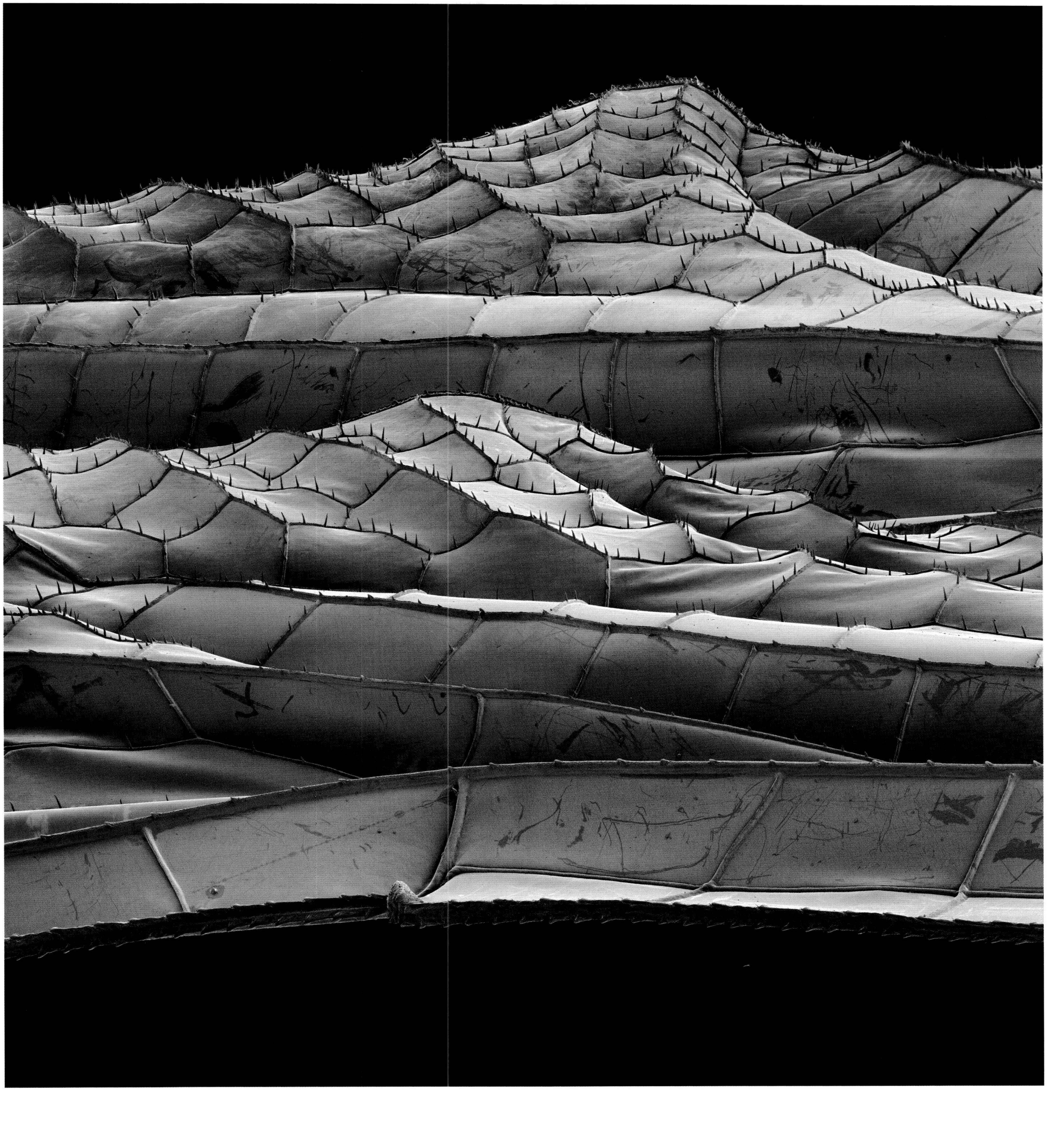

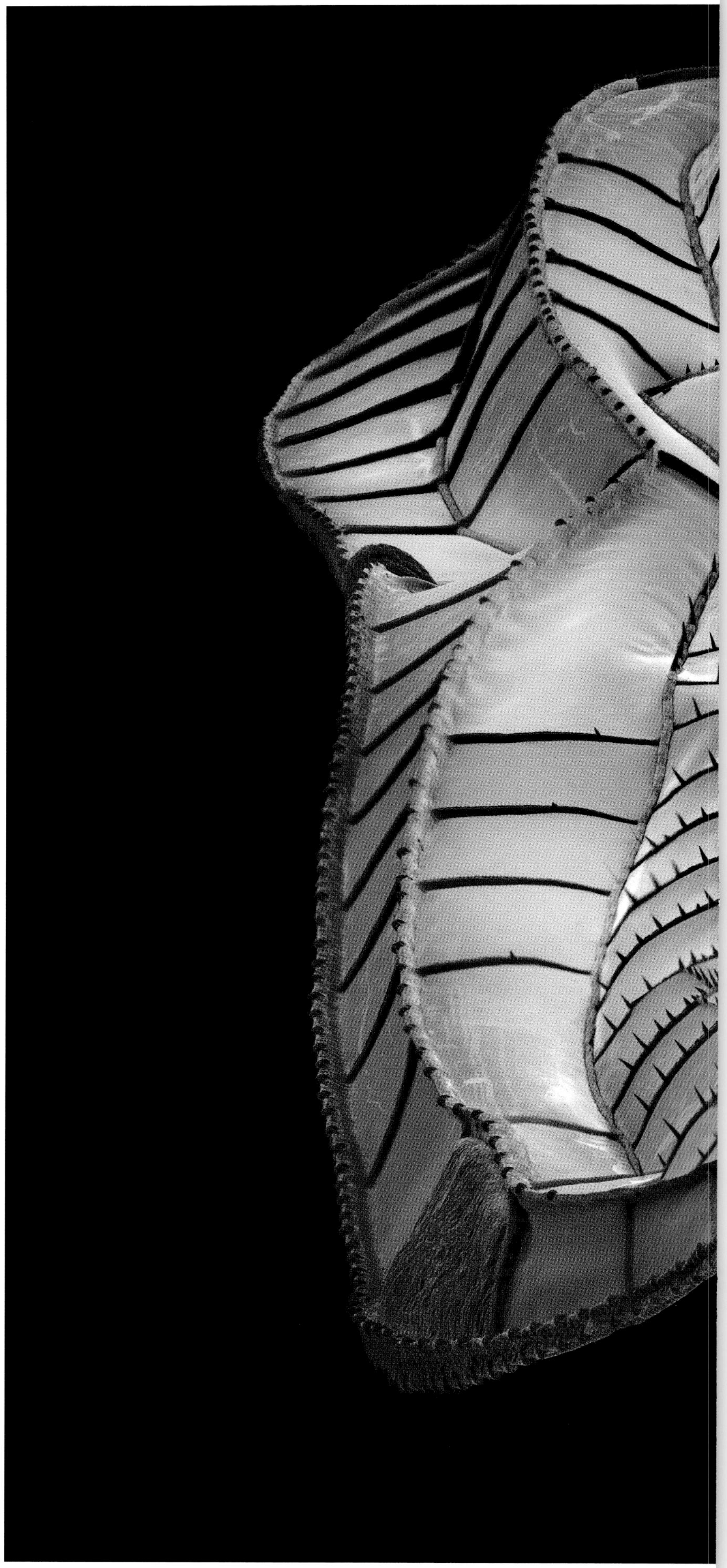

**Facing Page**

View from tip down leading wing of *Erythemis simplicicollis* dragonfly, Gatineau Park, Quebec. 106 x. Wing is about 3 millimeters wide

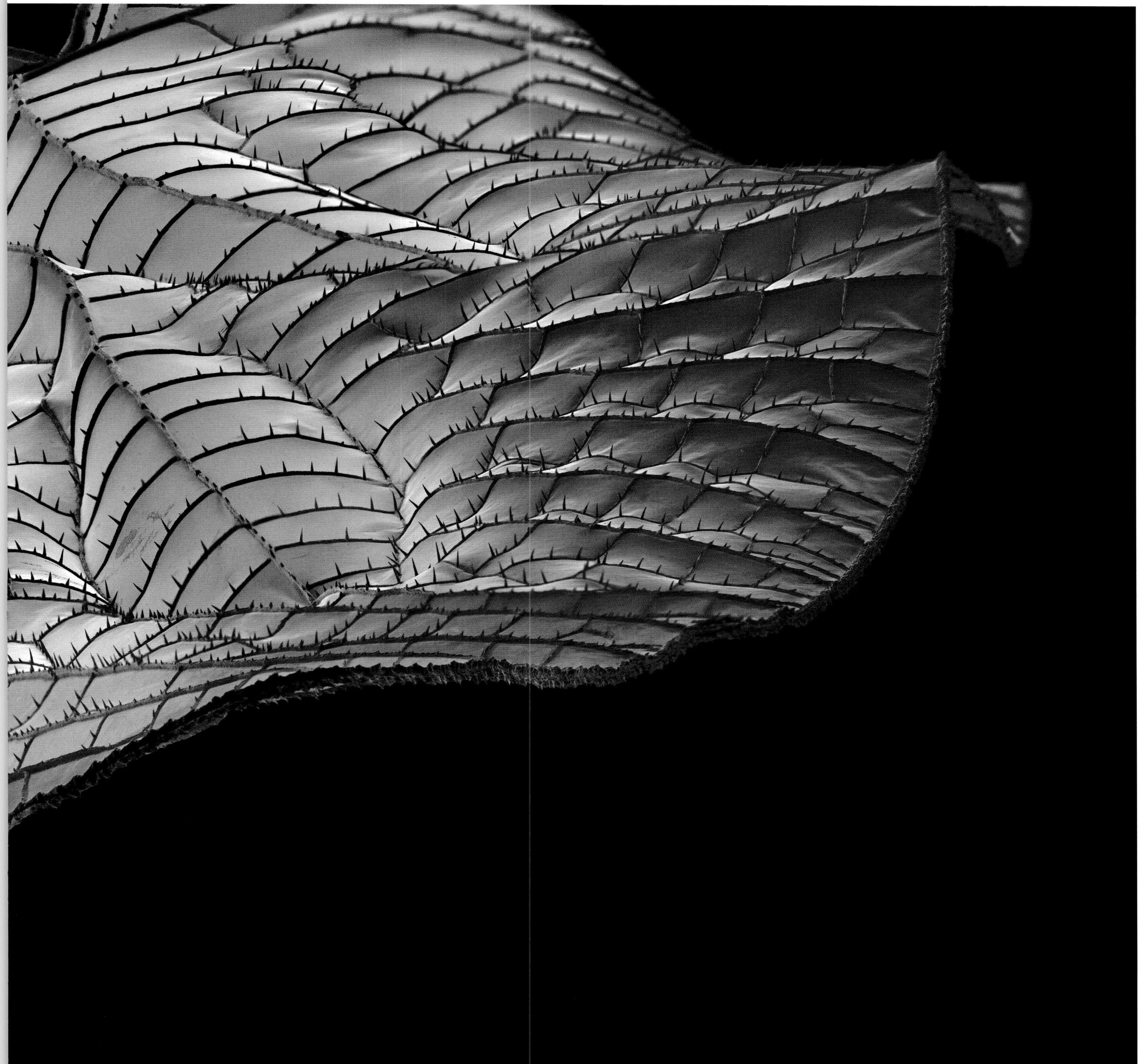

**Facing Page, Pages 282–283 →**

Stellar dendrite, Ottawa, Ontario.
83 x. 3.5 millimeters across

**Page 284 →**

Columnar snowflake with rime crystals, Ottawa, Ontario.
440 x. About half a millimeter wide

**Page 285 →**

Stellar dendrite with rime on one side, Ottawa, Ontario.
86 x. About 2.5 millimeters wide

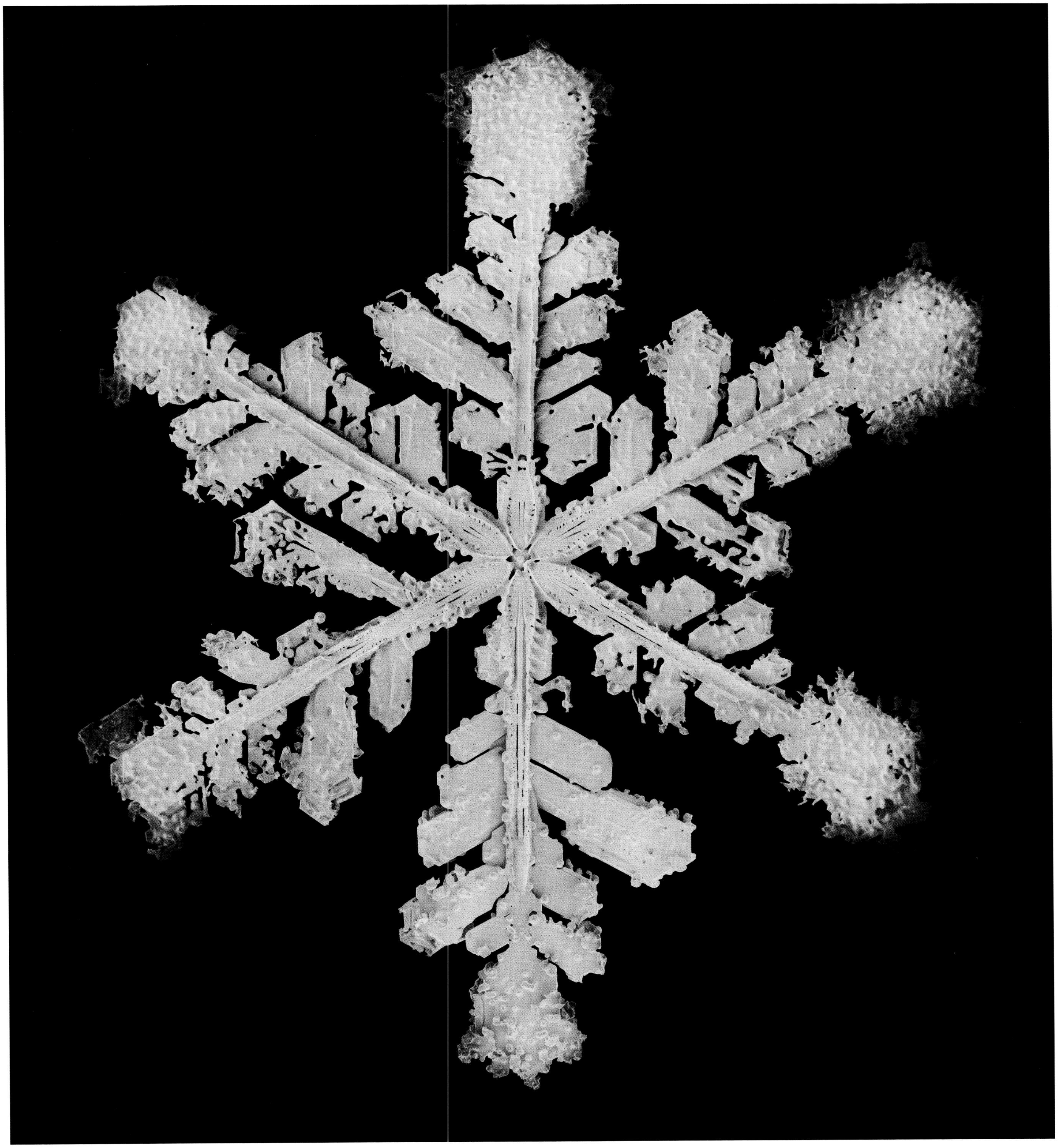

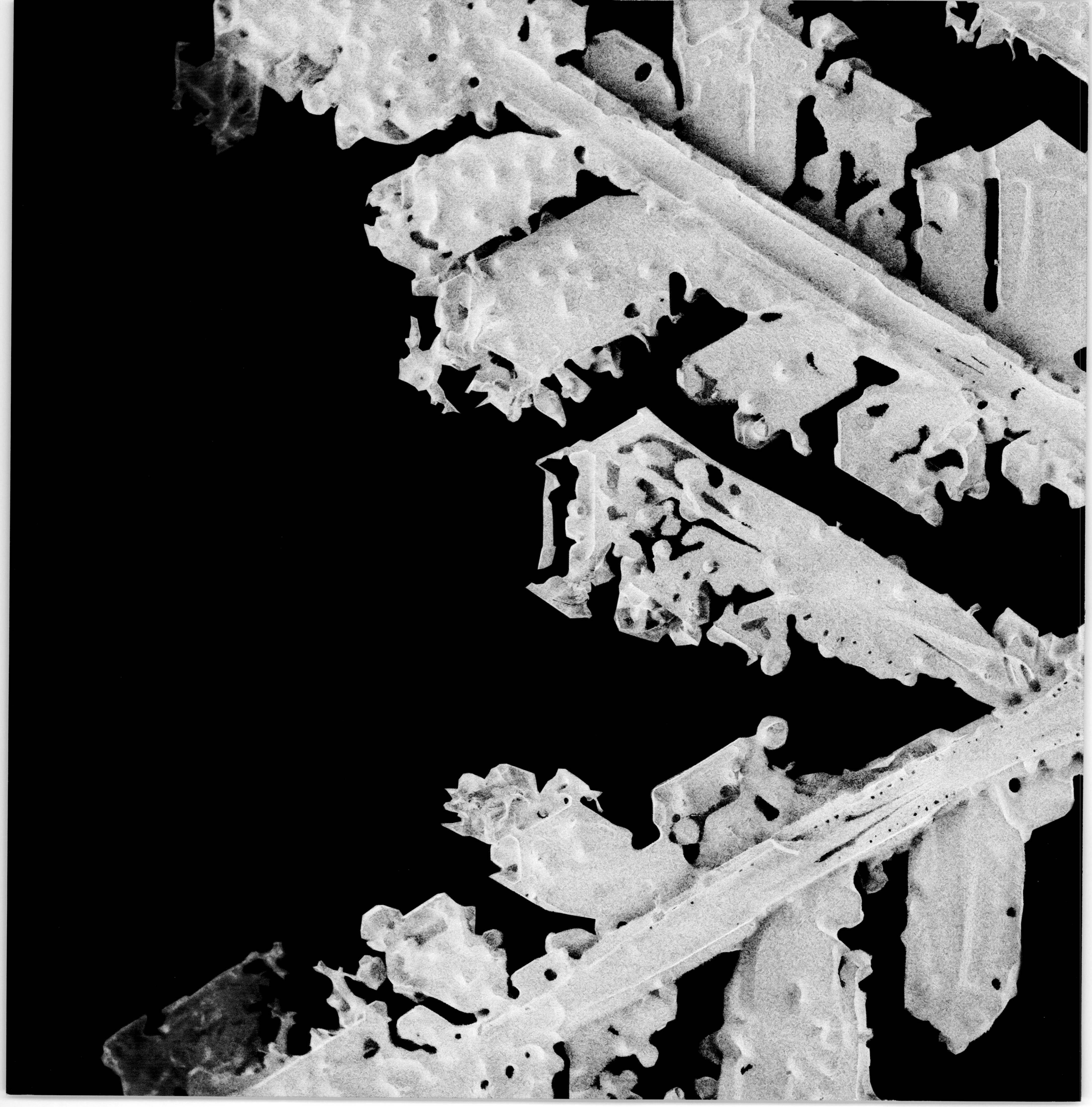

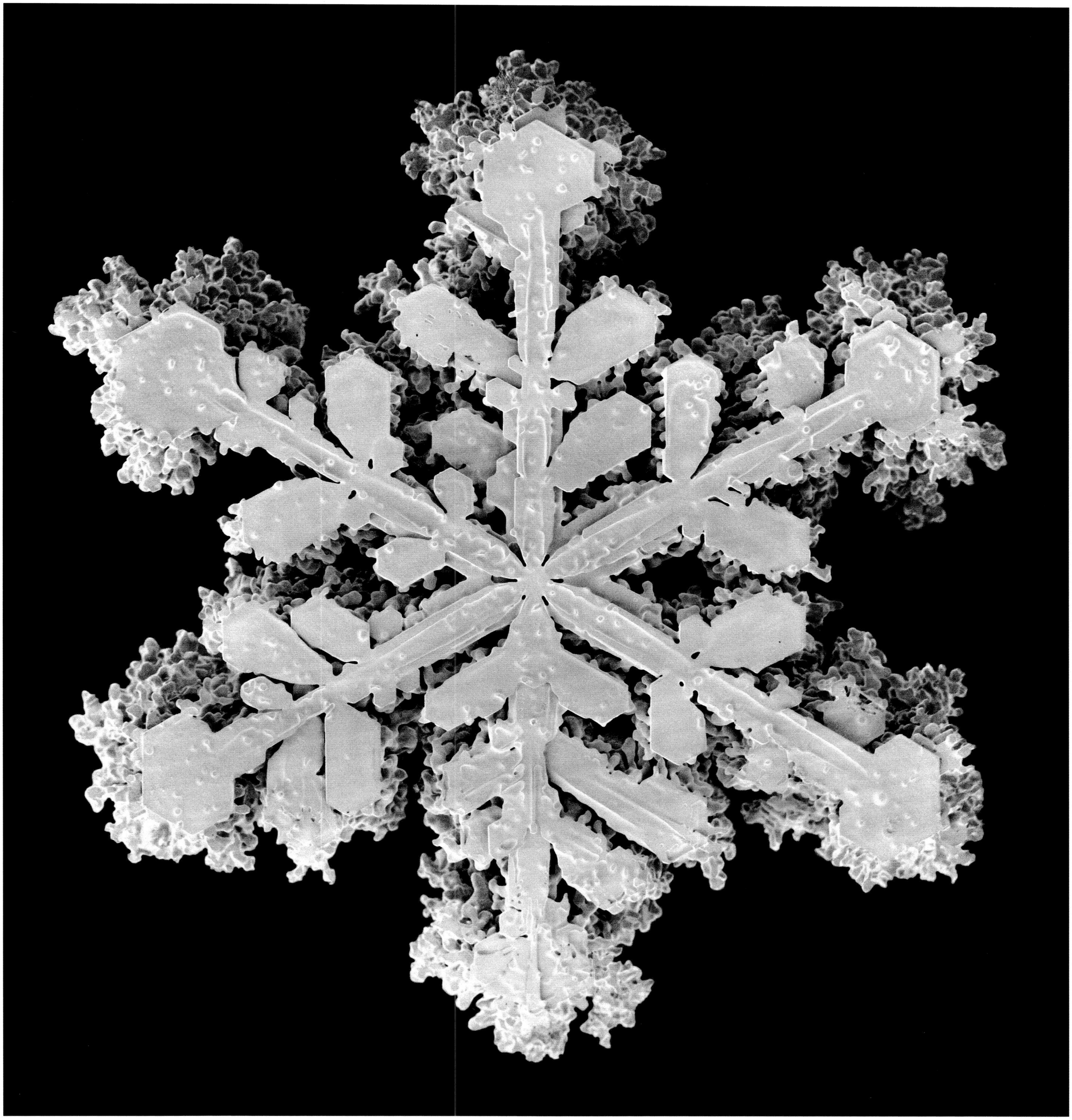

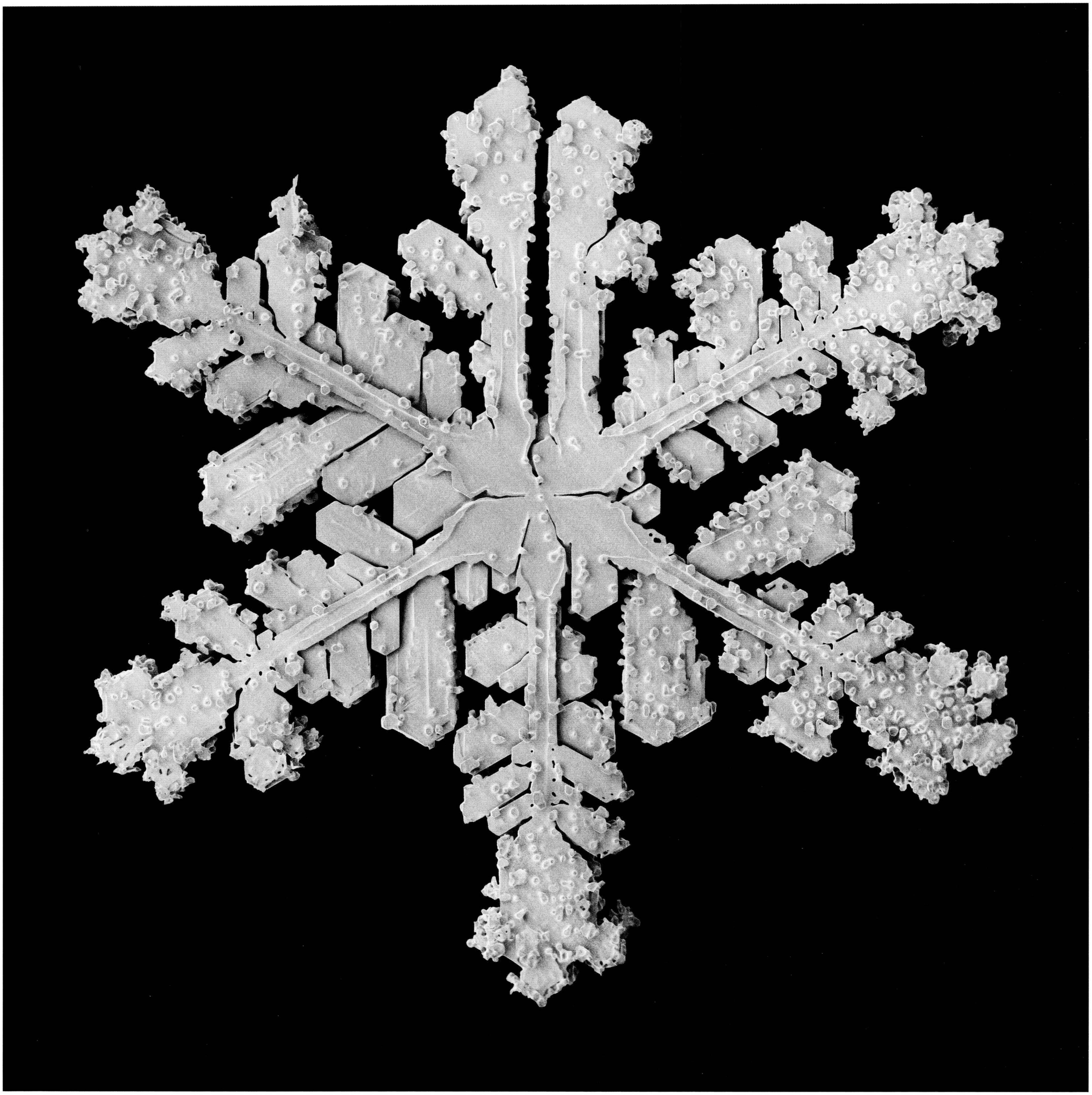

**Facing Page, Page 315**

Stellar dendrite with bisected arm, Ottawa, Ontario.
63 x. About 3.5 millimeters wide

# The Moon

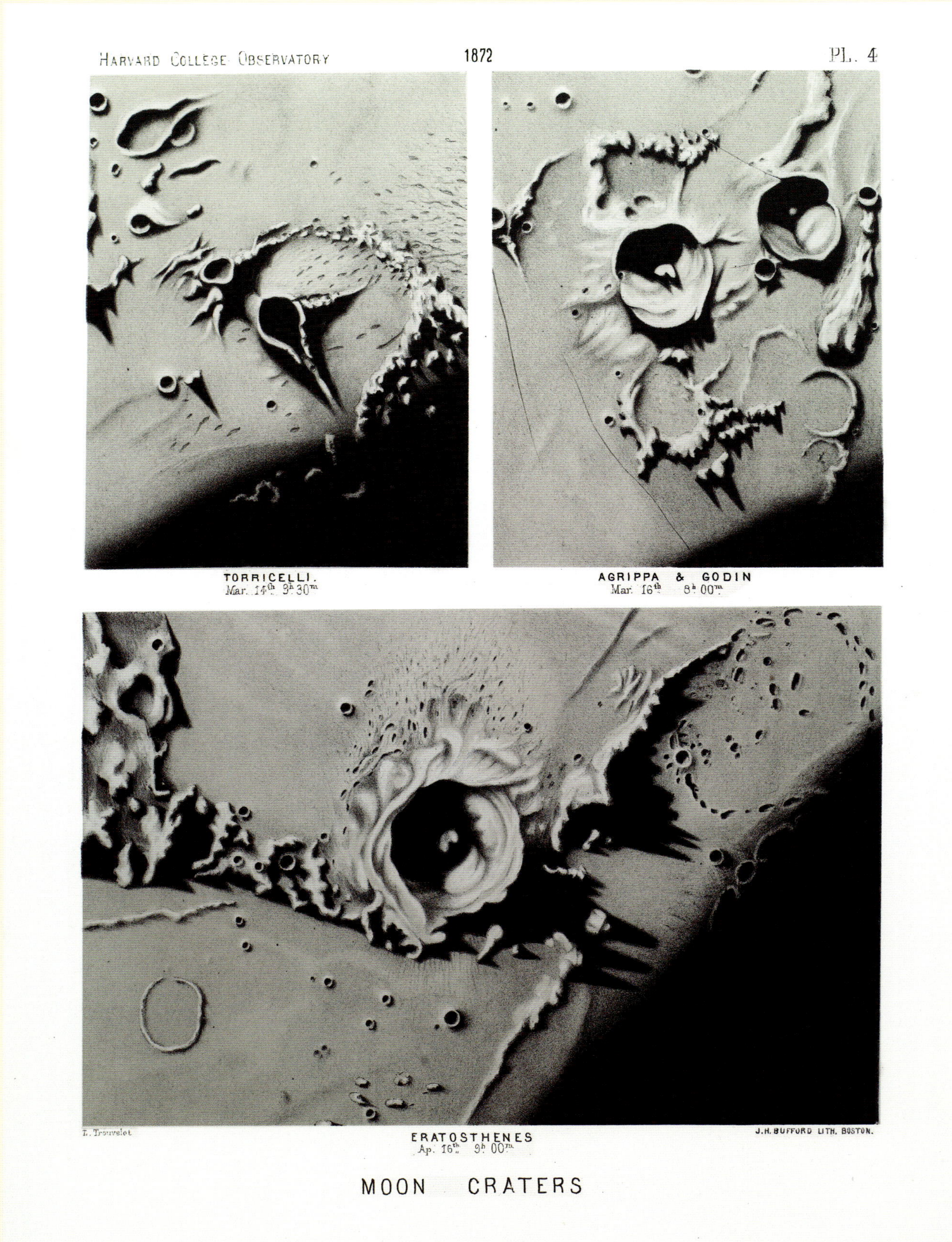
HARVARD COLLEGE OBSERVATORY
1872
PL. 4
TORRICELLI.
Mar. 14th 9h 30m
AGRIPPA & GODIN
Mar. 16th 8h 00m
ERATOSTHENES
Ap. 16th 9h 00m
J.H. BUFFORD LITH. BOSTON.
MOON CRATERS

Our planet's natural satellite was formed about 4.5 billion years ago, following a vast collision in which early Earth was impacted by another planetary body about the size of Mars. That epic encounter, certainly among the most violent events of the solar system's early days, abruptly excavated a Moon-sized spray of red-hot terrestrial rock. The debris wasn't thrown far enough to escape Earth's gravity, however, instead settling into orbit, where it gradually accreted into the Moon. Or so goes the prevailing theory.

Four and a half billion years being only about 100 million years after the solar system's formation from a saucer-shaped protoplanetary disc, the other planets were experiencing a similar pummeling. In effect, they were either merging with larger bodies (as with Earth and our Mars-sized interloper) or clearing orbital paths through the rubble left behind as the Sun's emergent system gradually collapsed down into innumerable planetesimals, or protoplanetary bodies.

Even as the young Moon assumed a spherical shape under the centralizing pull of its own gravity, successive major collisions continued to knock it about. To this day, the scars of that battering are clearly visible, with even the most cursory examination by low-powered telescope revealing the lunar surface's many craters. Even its so-called mare, or "seas"—actually dusty desiccated plains—were formed from colossal collisions somewhere between 3.2 and 4.2 billion years ago. Far from holding bodies of water, as mistakenly assumed by early astronomers, their dark tonalities are the result of molten rock that pooled and cooled within giant impact basins.

The Moon is proportionately the largest natural satellite in the solar system in relation to its planet, and its gravity gradually stabilized the Earth's spin. Our world had been jarred out of kilter by the impact, and that stabilization process unfolded partly through tidal interactions with Earth's oceans (that is, actual water)—a gravitational event that also resulted in the Moon's more massive face, the side with the lava-heavy "seas," becoming permanently oriented toward Earth. As a result, seasonal changes due to the planet's 23.4-degree axial tilt were moderated, producing milder and more predictable temperature variations than if the planet wasn't orbited by such a large satellite.

In this and other ways, the Moon's presence had a significant impact on the evolution of life. The increasingly dependable steadiness of our planet's seasons fostered the development of ever more complex biological forms. In effect, the Moon was functioning as a kind of outrigger, balancing its parent planet and eventually allowing for the emergence of creatures capable of mustering a creditable theory of its origins—something which might not have happened without it. As Blaise Pascal put it in the seventeenth century, "There is nothing so inconceivable as the idea that matter knows itself. We cannot possibly know how it could know itself."[1]

And yet evidently we do. Here we are—beings comprised of matter contemplating the role of matter in permitting the emergence of creatures capable of considering their own origins. And this in turn starts to answer a question that may already have occurred, namely: In a book dedicated to examining elements of natural design on Earth—a book predominantly focused on microscopic terrestrial life-forms—why start with the sterile, battered Moon?

I'm tempted to write: *Because it's there*. But the answer really is, to begin at the beginning. Moonless night in the small town. The stage, prior to the arrival of players. Genesis 7:20, if you prefer.

The most ancient lunar rocks were recovered by the last two Apollo missions, 16 and 17. Both flew in 1972, and both landed in the some of the Moon's most rugged regions.

When astronauts John Young and Charles Duke radioed a goodbye to Apollo 16 command module pilot Ken Mattingly on April 21, 1972, and descended toward the lunar surface, they thought they were going to a region that had once seen great

1 Blaise Pascal, *Pensées*, trans. W.F. Trotter (E. P. Dutton & Co., Inc., 1958).



**Left:** Étienne Trouvelot and J. H. Bufford, *Moon Craters*, 1872, lithograph volcanic activity. The heavily cratered Descartes Highlands, a region of disrupted, undulating terrain, would never have been chosen for the early Apollo missions. It was simply too rough, and therefore potentially dangerous to land on. But the Apollo program had grown bolder since the first landing four years earlier, with mission planners sending the later crews to destinations of greater geological interest, while also extending surface exploration times.

After setting down without major incident between North Ray and South Ray craters, each centered on prominent streaks of material ejected by the meteor strikes that made them, Young and Duke configured their lunar module for a three-day stay, squinted out at the brightly lit Cayley Plains, and tried to sleep for a few hours. Then they climbed into their space suits, depressurized the cabin, backed one by one down the ladder, and extracted their lunar rover from the lander's lower stage. Part of the purpose of the last missions—and Apollo 16 was the penultimate of six—was to better understand the Moon's early history. Descartes being quite a bit brighter than its surroundings, geologists believed that this mountainous moonscape must have been formed by volcanism.



It was in fact a primary reason for choosing these ancient highlands as a destination in the first place, and their landing site between two relatively young craters was deemed a perfect place to collect surface material. Such impacts were known to act as natural core samplers, drilling down to bedrock layers—in this case again presumed to be of volcanic origin. South Ray crater alone was 700 meters in diameter and 120 meters deep—picture about five big football stadiums—and had clearly been caused by an impact so violent that bright, fresh-looking streaks of ejected material undulated quite visibly across the Cayley Plains half a billion years later. Actually the astronauts quickly learned to navigate by these bright landmarks as they bounced across the lunar surface in their electric dune buggy.

What Young and Duke soon discovered at Descartes defied expectations, however. By the middle of their first excursion, geologists in Houston were already concluding that their volcanism theory was mistaken. What the astronauts were seeing were innumerable fragments of brecca—a type of rock formed by the heat and pressure of meteor impacts, rather than volcanism. Another visible manifestation of this were innumerable fragments of "impact glass," which glittered in the sun.

During that first rover excursion, the astronauts used an electric drill to penetrate three meters below the surface, after which they inserted a temperature sensor on a long wire—a "heat flow experiment." As they were doing this, one of them spotted a yellowish-green marble sitting on the ground nearby. Partially fractured on one side but otherwise intact, the 3.5 cm diameter glass sphere had probably been sitting in this same location on the lunar surface for around 500 million years. Somewhat distracted by their larger tasks, one of the moonwalkers dutifully bounced over and unceremoniously raked it into a sample bag.

That small globe of impact glass, forged from the intense heat of an ancient meteor strike, is now called Lunar Sample 60095. Following its return to Earth on April 27, 1972, four and a half billion years after the Moon's formation from molten terrestrial bedrock, it was stored in a nitrogen-filled vault at the NASA facility in Houston that is charged with preserving almost 400 kilograms of rocks and regolith retrieved by the Apollo program.

Sample 60095 provided every lunar landscape visible between pages 41–61 of this volume.

~

The Moon may be a chip off the Earth's proverbial block, but unlike its parent planet, whose own violent geological history has been largely covered over by ocean waves, plate tectonics, and the weather, our natural satellite has preserved a readable record of the early, shooting gallery solar system.

The oldest rocks on Earth date back about 4 billion years and are exceedingly rare. By contrast, the oldest lunar rocks are 4.5 billion years old—again, only 0.1 billion years younger than the solar system—and are thought to be parts of a surface crust that solidified soon after the impact that formed the Moon. Actually, the lunar origin theory outlined here is firmly rooted in Apollo samples, which contain the same isotopic signatures as terrestrial rocks—something virtually impossible without a shared origin.

Even at the level of fragments of a 3.5 cm-wide piece of glass, the lunar surface is a palimpsest—a landscape scribbled over many times by impacts, with the later inscriptions partially obscuring the older ones below. This is evident everywhere on the Moon, from the four-million-square-kilometer Oceanus Procellarium, the only lunar impact basin to rate "ocean" status, to the only slightly smaller mare or "seas," to the giant fifty-mile-wide craters like Tycho and Plato, and all the way down to comparative small fries like Apollo 16's North and South Ray craters.

And as can be seen in this book, that progression from large to small, from giant impacts to ever smaller ones, proceeds incrementally down to exceedingly small scales. Without an atmosphere capable of absorbing micrometeorites, the Moon has been exposed to a steady rain of tiny incoming fragments of minerals and metals for the duration of its existence. That unceasing bombardment, which in Earth's protective atmosphere would register as shooting stars, is what gave the Moon's mountains their gently rounded, sand-blasted appearance. While there is a lower limit to this micrometeoric pummeling, it's invisible to the unaided eye and can only be seen here due to the power of the scanning electron microscope.

And this brings us back to Pascal's observation about matter knowing itself. The samples brought to Earth by the Apollo program—prodigal material returned back to the world of its origin—have their own integrity as pieces of a whole. Like us, they're fragments of a larger story, and the point of presenting them here is to understand them both as such and also as artifacts.

They were forged, after all, by collisions with a body itself formed by collision, and they bear the marks of successive impacts. They are in fact nothing but impacts and material, extending all the way back to the original impacts and material that formed the Earth-Moon system—and, even prior to that, to the gradual, accretionary process that formed the solar system to begin with. It's turtles all the way down.

So what about that small blue-white sphere visible in the black sky above all those scattered, glittering mineral fragments? The exceedingly unlikely processes that unfolded on our world—an obscure proceeding eventually leading to life, intelligence, flight, space travel, and, eventually, that imperious flick of an Apollo 16 sample rake—are all miraculous. But so are these fragments of lunar surface considered separately. Like us, they are, as Pascal observed, "a nothing compared to the infinite, a whole compared to the nothing, a middle point between all and nothing." ■

# Radiolarians

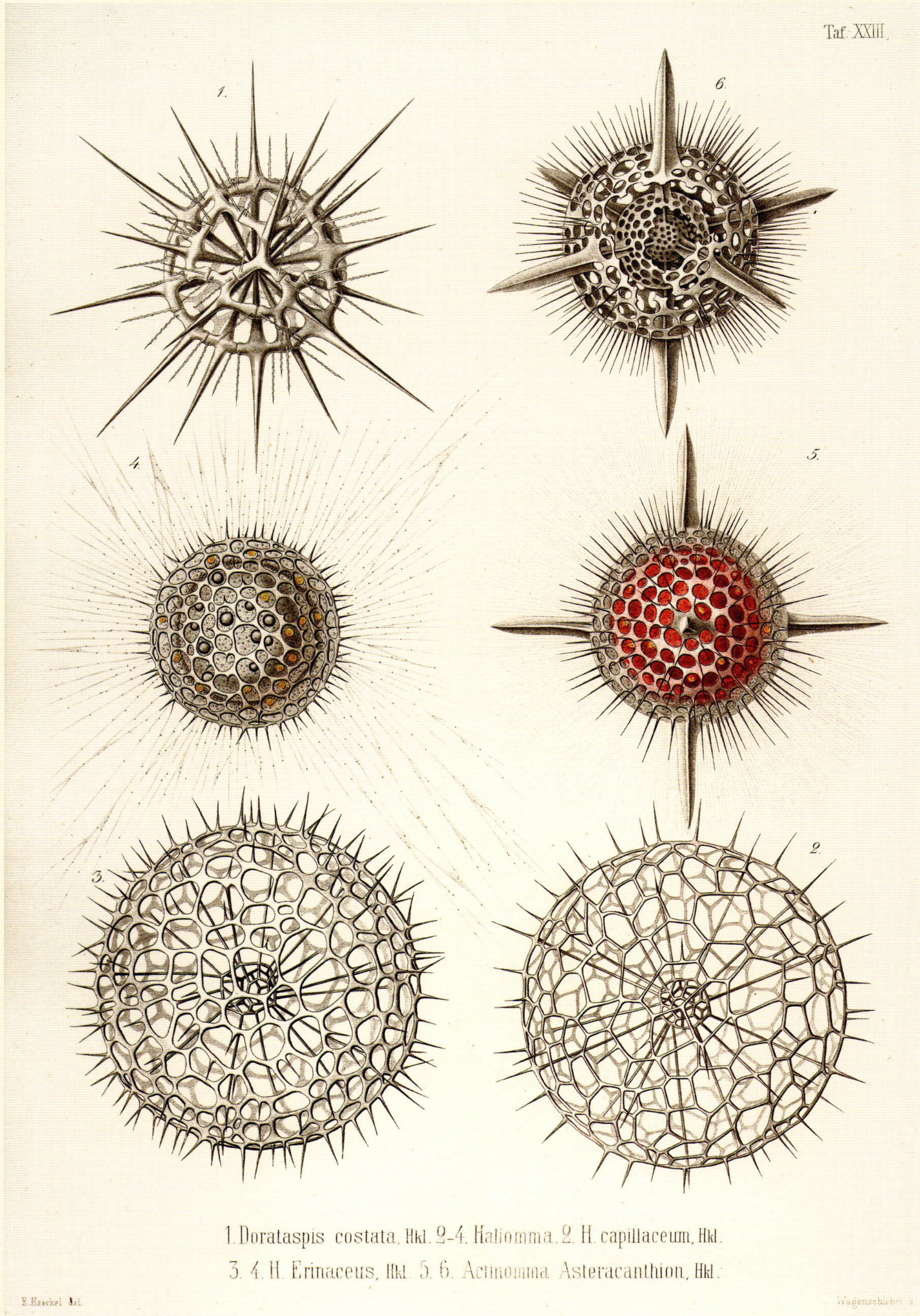

No creatures from the microscopic world have managed to captivate the human imagination as authoritatively as radiolarians. Since the late nineteenth century, they've made their existence known in art, architecture, design, and even literature. These single-celled marine organisms, which appear to have teleported straight from some rarified Euclidean space into all the world's oceans, seamlessly fuse geometry and biology. Their astonishingly complex glassine skeletons may serve to protect and regulate buoyancy for the cell inside, but they also seem a kind of physical embodiment in sculpted glass of what radiance means.

Radiolarians are adept at distilling silica, a form of quartz, from seawater, a process known as biomineralization. In this process, they generate a kind of mad geometer's vocabulary of forms—polyhedra, Archimedean solids, icosahedrons, dodecahedrons, and the like—tongue-twisting shapes perforated by radiating spines and microtubules. Their extraordinary morphologies serve to refute any suggestion that biological forms must simplify as they grow smaller. If anything, the opposite appears to be true. They can be found wherever salt water makes waves, but particularly in the warmer surface layers of the Indian and Pacific Oceans. No single species predominates.

Radiolarian skeletons are typically constructed of an orderly mesh of rigid polygons. These are frequently shaped into nested spheres—though there are many exceptions, including intricate axial symmetries. The central sphere contains the cell's endoplasm and nucleus, site of the greatest cellular activity. Surrounding that and protected by an outer shell is the cytoplasm, a frothy liquid of protein-synthesizing ribosomes, energy-producing mitochondria, energy-storing lipids, and the like.

Most radiolarians are heterotrophic, meaning they consume plant and animal matter, but many also possess photosynthetic abilities due to endosymbiosis (in which one organism, the photosynthetic one, lives within the larger radiolarian, to the benefit of both). This allows them to acquire energy from sunlight as well. Their characteristic projecting spines serve both as flotation modulators (via a complex process akin to raising and lowering protoplasmic sails) and as masts permitting the cell to extend itself outward into the surrounding water and feed. Complexity packaged within geometry is a radiolarian signature.

For all that, they only live for about two weeks. After which radiolarians sink, producing a steady snowfall of delicate glasswork onto the ocean floor. Depending on the depth, their skeletons can take up to fifty years to descend—a gentle rain continuously falling in all the world's oceans since at least the Middle Cambrian era, over 500 million years ago. The resulting carpet of fallen forms eventually compounds into a siliceous ooze which, on million-year timescales, eventually yields a particularly dense, fine-grained rock called radiolarite, aka chert or flint.

In this way, one of the first forms of life to have possessed a skeleton of any kind returns to the land. Chert was in turn used by some of the earliest protohumans to make tools and was thus a key material in the 3.4-million-year-long Stone Age. There are many ways to infiltrate human culture.

So how did radiolarians manage to do a bit more of that, in relatively recent history? Therein lies a tale.



In 1859, a young German physician with a newly minted license to practice surgery and obstetrics decided to go on a trip to Italy. Ernst Haeckel had never been entirely convinced that medicine was his calling. He'd gone into the field under parental pressure and had been repelled by the surgical gore of medical school. But Haeckel had a decidedly research-oriented side and had been supported in his interest in the natural sciences by a distinguished professor at the University of Wurzburg, anatomist Albert von Kölliker. Deploying the newest compound microscopes then being produced by Carl Zeiss in nearby Jena, von Kölliker had in effect moved from the dissection table to the lab, thereby founding a new medical field: microscopic anatomy.

Haeckel's Italian trip was part of a surreptitious alternate life plan. While visiting a friend at the University of Jena, he had been

**Left:** Ernst Haeckel and Adolf Giltsch, *Radiolarians*, 1887, lithograph offered the prospect of an academic career—something he chose to keep confidential. The friend, Carl Gegenbaur, was already a professor at the university. He suggested Italy as an ideal educational destination for young Ernst in preparation for a possible position in Jena.

At the outset of his trip Haeckel was still shaken from the death, probably by suicide, of one of his professors, the esteemed physician and marine biologist Johannes Müller. He was determined to continue research into a type of marine organism, the radiolaria, that Müller had initiated. Considered one of the greatest natural philosophers of the nineteenth century, Müller had been among the first to describe this group of single-celled plankton with intricate skeletal structures. He had also promoted the term radiolarian, a derivative from the Latin *radiolus*, which referred to their characteristic radiating points. Working primarily in the North Sea, Müller had pioneered the use of extremely fine nets to collect these and other tiny marine organisms.



On an extended trip to Messina, Haeckel set about resuming Müller's radiolarian work. He commissioned an elderly fisherman in the Sicilian harbor city, Domenico Nina, to row him around while he sieved the sea with nets based on Müller's design. Examining what he brought back with a growing fascination, he quickly realized that the rich waters of the Strait of Messina—a unique Mediterranean environment where swift currents funnel nutrients through a constricted space—was yielding dozens of previously undiscovered species. Just as von Kölliker's use of the microscope in anatomy had created a new field, now the combination of Müller's nets, Haeckel's artistic abilities, and a specialized feature of the microscope he had brought from Germany, produced by the Berlin company Schiek, enabled a new chapter in marine biology—one that he had the opportunity to write.

Amazed by their radial symmetry and jewel-like skeletons, Haeckel realized that the only way to do radiolarians justice was to draw them with as much precision as possible. To accomplish this, he used his microscope's camera lucida attachment—an advanced feature which enabled him to project his marine samples directly onto paper, where they could be traced by his pen. The results were unprecedentedly exact renderings.

By early 1860, Haeckel had discovered over a hundred new radiolarian species. In harnessing his artistic abilities to what was clearly a creative scientific drive, he'd also found his calling. By 1862 he had published an illustrated monograph, *The Radiolarians*, which he dedicated to Johannes Müller. It contained ten sumptuous color plates, with each depicting multiple species in detail. (One example is reproduced on page 292.) Nothing quite like it had been seen before, and when Haeckel sent a copy to Charles Darwin, whose *On the Origin of Species* had been published just three years before, he soon received an enthusiastic response. "It is one of the most magnificent works which I have ever seen, & I am proud to possess a copy from the author," the great naturalist wrote. "It is very interesting & instructive to study your admirably executed drawings; for I had no idea that animals of such low organization could develop such extremely beautiful structures."[1]

Having established himself as a marine biologist and initiated a link with the greatest scientist of his time, Haeckel rapidly ascended to international prominence. Successive illustrated monographs on sea sponges, jellyfish, and zoology earned him a full professorship in Jena, and he became the leading proponent of Darwin's evolutionary theories in Germany. By the late 1870s, he was asked to take part in the gargantuan task of evaluating and cataloguing the results of the five-year *Challenger* expedition—an ambitious British global research voyage, the first ever dedicated purely to science. Haeckel's contribution to the final 50-volume *Report of the Voyage of H.M.S. Challenger* took a decade to complete and spanned three volumes, 2,750 pages, and 130 plates.

The latter were produced with talented master lithographer Adolf Giltsch, a lifelong collaborator in Jena who possessed "the knowledge-seeking drive of a true naturalist," as Haeckel put it. In his *Challenger* work alone, Haeckel described over 4,000 new

1 Charles Darwin's letter to Ernst Haeckel, March 3, 1864. darwinproject.ac.uk/letter/?docId=letters/DCP-LETT-4419.xml.

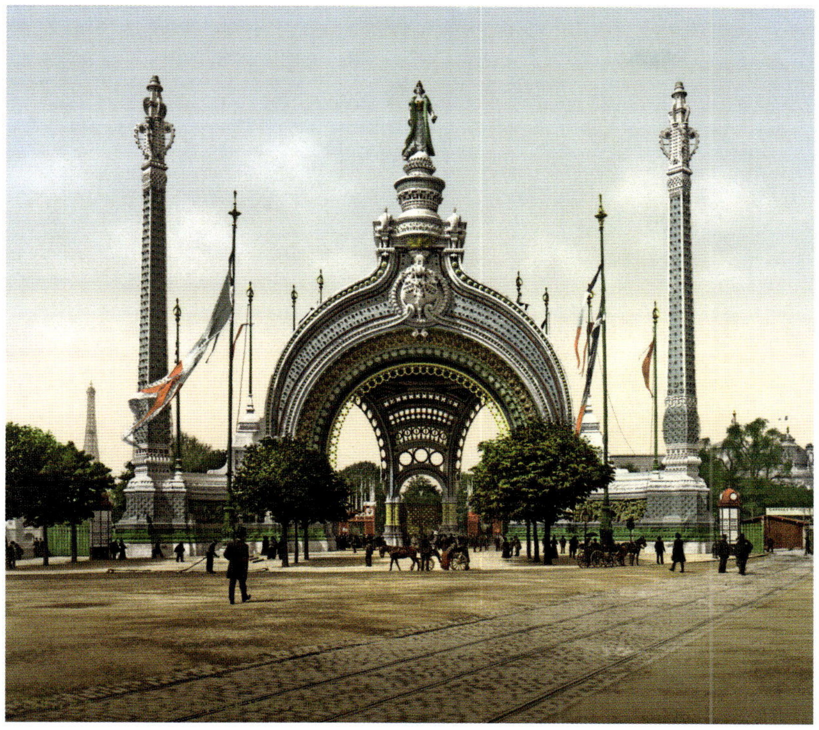

Unknown photographer, *René Binet's Porte Monumentale at the 1900 Paris Exposition*, circa 1889–1900, photolithograph

radiolarian species. He had become by far the leading authority on the subject.

By the turn of the century, Ernst Haeckel had amassed enough drawings to produce his great masterpiece and the single work most associated with his name, *Kunstformen der Natur* (*Art Forms in Nature*). Once again, Giltsch's lithographs turned Haeckel's drawings into finely tuned renderings suitable for mass reproduction. Radiolarians were visible throughout the book, which was published as separate volumes between 1899 and 1904 and finally as a complete edition in 1904 with one hundred full-page lithographs.[2] An immediate commercial success, it has never gone out of print.

As already touched on in the Introduction, *Art Forms in Nature* almost immediately impacted the visual culture of the early twentieth century. Artists of the art nouveau and German Jugendstil movements channeled Haeckel's discoveries into their work, as did such leading architects and designers as Antoni Gaudí and Louis Sullivan. In 1898, architect René Binet unveiled his design for the Porte Monumentale, a grand three-legged archway intended to serve as the entrance portal to the upcoming 1900 Paris Exposition. The massive structure was based on the radiolarians that Binet saw in Haeckel's *Challenger* plates, a fact which he made clear in letters to the marine biologist, now in his late sixties.

Binet's Porte was duly constructed late the following year and marked a kind of apotheosis of natural design incorporated within architecture. Comparable in size to the Arc de Triomphe, the Porte Monumentale also contained surface ornamentation based on Haeckel's exacting lifelong documentation of organisms to be found in the world's oceans.[3] When the Exposition closed that November, however, it was unceremoniously disassembled, never to be seen again.[4]

By contrast, Ernst Haeckel's *Art Forms in Nature* continues to be among the most influential illustrated books ever published. Ranking as the first international image-based arts-science bestseller, it brought the enduring mystery of nature's extraordinary ability to forge dazzlingly intricate structures to the public. ■



2 "Obituary of Adolf Giltsch by Ernst Haeckel," Kunstformen der Natur, Mediamatic, mediamatic.net/en/page/369299/obituary-of-adolf-giltsch-by-ernst-haeckel; see also "Plates," Kunstformen der Nature, Mediamatic, mediamatic.net/en/page/224004/plates.

3 Robert Proctor, "A World of Things in Emergence and Growth: René Bidet's Porte Monumentale at the 1900 Paris Exposition," in *Symbolist Objects: Materiality and Subjectivity at the Fin-de-Siècle,* ed. Claire I. R. O'Mahony (Rivendale Press, 2009), 220–244

4 "Christina Svendsen, "Ephemeral Buildings: Threshold Experiences of Memory in Paris, 1889 and 1900," Amodern, amodern.net/article/ephemeral-buildings.

# Diatoms

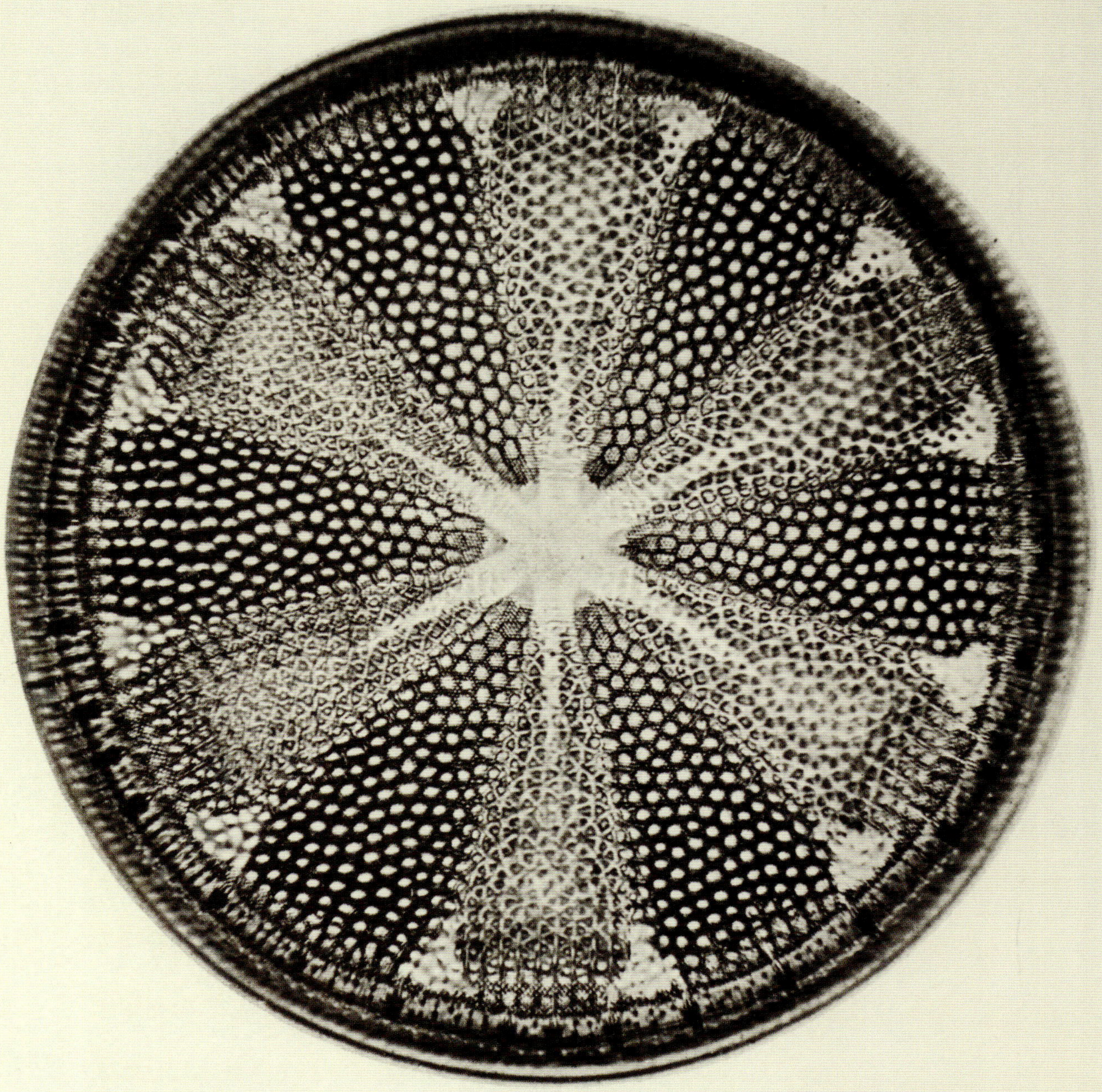

FIG. 3.
DIATOM, HELIOPELTA METII.
× 350.

On first observing diatoms through his achromatic microscope—an impressive instrument that permitted him to magnify specimens over 1,300 times—Charles Darwin was amazed. What could possibly account for the exquisite complexity of these organisms? What natural evolutionary processes might produce such intricate precision of design, to say nothing of their wildly different forms?

He exchanged views on the subject with Ceylon-based botanist G. H. K. Thwaites, who was engaged in trying to prove that diatoms weren't animals at all, but algae. They discussed how diatoms distilled their glass-like shells, or frustules, out of silica dissolved in water. In his writings, Thwaites had put forward a provocative philosophical question: Was it possible that such beautiful organisms had been created specifically so that one day they could be admired by humanity? Did they exist in order to justify the invention of the microscope?

In *On the Origin of Species*, Darwin pushed back against such anthropocentric ideas. Referencing diatoms specifically ("Few objects are more beautiful than the minute siliceous cases of the diatomaceae"), he addressed larger questions concerning the symmetrical and seemingly harmonious forms produced by evolutionary processes. Darwin wrote that some of his fellow naturalists were of the opinion "that very many structures have been created for beauty in the eyes of man, or for mere variety. This doctrine, if true, would be absolutely fatal to my theory." In a letter to Thwaites in 1860, he meditated further on the question: "I should attribute most of such structures to quite unknown laws of growth; & mere repetition of parts is to our eyes one main element of beauty. When any structure is of use (& I can show what curiously minute particulars are often of highest use) I can see with my prejudiced eyes no limit to the perfection of the co-adaptations which could be effected by natural selection."

Although he clearly suspected that the bilateral symmetries, neatly arranged grill-like pores, and extraordinarily intricate ribs characteristic of diatom shells had a purpose—a "*highest* use"—Darwin could not yet know that these marvels of microscopic engineering, which look so much like miniaturized technology of the highest order, are responsible for the production of between 20 and 50 percent of all the oxygen infused into Earth's atmosphere each year. Diatoms are the biggest component of the total oxygen production of all marine photosynthetic organisms, which collectively produce about 70 percent of the planet's oxygen—a figure neatly in line with how much of Earth's surface is covered by ocean. (By contrast, trees and other land plants produce about a third.) They also absorb between ten and twenty billion metric tons of carbon dioxide annually—a figure roughly double the net figure of carbon captured by all of the planet's trees. Unfortunately, as with so many other taxonomic families, diatom populations are in steep decline, thereby reducing the amount of carbon absorbed from the atmosphere and injected safely into deep ocean storage.

The term "diatom" covers an exceedingly large group of single-celled creatures with clear, glass-like shells. These function a bit like microscopic petri dishes, simultaneously protecting and revealing the cells within. Diatoms are in fact now considered microscopic algae, a type of phytoplankton, and can be found in both salt water and fresh water, as well as damp soil. Their frustules can even be detected within airborne dust—as Darwin himself determined when he sent air samples he'd collected in the Cape Verde Islands in 1832 to another natural philosopher, Christian Ehrenberg. The German zoologist confirmed that the crystalline forms Darwin had observed by microscope on board the *Beagle* were silica diatom shells. Unlike Thwaites, Ehrenberg, who specialized in microscopy and microscopic organisms, was of the belief that diatoms were animals, not plants.

The specimens Darwin had recovered from the thin air of an island group over 300 nautical miles from Africa were part of the vast annual intercontinental fertilization of the Amazon basin by diatom shell dust, an estimated thirty million metric tons of which is transported by trade winds annually from storms in the Sahara to the tropical rainforests of South America. Much of it comes from the Bodélé Depression, where a vast system of

**Left:** Arthur E. Smith, *Diatom*, 1909, photomicrograph

freshwater lakes existed for millennia before gradually drying up. Many of the diatoms visible in this book come from freshwater sources. These include Siberia's Lake Baikal—the largest freshwater lake by volume in the world.

Apart from proliferating in rivers and lakes, suffusing the upper regions of the sea from pole to pole, and wafting about on the wind, diatom residue can be found in thick sedimentary layers half a mile deep on the ocean floor. (They are joined by radiolarian silica.) In fact, diatoms constitute up to half of all the organic material found in oceans, meaning that creatures invisible to the naked eye comprise an extraordinarily large portion of Earth's biomass. Diatoms absorb an estimated 6.7 billion metric tons of silicon from the waters where they live every year—with silicon's mineral form, quartz, being the second most abundant element of Earth's crust after oxygen, and readily dissolvable in water.

Like plants, diatoms use photosynthesis to convert energy from light into chemical energy. But in practice, Ehrenberg, who thought they were animals, and Thwaites, who argued that they were plants, were both wrong. Diatoms and other protists—a biological kingdom comprising mostly single-celled organisms—are somewhere in between. Diatoms are roughly divided between mostly disc-shaped planktonic forms, which are lighter and float freely in the water column, and bottom-dwelling or benthic species, which tend to exhibit an elongated bilateral symmetry and are heavier. The former have evolved various strategies for remaining at sunny depths, while the latter are capable of moving along ocean or lake floors, much like tiny snails. Bottom-dwelling diatoms invariably exhibit paired slits in their shells, called raphes, which enable them to extrude an adhesive cytoplasm and creep about. One leading diatom researcher, Paul Hamilton of the Canadian Museum of Nature, has frequently observed bottom-dwelling varieties moving quickly to escape his intrusive micro-pipette as he attempted to vacuum them up for study—an evasive action which didn't prevent his discovery of over seventy new species.

The fact that single-celled organisms invisible to the naked eye produce almost half the oxygen we breathe, not to mention absorbing far more carbon dioxide than all land plants put together, is thought provoking to say the least. It serves as more than confirmation of Darwin's view that their intricacy had a purpose beyond ornamentation. It also could be considered as substiantial empirical evidence of the Gaia theory, which proposes that all living beings interact with their surroundings to form a self-regulating whole capable of perpetuating life on Earth.

In fact, the precision and seemingly machined quality of the diatoms' silica shells, which look extraordinarily similar to examples of human technology, even call into question what is meant by design. Is design always something produced by human beings, or is this yet more anthropocentrism? After all, as already mentioned in the Introduction, the origin of the word "technology" is a compound of the ancient Greek word *tekhne*, for "craft, art," and *logia*, for "study, knowledge." And as Aristotle wrote in *Physics*, "When a thing is produced by nature, the earlier stages in every case lead to the final development in the same way as in the operation of art, *and vice versa*." [Emphasis added.]

Aristotle's formulation equating the productions of humanity and nature may have been known to Darwin, whose *On the Origin of Species* came out a mere twenty-three centuries later. In any case, the clear inference here is that complexity and refinement emerge from evolution—something inarguable when studying natural selection's 3.5 billion year sweep. And certainly, for much of human history, our imitations of nature's designs were thought to represent ultimate craftsmanship. Even if we discard the idea of a divine maker, we're left with diatoms as a stunning example of natural design, their silica shells the result of upward of 200 million years of incremental refinement. Call it nanotechnology long before we managed to get there—long before we even arrived on the scene, in fact. And it's capable of providing half the oxygen we breathe.

No surprise, then, that diatoms are being studied for their potential use in exactly that: nanotechnology. So-called diatomaceous earth is already used as a fine abrasive for the polishing of various materials and also in water filtration. But as part of their inscrutable purposes, these unicellular organisms also seemingly specialize in the production of exceedingly tiny valves of various complex shapes—an ability that could permit their deployment in the production of micro-scale structures usable across a wide range of potential industries. These might eventually include diatom-powered semiconductor nano-lithography, the use of diatoms as tiny vehicles for drug delivery, and even the cultivation of diatoms for use in the mass production of nanoscale components.

Of course, to do that we would need to divert at least some of these creatures from their critical role in the carbon and oxygen cycles so integral to the maintenance of life on Earth—a fact that should give us pause, but probably won't. ■

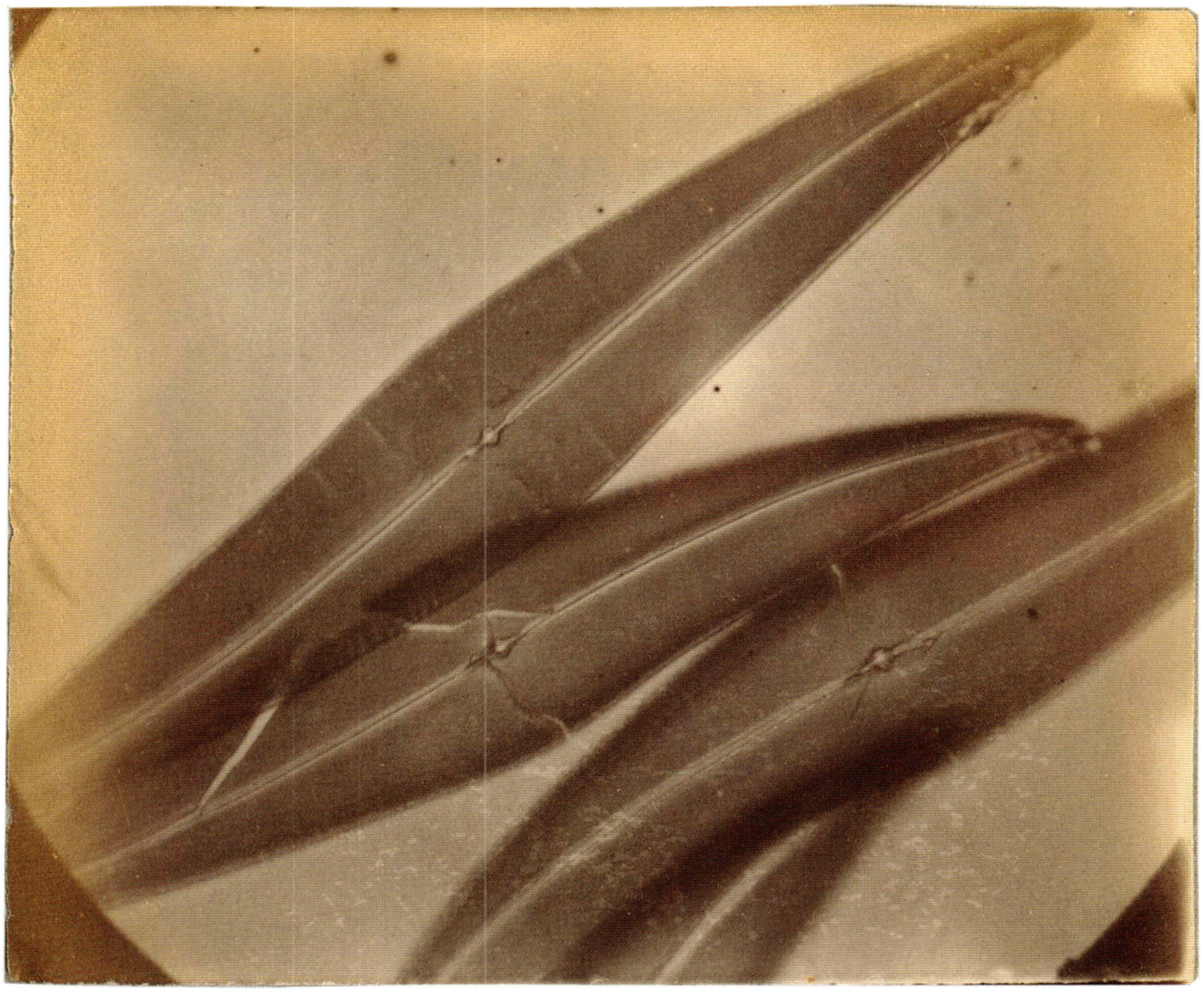

Marianus Piter Filbri, *Microscopic Image of Diatoms*, circa 1887–1888, albumen print

# Dinoflagellates

*Haeckel, Kunstformen der Natur.* Tafel 14 — *Peridinium.*

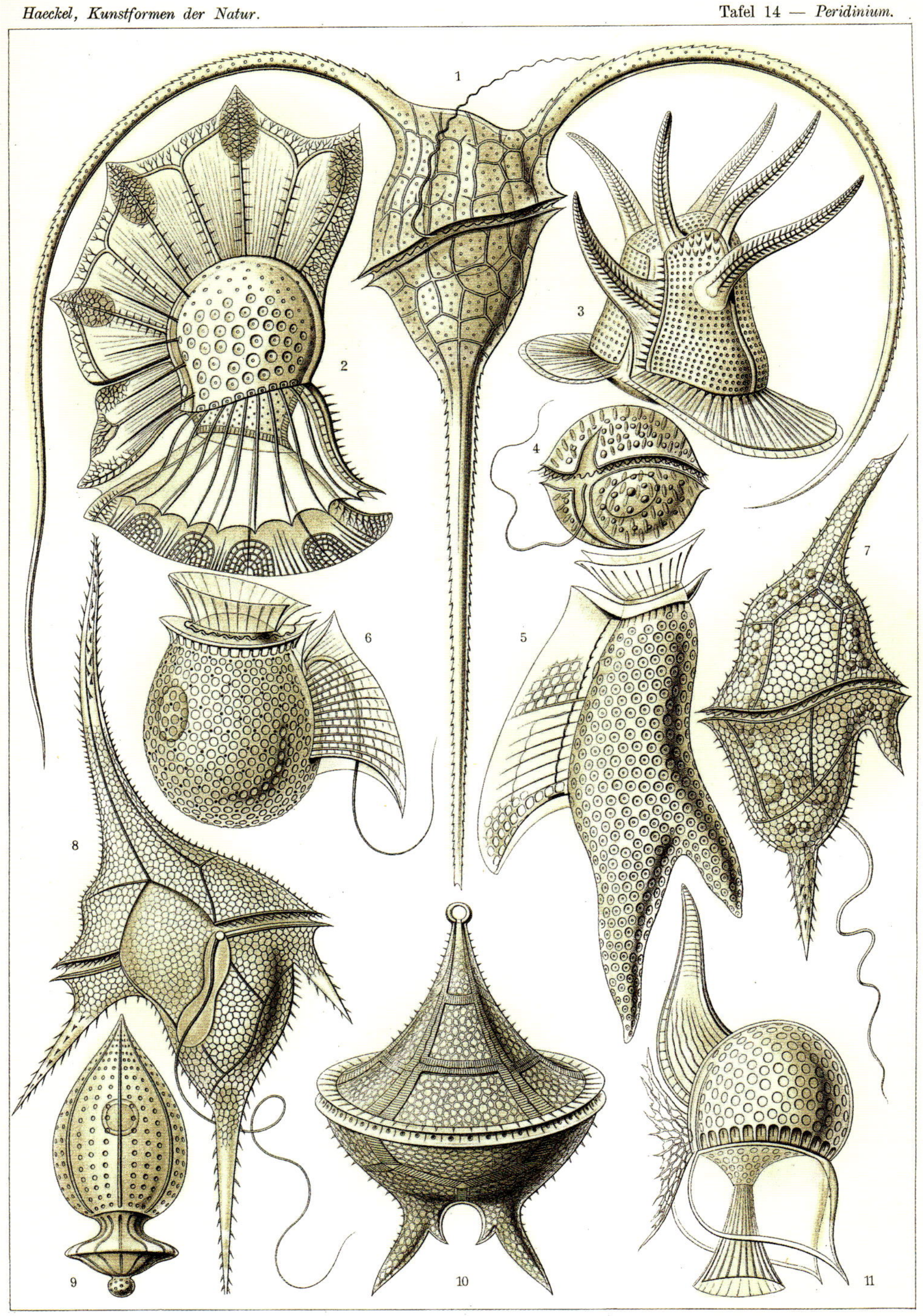


Peridinea. — Geißelhütchen.

FEW THINGS BETTER EMBODY THE CRYPTIC, kinetic qualities of life within Earth's vast watery realms at microscopic scales than dinoflagellates. A close examination of their external topographies in the SEM, an instrument capable of revealing every complex ridge and fissure, can produce a profound, even metaphysical sense of appreciation for the alchemical ways in which living matter is sculpted by time. In appearance, dinoflagellates reveal both a fiercely embodied will to prosper—to go forth and multiply—and an immutable linkage to an epic webwork encompassing all living organisms—the vast nexus conjoining us all. Although seemingly alien in appearance and function, they are very much of this world, an extraordinary ongoing expression of life's fundamentally mysterious workings.

Diverse in their shape, behavior, and living environments, these single-celled aquatic microorganisms live in fresh water, salt water, and all gradations in between, up to and including ice, snow, and damp soil. In this they are quite different from radiolarians, which can be found only in salt water. Their largest fields of play are the oceans, however, where over 1,500 known species whirl in their inimitable gyrating way, each propelled by a pair of flagella—whip-like motors of motion. They constitute one of the largest groups of eukaryotes, a domain including all plants, fungi, and animals. After diatoms, the subject of the next chapter, dinoflagellates are among the most prolific marine plankton groups.

It's difficult to study the complex hydrodynamical exteriors of dinoflagellates without feeling a sense of wonder at what an estimated 250 to 550 million years of evolutionary optimization has wrought. (These numbers are hard to pin down. Dinoflagellates have been found in fossil records dating back as far as 240 million years, but evidence suggests they originated earlier, in the Paleozoic era.) Thecate or armored dinoflagellates sometimes resemble medieval battlements or weapons. Their armor is composed of complex arrangements of plates, called theca, joined together by undulant ribbon-like fissures called sutures. By contrast, athecate or soft-sided dinos can have a plush, upholstered appearance, sometimes with fleshy fins, sails, and even umbrella-like crowns.

Despite this diversity, their morphology is always defined by paired latitudinal and longitudinal grooves, with each housing a threadlike flagellum. The latitudinal groove typically runs along the entire length of the creature's equator. The longer transverse flagellum typically has a ribbon-like quality and is used to power rotation around a longitudinal axis, producing the characteristic spinning motion of almost all dinoflagellates. The lower longitudinal flagellum is shorter and beats more quickly, serving as a kind of propeller generating forward motion—though actually, both flagella are used for propulsion and steering.

Like diatoms and radiolarians, dinoflagellates are usually (but not universally) considered protists—organisms possessing attributes of both plants and animals. While about half are photosynthetic, the rest are mixotrophic, meaning they use the energy provided by their ability to produce sugar from sunlight, water, and carbon dioxide to scoot off and hunt prey (with the latter, of course, being another source of energy). Typically, they're chasing diatoms, algae, ciliates (like paramecia), and also other dinoflagellates. But even the purely photosynthetic kind, the ones without the hunter-killer instinct, are highly mobile and buzz through the water like spinning tops. If they sometimes appear to possess a sense of self when doing so, it's not because they're philosophically inclined or prone to doubt, but rather because they seem to exude a palpable sense of purpose. For example, like diatoms, the subject of the next chapter, dinoflagellates will do their best to escape the giant vacuuming snout of a laboratory pipette, and can be observed literally diving for cover within a petri dish. Such behavior is more generally associated with animals than plants.

Although far too small to be readily differentiated with the naked eye normally, dinoflagellates are in fact unusually discernible by humans, both individually and in aggregate. This is for two reasons. Perhaps most magically, about eighteen of the one hundred twenty-five known dinoflagellate genera possess bioluminescent properties—meaning they glow in the dark when agitated. ("Genera" being a taxonomic category above individual species but below "family.")



**Left:** Ernst Haeckel and Adolf Giltsch, *Dinoflagellates*, 1899, lithograph

That blue-green bioluminescence is what causes the twinkling flashes in boat wakes at night, and can sometimes be seen when waves break as well. (When British naturalist-poet Henry Baker discovered dinoflagellates in 1753, he described them as "Animalcules which cause the Sparking Light in Sea Water.") Dinoflagellates' seemingly preternatural ability to shine when agitated is thought to have arisen as a response to predation. One theory holds that while it may serve to give away the position of panicky individuals to predators like copepods or fish larvae, to their detriment, the general glow also attracts the attention of larger hunters like adult fish, which proceed to graze on the grazers—a net gain for our tiny protagonists.

Along with presumably saving many dinoflagellate lives, their bioluminescent early-warning system did much the same for US Navy fighter pilot Jim Lovell at the midpoint of the last century. Lovell was on a night training mission off the coast of Japan in 1954 when the navigation systems in his F2H Banshee jet failed, leaving him unable to locate his aircraft carrier, the USS *Shangri-La*. With his ability to find his ship suddenly very much in doubt, a disoriented Lovell instinctively switched off his cockpit lights. As his eyes slowly adjusted, he squinted into the darkness—and gradually discerned a nebulous blue-green glow within the waves below his plane. It inscribed a phosphorescent, undulating ribbon extending to the horizon.

That ghostly light was due to the vigorous thrashing of thousands of dinoflagellates by the distant carrier's powerful propellers. The survival strategy of a profoundly alien species, it pointed the way to Jim Lovell's survival. A decade after he located and landed on the *Shangri-La* without further incident, Lovell was accepted as an astronaut trainee. His grace under pressure subsequently served him well as commander of Apollo 13, which suffered a crippling explosion on its way to the moon in 1969. As of this writing, he's the oldest living astronaut and the only human being who has flown to the moon twice.

The second leading reason why dinoflagellates can become visible to the naked eye is their tendency to produce toxic blooms in the water —so-called red tides. Another historical luminary, one slightly more prominent than Jim Lovell, was thought to have seen the results of this kind of dinoflagellate activity: Moses, the foundling of the Nile. From *Exodus* 7:20-1:

> And Moses and Aaron did so, as the LORD commanded; and he lifted up the rod, and smote the waters that were in the river, in the sight of Pharaoh, and in the sight of his servants; and all the waters that were in the river were turned to blood. And the fish that was in the river died, and the river stank, and the Egyptians could not drink of the water in the river; and there was blood throughout the land of Egypt.

In fact, the phenomenon described, one of seven biblical plagues that struck Egypt, almost certainly constituted a classic red tide. Caused by vast uncontrolled blooms of dinoflagellates and diatoms, such tides suck oxygen out of the water, killing almost all the fish and other aquatic life and producing a characteristic crimson tint. (Blooms like these almost certainly provided the name of that other Exodus stage set, the Red Sea.)

Dinoflagellates have other important, sometimes mystifying characteristics. For example, each packs up to ninety times more DNA than an individual human cell—a genetic archive so tightly squeezed into a special nucleus called a dinokaryon that researchers don't understand how it could possibly be used for transcription purposes by the host cell. (Recall that the genetic information carried by DNA needs to be accessible by a complementary nucleic acid, RNA, in order to direct protein production within cells.) While it might seem difficult to use given how packed-in all that DNA is, it's possible that their seeming overload of genetic material is what allows dinoflagellates to survive in such disparate environments in the first place.

And it may also be behind another survival strategy designed to keep predators at bay. About a hundred of the 2,500 known species of dinoflagellates produce dangerous neurotoxins, which is one reason why red tides kill fish. One of them, saxitoxin, is among the strongest known natural poisons. A thousand times more potent than cyanide, it is produced by representatives of the dinoflagellate genus *Gymnodinium*. Just five hundred micrograms of the stuff can paralyze and then kill a human being. That's 0.0005 of a gram—the merest pinpoint, but one easily accumulated by only a dozen or so filter-feeding mussels or clams.

It's something several members of English explorer George Vancouver's crew discovered to their detriment in June 1793, soon after eating about that many shellfish each on British Columbia's central shore. In this incident—the first documented case of paralytic shellfish poisoning on North America's west coast—one sailor died, and four others suffered severe illness. While the friendly Indians they had encountered along the way knew very well to stay away from shellfish during red tides, several of which had been observed that June, the Royal Navy wasn't exactly in the habit of asking them for advice. In any case, is there a better proof of the conjoined relationships of all life on Earth than the association of this tiny whirling organism with human death and survival throughout recorded history? ■

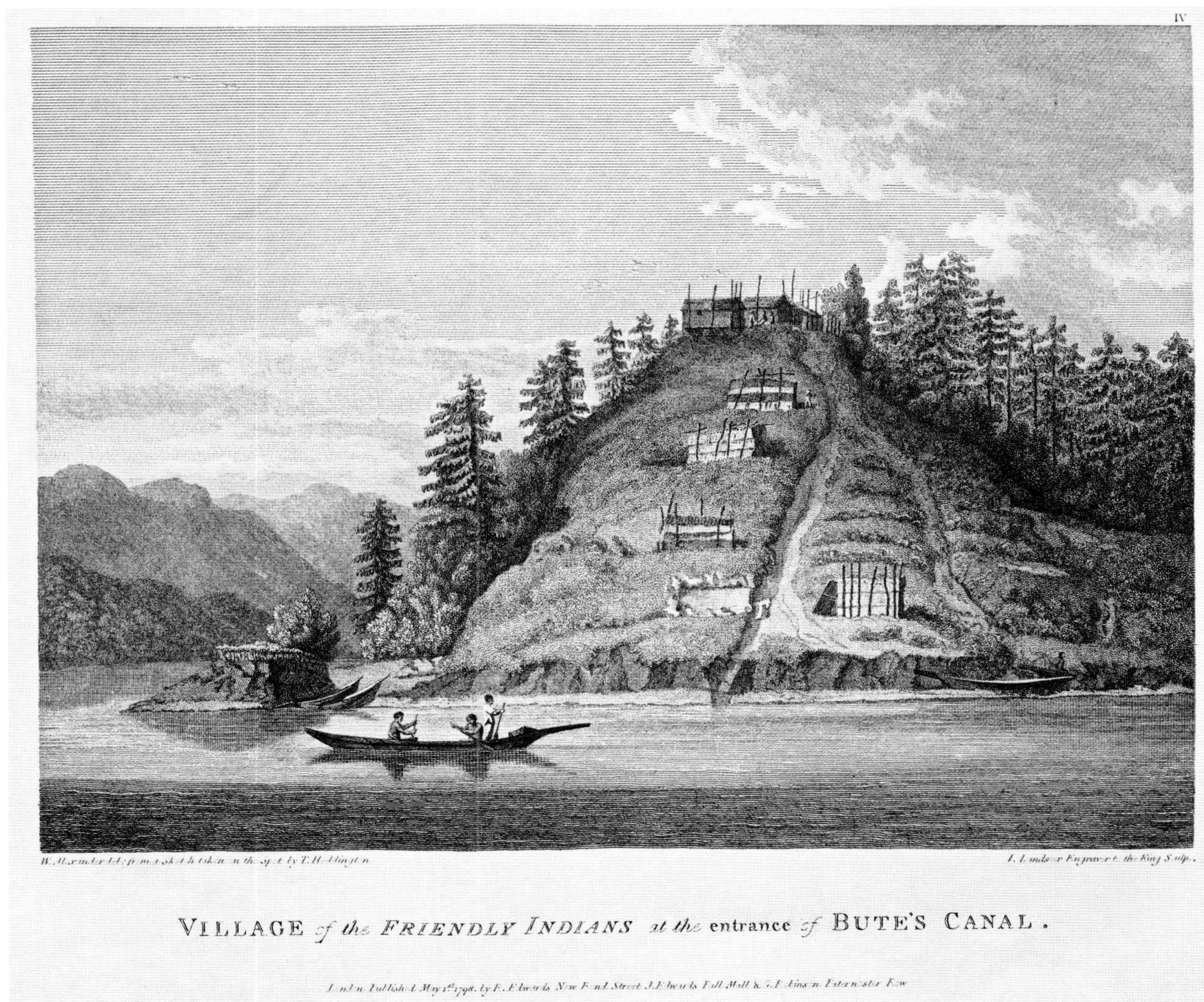


W. Alexander from sketch by T. H. Heddington, *Village of the Friendly Indians at the entrance of Bute's Canal*, 1798, engraving. Bute's Canal, now Bute's Inlet, is on the southwest coast of British Columbia.

# Aphids, Ants, and Plants

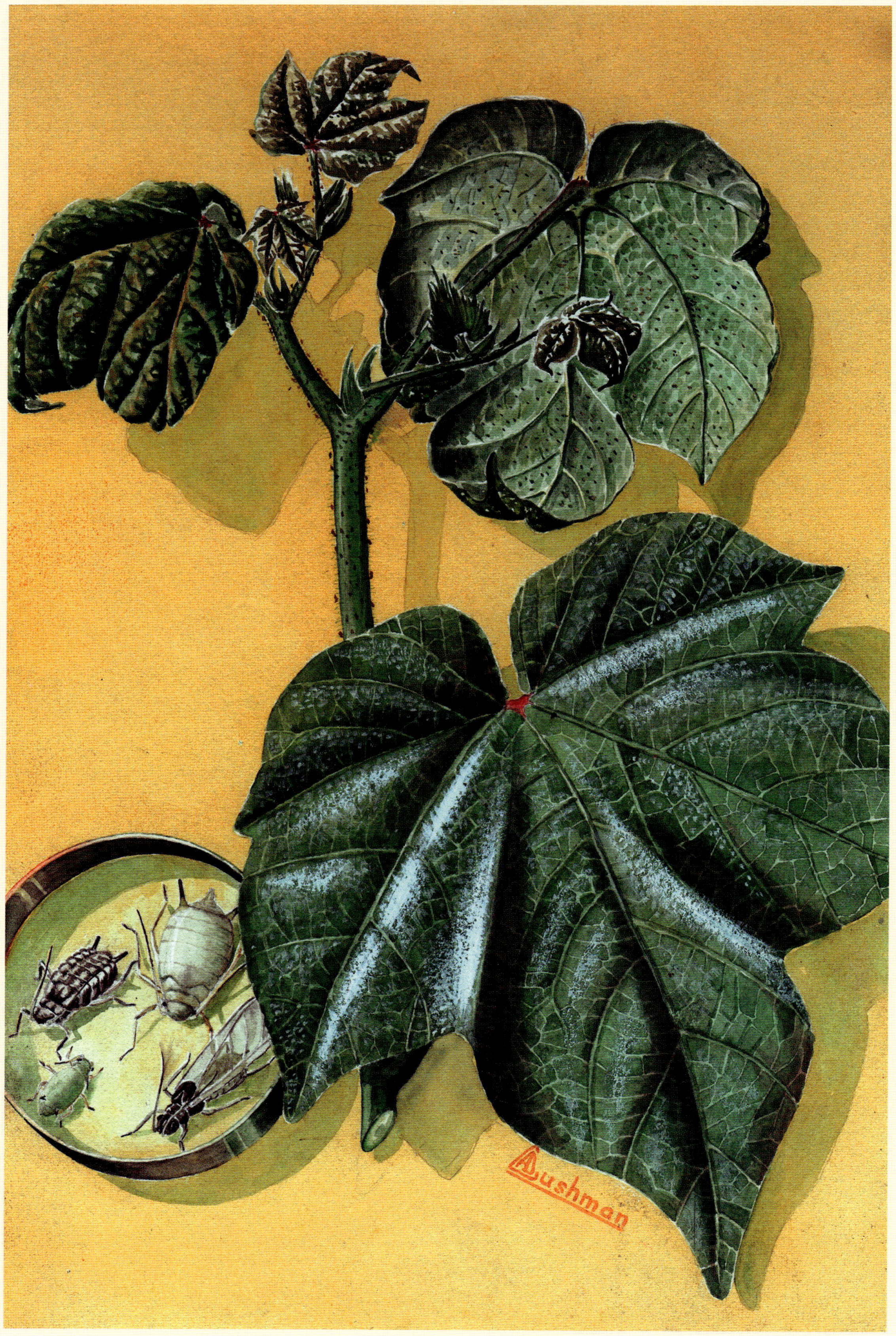

Behold the aphid—sap-sucking nuisance, cash cow of ants. Glorious result of 280 million years of evolution, ignominious pest to gardeners worldwide, aphids are superbly designed for exactly their ecological niche. Appearing to be stitched together from castoff burlap, these tiny bugs have corrugated, soft-sided bodies, exhibiting none of the fancy exoskeletal structures that make so many insects fascinating under the scanning electron beam. Instead, there's an endearingly elephantine quality to them, with their wrinkled skin and protruding, trunk-like abdominal cornicles—an extraordinary quality in something so small.

Aphids make their living by plunging their sucking mouthparts, called stylets, into shrubs, herbs, ferns, and flowers. Encased in a dangling, phallic-looking assembly called a rostrum, the stylet is an evolutionary development of the standard insectoid mandible and maxilla toolkit. Once it has reached the phloem—a layer of tissue containing the sugar-bearing "blood" of a plant, part of a circulatory system transporting liquid organic compounds produced from photosynthesis—the aphid taps this liquid, which courses through its body under high pressure, eventually blasting out of its anus after being transformed into a sticky, sugar-rich secretion called "honeydew."

This is, of course, kind of hilarious. It's no accident that aphids have a sac-like quality; they're designed to expand and contract under pressure. Several of the five hundred or so aphid species even manage to geyser out their full body weight in honeydew *every hour*.

Some aphids specialize in attacking a single plant species. They methodically suck their increasingly dehydrated hosts dry, all the while drenching them in sticky residue—a process frequently resulting in the dark mold familiar to many an exasperated gardener. Because the explosive byproduct of an aphid's feeding constitutes more than 90 percent sugar when dry, with much of the rest being amino acids, vitamins, and minerals, honeydew is an ideal food source for ants, which ingest the fluid, return to their nests, and regurgitate it for the benefit of the colony. This has resulted in a symbiotic relationship between ants, typically the dominant invertebrate in an ecosystem, and aphids. Scampering about energetically, powered by honeydew sugars, the ants tend to large herds of aphids, protecting them from predators and parasites and sometimes physically carrying them onwards to greener pastures. Bees also collect the sweet residue left behind by aphids for processing into a particularly strong, dark honey.[1]

One might think all that honeydew would be more than enough liquid passthrough for any self-respecting phloem processor, but when we turn the bug around and examine its backside, another aspect of aphid architecture reveals itself: paired cornicles. These backward-facing tubes were once erroneously thought to be the honeydew source, a mistake even the esteemed eighteenth-century Swedish zoologist-taxonomist Carl Linnaeus made. In fact, the cornicles exude a much less sugary liquid—a fast-drying, fat-bearing wax. Some have suggested that the wax has a defensive function and is used to trap and immobilize parasites. Contrarian entomologists contend that the excreted substance is simply how aphids rid themselves of excess fats.

Hosting the aphids visible in this volume are flowering plants of a variety you would barely notice, due to their size. In fact, the entire plant pictured on page 217 is only about a centimeter wide, with the aphid itself no more than two millimeters long. It was plucked from an Ontario lawn immediately prior to the passage of a mower so monstrous that its tiny passenger would have had no chance of exuding alarm pheromones. As for the flora that surrounds it—a subspecies within the verbena family—what you see here is a classic example of inflorescence, meaning it exhibits the complete head of a multi-flower plant, including pedicels (stems), bracts (modified small leaves centering on the flowers), and the flowers themselves.

A similar plant-aphid pairing can be seen starting on page 221, in this case in an image captured from the shore of the Croatian Adriatic. Flowering plants, also known as angiosperms, are the

1 The origin of the term "honeydew," however, apparently predates the realization that bees sometimes make use of it, and originates both in its resemblance to dew and its sweet taste.

**Left:** Arthur Cushman, *Cotton Aphids, Aphis gossypii, and Cotton Leaves*, 1952, illustration most diverse on Earth by far, comprising about 90 percent of all species. There are more than 300,000 kinds, including most trees (but not conifers, which are called gymnosperms[2]). When viewed up close, both of the aphid-bearing flowers seen here exhibit the complete angiosperm reproduction vocabulary. (As does the one with a weevil in it, for that matter, on pages 203-205, and the one wth the ant in it on pages 227-229.)

Their reproductive systems include the stamen—the male part—comprising stem-like filaments and fecund-looking anthers, with the latter producing pollen, whose tiny hard shells protect the sperm encased within during their journey to another plant, be it by wind or insect. Also visible are numerous pistils, or female reproductive organs, the protruding Y-shaped structures. These comprise stigma held aloft by the stalk-like style, with the ovaries of the flower hidden from view below. The sticky stigma collects and holds the pollen, rehydrating it after its journey. It's no accident that most flowers have open, welcoming forms. They are designed to maximize the chances that incoming pollen from other plants of the same species will be received and retained.

They're also frequently adapted to serve as perfect targets and landing grounds for winged insects of the kind to be seen starting on page 249. So what's keeping these perfect hermaphrodites from self-fertilizing—and thus radically reducing the beneficial effects of natural selection and genetic diversification? While that does sometimes happen, in most cases, flowering plants have developed various mechanisms designed to reduce such self-fertilization. Some flowers have longer stamens than pistils, or vice versa, thus minimizing the probability that pollinating insects will transfer pollen grains inappropriately. Or the stamen and pistils simply grow away from each other, as can be between pages 208-211. And finally, many flowering species have biochemical mechanisms designed to reject fertilization by pollen grains from the same plant.

Meanwhile, our rostrum-bearing, honeydew-geysering aphids benefit greatly from all the sappy activity on view. Far from being beneficial pollinators, however, they're a serious nuisance to their flowery targets, even while supplying sugars, nutrients, and vitamins to their ant tenders. ■



2 Gymnosperms reproduce via unenclosed seeds, as opposed to the ovary-encased seeds of the angiosperms, or flowering plants. In pine trees, gymnosperms are visible as cones.

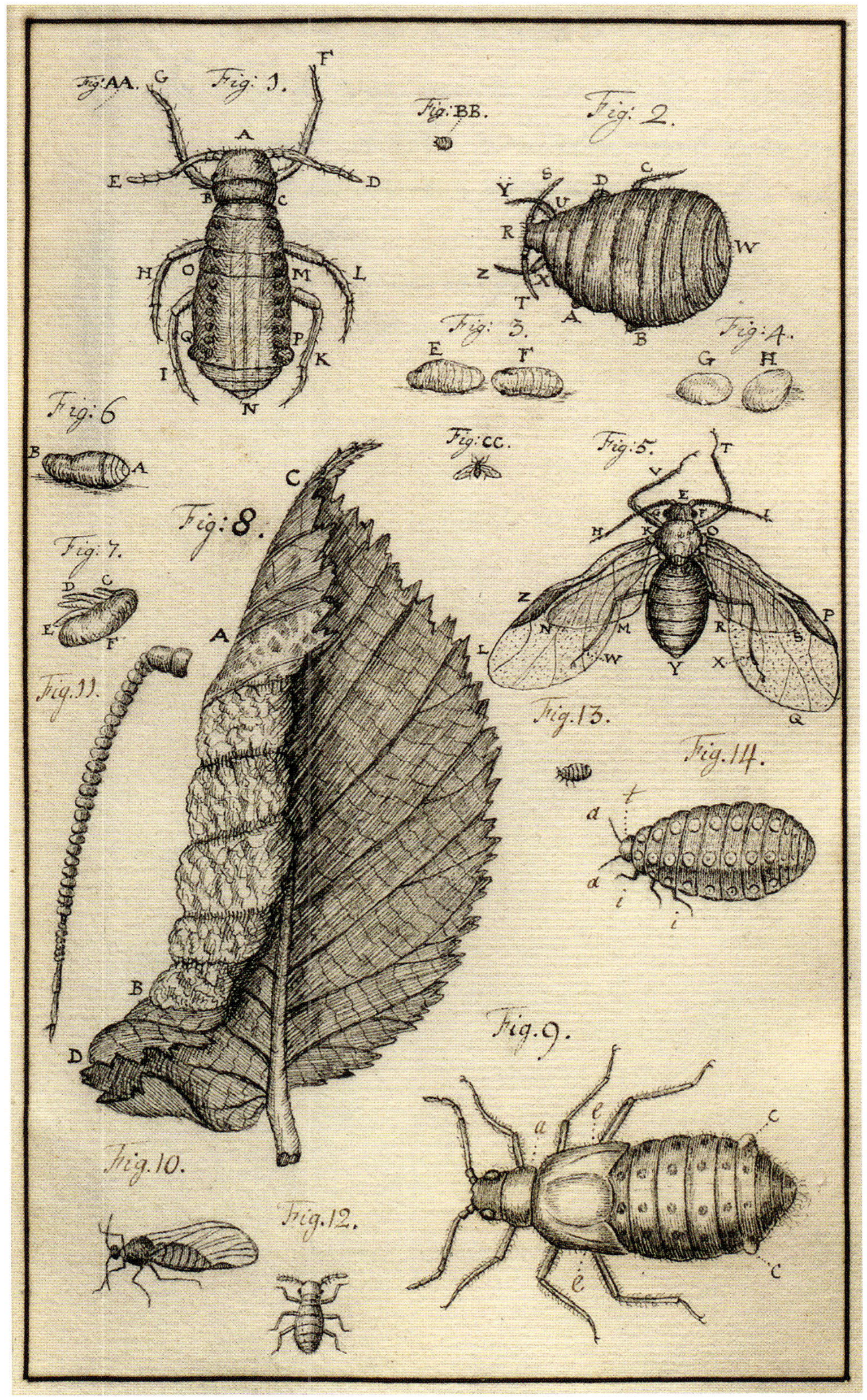


Charles De Geer, aphid life cycle, *Insekt illustrationer utförda med svart tusch*, 1778, ink on paper

# Beetles

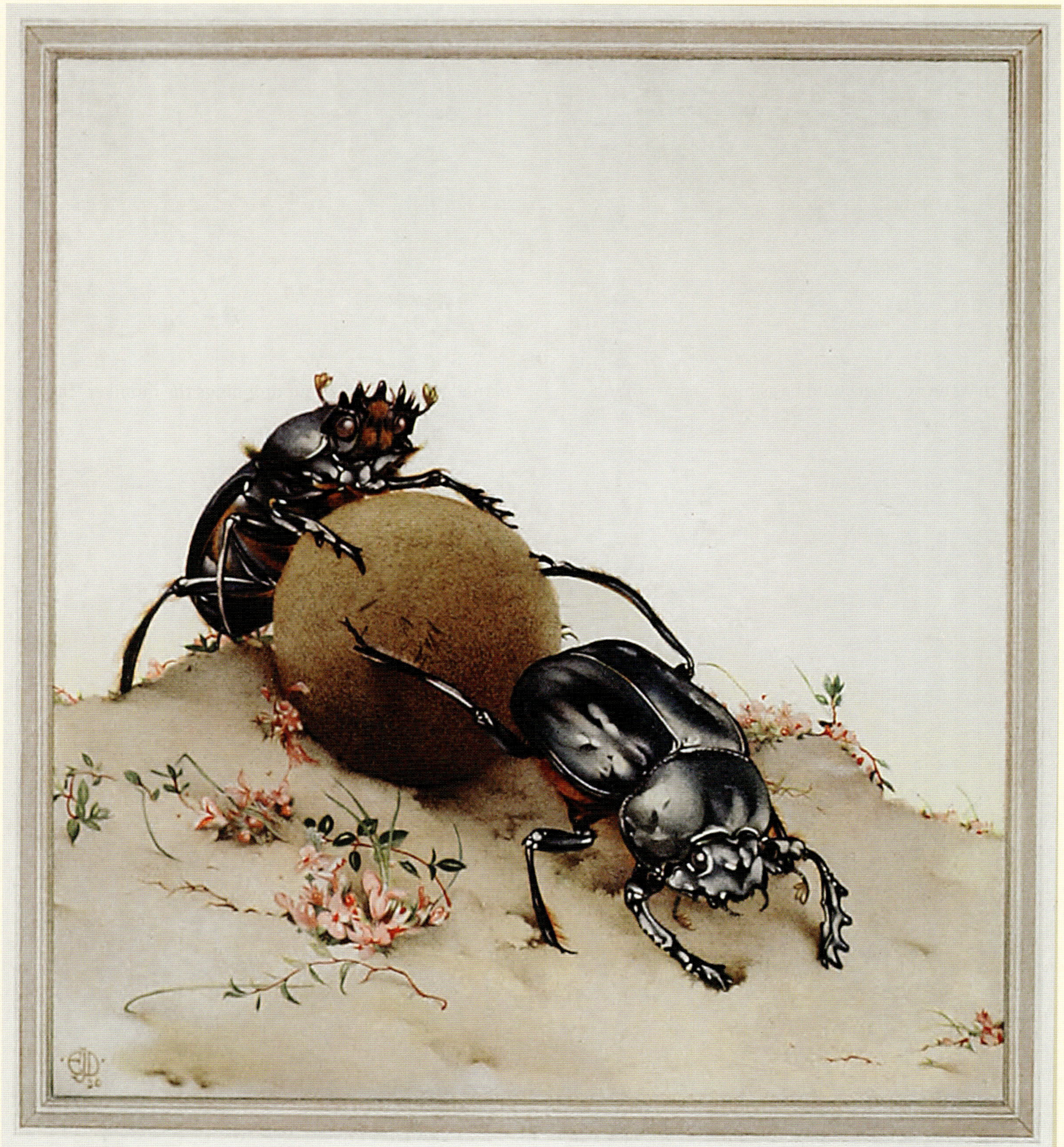

British geneticist and evolutionary biologist J. B. S. Haldane is probably best remembered today as originator of the so-called primordial soup theory, in which life is thought to have arisen from an abiotic fusion of organic compounds in early Earth's primordial ocean. Among most brilliant scientists of his time, Haldane was widely known for his quick wit. Asked by a group of theologians what he could intuit about the Creator from his long study of natural history, Haldane reportedly quipped, "that God has an inordinate fondness for beetles"—a reference to the extraordinary proliferation of beetle species worldwide.[1]

Whatever the deity's predilections, beetles can be found everywhere on Earth other than Antarctica. About 400,000 species have been documented, but estimates hold that somewhere between one and two million actually exist, with some entomologists maintaining that the number could be as high as five million. Even at the currently accepted figure, beetles constitute close to 40 percent of all documented insects, making them an astonishing 25 percent of known animal species.

Let that sink in for a second. *One out of four* of all animals that walk, fly, swim, squirm, or otherwise locomote within the confines of our planetary sphere are beetles. Ladybugs are beetles. Fireflies are beetles. Weevils are beetles, too—all 97,000 described species. Which in turn makes these snout-nosed herbivores just under a quarter of all insects.

Furthermore, beetles "represent a hyperdiverse monophyletic clade in the tree of life," as entomologist Arthur V. Evans describes them.[2] Translation: an extremely diverse group of organisms that have descended from a common ancestor or evolutionary group. One marker of their evolutionary success is that all four of the extant suborders of beetles to be found today already existed during the middle of the Triassic, 250 million years ago.

In short, beetles are incredibly ancient. They probably owe their survival through multiple planetary mass extinction events to their innate ability to burrow deep and lay low.

One way we know this is by examining the fossil record. Unfortunately, insect fossils preserved in stone tend to be mashed otherwise busted up. But another promising recent method of detecting the presence of ancient beetles deploys synchrotron X-ray tomography to scan petrified dinosaur dung. Such scans enable highly detailed three-dimensional reconstructions of whatever may have been enjoying a manure bath at the time. They've turned up numerous perfectly preserved beetles—an invaluable resource for the entomologically-minded paleontologist. Apart from highlighting an ancient scatological connection that we'll explore in more detail later, scans of this kind have also earned fossilized dinosaur droppings the moniker "the new amber." They are particularly useful because dinosaur dung comes from geological periods long before amber, which is the fossilized sap of coniferous trees that only arose in the Cenozoic, 65 million years ago.

Perhaps unsurprisingly for such a dominant order of insects, beetles have made a significant mark on our study of the structures and behaviors of the natural world. Aristotle first recognized them as a group circa 350 BCE, naming them *Coleoptera*. Pliny the Elder—the scholarly Roman author of a landmark encyclopedia of natural history who died in the eruption of Vesuvius in 79 CE—documented dozens of beetle types, including swimmers, borers, rovers, glowers-in-the-dark, and dung rollers. The important eighteenth-century Swedish biologist-taxonomist Carl Linnaeus, author of the binomial system for naming organisms that we still use, retained Aristotle's Coleoptera designation, now known to be the largest order of the insects. And both of the distinguished nineteenth-century English biologists who independently arrived at theories of natural selection, Charles Darwin and Alfred Russel Wallace, relied in part on observations of tropical beetles to substantiate their studies.

1 Haldane's epigram has ricocheted back from apocryphal status to being ostensibly confirmed. The question was finally resolved, at least to my satisfaction, by Stephen Jay Gould in his January 1993 article "A Special Fondness for Beetles" in *Natural History*. Not only did Haldane say it, but he was pleased to repeat it, Gould reported, sometimes with the variation "God has an inordinate fondness for stars and beetles."

2 Arthur V. Evans, *The Little Book of Beetles,* Princeton University Press, 2024.

**Left:** E. J. Detmold, *Sometimes the Scarab Seems to Enter into Partnership with a Friend*, 1921, lithograph

Beetles are different from most other insects in that their front wings have hardened into shell-like cases, called elytra. This is already visible in the earliest known example, *Coleopsis archaica*, which was discovered fossilized in 295-million-year-old rocks near Germany's Saar River in 2013. The hardened forewings serve as a protective carapace for the transparent hindwings, with only the latter used for flight. (It's how Aristotle arrived at "Coleoptera," *koleós* being ancient Greek for a leather sword cover, and *pterón* referring to wings.) The elytra give beetles their strange supercar aspect as a pair of doors seemingly rise on each side of their bodies just before flight, after which their larger hindwings unfold and they take to the air. The whole assembly is then rapidly refolded on landing. (The gleaming visual impression of their folded elytra, meanwhile, is how the VW Beetle earned its name, along with that car's overall resemblance to a ladybug.)

The elytra are considered an important factor in the Coleoptera's remarkable evolutionary success. They protect the fragile flight wings and delicate internal organs. The oblong protected space created by the folded elytra also allows desert beetles to conserve drinkable moisture (while simultaneously regulating their internal temperature), and it gives aquatic beetles a kind of readymade scuba tank to fill with breathable air before diving (while simultaneously regulating their buoyancy). And the elytra also enable these insects to burrow through abrasive materials, surface unscathed, and then take wing to escape predators on the other side.



Apart from catching the eye of some of our greatest scientists, beetles have had a significant impact on human cultural and economic history. Pests such as the boll weevil, which feeds on cotton bolls, the Colorado potato beetle, which prefers spuds, and the emerald ash borer, an Asian invasive species responsible for decimating ash forests across North America, have crashed agricultural production worldwide.

But that's taking an anthropocentric view. From their perspective, each of these species has enjoyed periods of great bounty—until pesticides such as DDT, calcium arsenate, and malathion descended from on high, spoiling the party. Typically only momentarily: these resilient insects tend to bounce back.

Actually, most beetle species are highly beneficial to the environment. They process plant and animal debris, eat agricultural pests such as aphids, and in general play a useful role in cycling decaying organic matter, typically a primary element of their diet. They are prolific pollinators as well, although not quite as important as bees or butterflies. Beetles were already well established when flowering plants evolved, however, and they may have been the first insects to shuttle pollen grains between them.

Evidence of this can be found in an ancient piece of amber discovered in northern Myanmar. Encased within it is a beetle scientists dated to 99 million years ago—the mid-Cretaceous period. Examined closely, this previously unknown species, now named *Angimordella burmitina*, proved to be covered in yellow grains. It also had pollen-collecting appendages near the mouth, comparable to those found in contemporary beetles. This 2019 discovery demonstrated that insect assistance of flowering plant reproduction was already well developed 99 million years ago, pushing the history of that insect-plant relationship back by about 50 million years.[3]

As already mentioned, about a quarter of all beetles are weevils—well over 60,000 recorded species. Seen via electron beam, these tiny creatures with their elongated noses have a profoundly alien and yet somehow benevolent appearance. (See pages 232–232.) The recently discovered iridescent jewel weevils of the Philippines' Mindanao islands glitter in dappled sunlight high up in the rainforest canopy. Their vivid markings are a signal to predators to beware: if eaten, they will cause severe indigestion, as their bodies are virtually indestructible. When chewed, their elytra tend to shatter, with the shards then working their way, much like broken glass, through the digestive systems of any creature foolish enough to eat them.

---

3 Stephanie Livingston, "This amber-encased beetle may have been one of the first insects to pollinate flowers," *Science,* November 11, 2019, science.org/content/article/amber-encased-beetle-may-have-been-one-first-insects-pollinate-flowers.

But if their brilliant appearance alone isn't enough to deter predators, the jewel weevils have perfected a particularly effective escape strategy. All four hundred or so species are incredibly sensitive to vibrations. The second anything makes the slightest bump—for example, into the branch they happen to be clinging to—they immediately drop to the forest floor, falling belly-up. Their dark undersides are the exact opposite of their vivid top halves, providing the perfect camouflage among dry leaves and tree litter. With their limbs pulled in, they look like innocuous seeds, if they are discernible at all.[4]

As for their impact on culture, no insect has played a larger role within human mythological constructs than a particular species of Mediterranean dung beetle, *Scarabaeus sacer*. Venerated by the ancient Egyptians, who identified it with the deity Khepri—a name meaning either "the one that arose by itself" or "to be born"—it was symbolically connected to the Sun. Usually depicted in wall art as a scarab hieroglyph, and sometimes as a human body with a beetle face, Khepri represented the end of the Sun's nightly journey through the underworld, the Sun God Ra's reappearance, and the beginning of the Sun's daily course through the desert sky.

Here, the importance of Ra in the ancient Egyptian belief system comes into play. The Sun God was seen as the creator of all life and the ruler of the earth, the heavens, and the underworld. Given these facts, even a cursory look at *Scarabaeus sacer*'s shape and behavior gives clues as to why this insect managed to achieve such an exalted status. Its head features six well-spaced projections, resembling rays. These link visually with four more horns on each of two large front legs, making fourteen evenly-spaced "solar rays" at the scarab's front end.

As to its behavior, like most dung beetles, *Scarabaeus sacer* collects fresh excrement, sculpts it into a ball larger than itself, and rolls it laboriously to a safe location for burial—where it remains stored for quite some time, a fragrant source of future nutrition. Female scarabs, meanwhile, do much the same while also laying a single egg within each ball, thereby providing an immediate food source for the resulting larva. For the Egyptians, this activity was seen as the veritable embodment of Khepri, who was often depicted rolling the sun across the firmament.

Along with the radiant front end of these beetles and the extraordinary resourcefulness necessary to form, roll, and bury spheres of dung many times their own size, the eventual emergence of a young scarab from within—seemingly without an act of procreation—further suggested divine provenance to the ancient Egyptians. The scatological element of this story, clearly, was not a problem.

In any case, rarely will you find a better illustration of evolution's ability to refine natural designs into ever more constructivist forms than the blunt, take-no-prisoners design of the dung beetle. There's no mistaking the compressed power visible in the limbs and body shape of the many varieties worldwide (and there are over 5,000 species). There something implacable about these industrious creatures. Their spade-like faces and recessed eyes resemble medieval Samurai battle helmets. One easily imagines the sheer elastic tensile strength of the muscles inside, anchored to reinforced corners of their bulldozer-like exoskeletons.

The very idea that these little dynamos can also fly is wondrous—a bit like seeing an Abrams tank lift off. One nocturnal species has acquired the ability to navigate by the light of the Milky Way in flight—the only insect known to have such a capability. Some estimates hold that the daily droppings of a single African elephant can support a staggering two million dung beetles. Individual specimen have been observed sculpting and rolling quantities of dung two-hundred fifty times their own body mass in a single night—a superpower worthy of the Marvel Universe. ■

4 Jason Bittel, "Dazzling new mimic beetles found, may already be under threat," *National Geographic*, August 1, 2019, nationalgeographic.com/animals/article/jewel-weevils-philippines-beetles-discovery.

# Snow

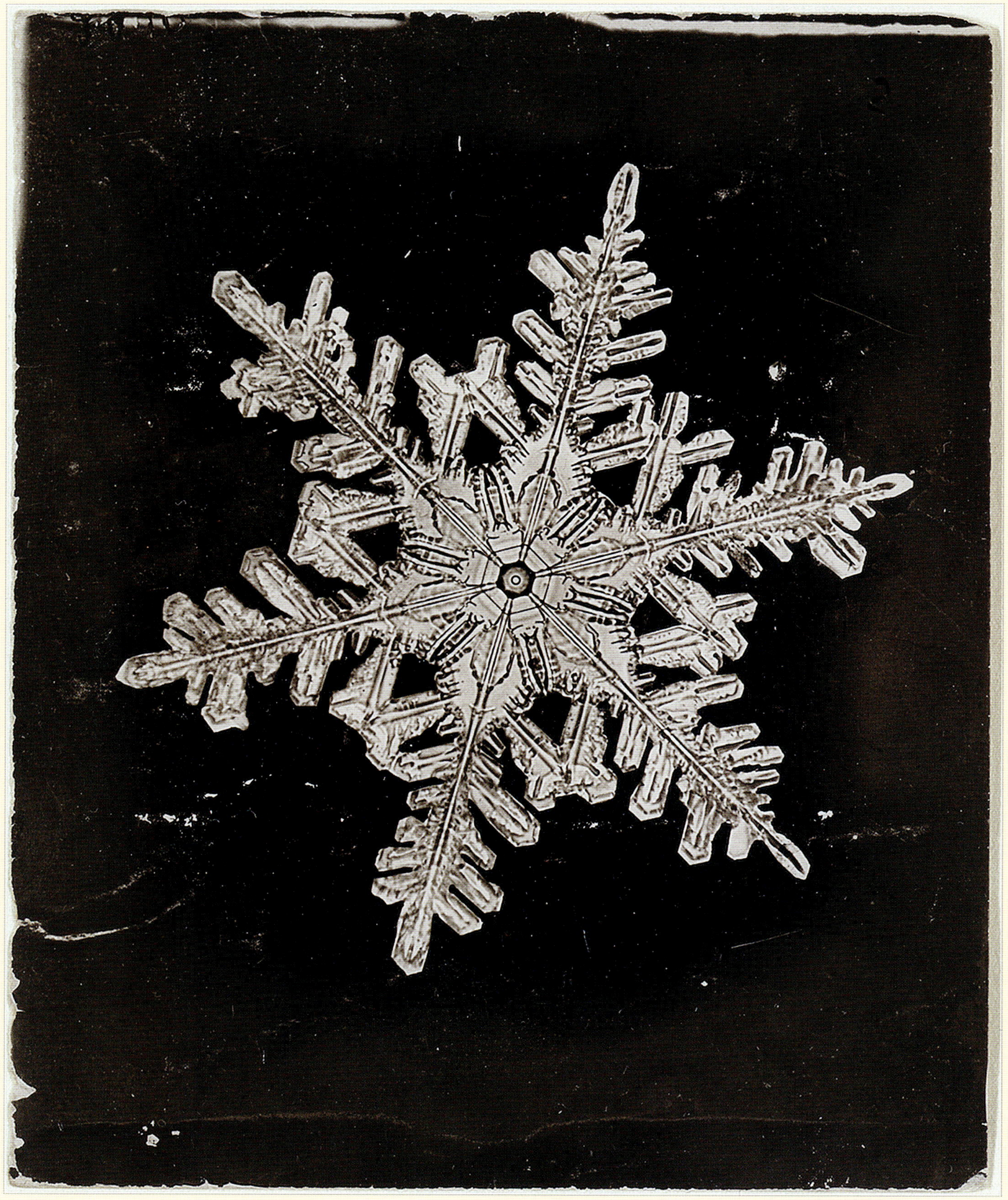

IT MIGHT SEEM COUNTERINTUITIVE TO consider ice a mineral. Be it transparent, opaque, solid, or a concatenation of loose crystals, water in its frozen form constitutes one-third of a tripartite terrestrial cycle in which $H_2O$ morphs through time, shapeshifting seamlessly from liquid to vapor to the subject of this chapter, depending on the conditions at hand. Whereas minerals generally require forcible extraction from the ground. So how could ice be a mineral?

Let's set its polymorphous qualities aside for the moment. Water ice meets all the requirements. It's an inorganic solid. It has an organized crystalline structure, with its molecules arranged in neatly hexagonal lattices. It occurs naturally. And it has a well-defined chemical composition. Furthermore, when far enough below the freezing point, it's hard as rock.

In short, it's a mineral.

Seen from space, Earth appears to have an entire continent of the stuff: Antarctica, all 5,500,000 square miles of it. The north pole too is still capped by ice every winter—though that's growing ever more irregular as global temperatures rise. Wherever it is, ice in terrestrial conditions tends to hover relatively close to zero Celsius, or 32 degrees Fahrenheit, meaning it's forever on the verge of either melting or sublimating (that is, going from solid to vapor without an intermediate liquid stage). Farther away from the Sun, however, many of the moons orbiting gas giant planets such as Jupiter and Saturn are composed mostly of water ice. Given prevailing temperatures of anywhere between -160 to -330 degrees Celsius, or -256 to -562 Fahrenheit, their ice tends to remain more or less definitively in its mineral stage, at least on the surfaces of those satellites—though even such frigid places as Europa and Ganymede, two of Jupiter's largest moons, and Saturn's diminutive satellite Enceladus, are also known to possess subsurface liquid oceans beneath their ice crusts. (Of course, Earth also has liquid minerals under its surface, some of which occasionally emerge. We call it lava.)

On Earth, the most visually complex and stunning manifestation of water ice appears when it floats down from the sky in the form of flakes. As individual forms, these comprise single ice crystals that have coalesced around particles of dust in the air. As each flake wafts about, encountering differing temperatures and levels of humidity in the atmosphere, its crystalline structure shifts, expanding, contracting, and growing ever more complex. Once an initial droplet has frozen around that microscopic nucleus, it attracts other water molecules, which transition from airborne vapor to solid, expanding the size of the nascent flake until it reaches hundreds of microns up tpo several millimeters in size. By depleting atmospheric vapor and emerging as mineral, the snowflake is suddenly susceptible to gravity, becoming precipitate—a word signifying both the creation of a solid from a liquid and the fall of either from the sky. In the case of snow, that gravitational susceptibility is reinforced by the tendency of individual flakes to clump together.

Snowflakes transition from vapor into a complex variety of forms, morphing from molecular water into both crystalline lattice and language, becoming needle, plate, column, capped column, rime, cup, rosette, crystal, stellar dendrite, and needle cluster—themselves constituent parts of approximately eighty snowflake variants organized into eight classification schemes. As they fall and grow, or fall and diminish, or waft through varying levels of humidity while exhibiting ever more variegated shapes, the flakes reflect light through their crystalline facets. Individually they are clear, but collectively they diffuse light such that they appear white. As the driven snow.[1]

Capturing good images of snowflakes in the scanning electron microscope is a fraught process. It requires speed, precision, the right pressure in the vacuum chamber (not too low), the right beam strength (not too strong), patience, and a generous supply of LN2, or liquid nitrogen—the latter being a fractional distillation of liquid air, appropriately enough. It can also require a skilled collaborator, and all the snowflakes seen here were made with the help of Dr. Jianqun Wang in Ontario, a Canadian province known

1 *See in particular* Margaret Wertheim, "Building a Better Snowflake: An Interview with Kenneth Libbrecht," *Cabinet Magazine*, Spring 2008.

**Left:** Wilson Bentley, *Photograph of Snowflake*, circa 1910, gelatin silver print for seven-month stretches of near-continuous snowflake production annually. The liquid nitrogen enables the cooling of a dime-sized sample stub to circa -196 Celsius, or -320 Fahrenheit. Of course, any successful SEM snowflake hunt also requires the right conditions *outside* the lab, meaning anywhere from -5 Celsius, or -23 degrees Fahrenheit, to much colder at ground level—though the flakes typically fall through lower temperatures still during their complex atmospheric trajectories.

Even when not pinioned by electron beams, snow transforms almost continuously. When it's not melting due to temperature rise, it's sublimating due to low humidity—as happens all the time in the arid conditions of western North America, for example, where newly fallen snow can vanish within days even when temperatures stay well below freezing. But this process is radically accelerated in the SEM, where the vacuum chamber's artifically low pressure atmospheric combines with the beam's energy to limit snowflake life to minutes. Still, these forces are countervailed in part by the liquid nitrogen and cryogenic "stage"—the latter being a supercooled metal block that keeps the sample stub temperature low.



The conceits that snowflakes are "perfect" and that no two are ever alike appear to stem from photographer Wilson Bentley, a native of Vermont who started capturing them through a microscope on glass plate negatives in January 1885—early enough that he's considered the first to have documented snowflakes photographically. Bentley proceeded to record over 5,000 flakes against black velvet until his death in 1931, which was also the year he published *Snow Crystals*, the 227-page monograph bearing about half his life's work. In the process he became one of Vermont's famous sons, and to this day it's likely that if you've seen a photograph of a snowflake, it was taken by Bentley in a 90-second exposure outside of his barn.

Apart from printing his photographs, which focused largely on symmetrical six-sided flakes rather than the many irregular forms, Bentley occasionally meditated on his lifelong calling. "Every crystal was a masterpiece of design and no one design was ever repeated," he wrote—an observation that captured the popular imagination and proved correct: no two flakes are ever exactly the same, and even in laboratory conditions it's difficult to make them appear so. As to their perfection, an attribute notoriously difficult to quantify in either nature or human affairs, whatever references Bentley made were largely rhetorical and couched in the utopian language of nineteenth-century naturalism. Snowflakes, he wrote, "come to us not only to reveal the wondrous beauty of the minute in Nature, but to teach us that all earthly beauty is transient and must soon fade away."[2]

Like many an image maker, Bentley tended to ignore forms he deemed less photogenic, instead shooting what he liked. In 1882, German meteorologist Georg Gustav Hellmann saw reproductions of Bentley's work in Prussian Silesia and sensed something was off, so he commissioned local photographer Richard Neuhauss to conduct his own study. On examining Neuhauss's work, Hellmann noted many irregular flakes and a variety of crystalline shapes not to be found in Bentley's pictures. Although Bentley had never claimed that his pictures represented the full scope of snowflake morphologies, Hellmann accused him of a grave scientific crime: the falsification of data.

The resulting battle lasted for decades, with Bentley eventually admitting to having made direct interventions onto his plates to remove imperfections such as dust particles—but only to keep his photographs "as true to nature as possible, and if retouching will help in this respect, then it is fully justified."[3] Although their bitter public argument was never really resolved, by carping on nature's imperfections, Dr. Hellmann ultimately helped his adversary win, at least in the court of public opinion, and Bentley's pristine six-sided flakes continue to be what we picture when we consider the snow. ■

2 Andrea Shea, "How Vermont's 'snowflake man' created the first-ever images of fleeting snow crystals," WBUR, February 25, 2022, wbur.org/news/2022/02/25/snow-flake-man-photographs-wilson-bentley.

3 Helen Pilcher, "The Great White Lie, What Snowflakes Really look Like," *New Scientist*, December 17, 2013.

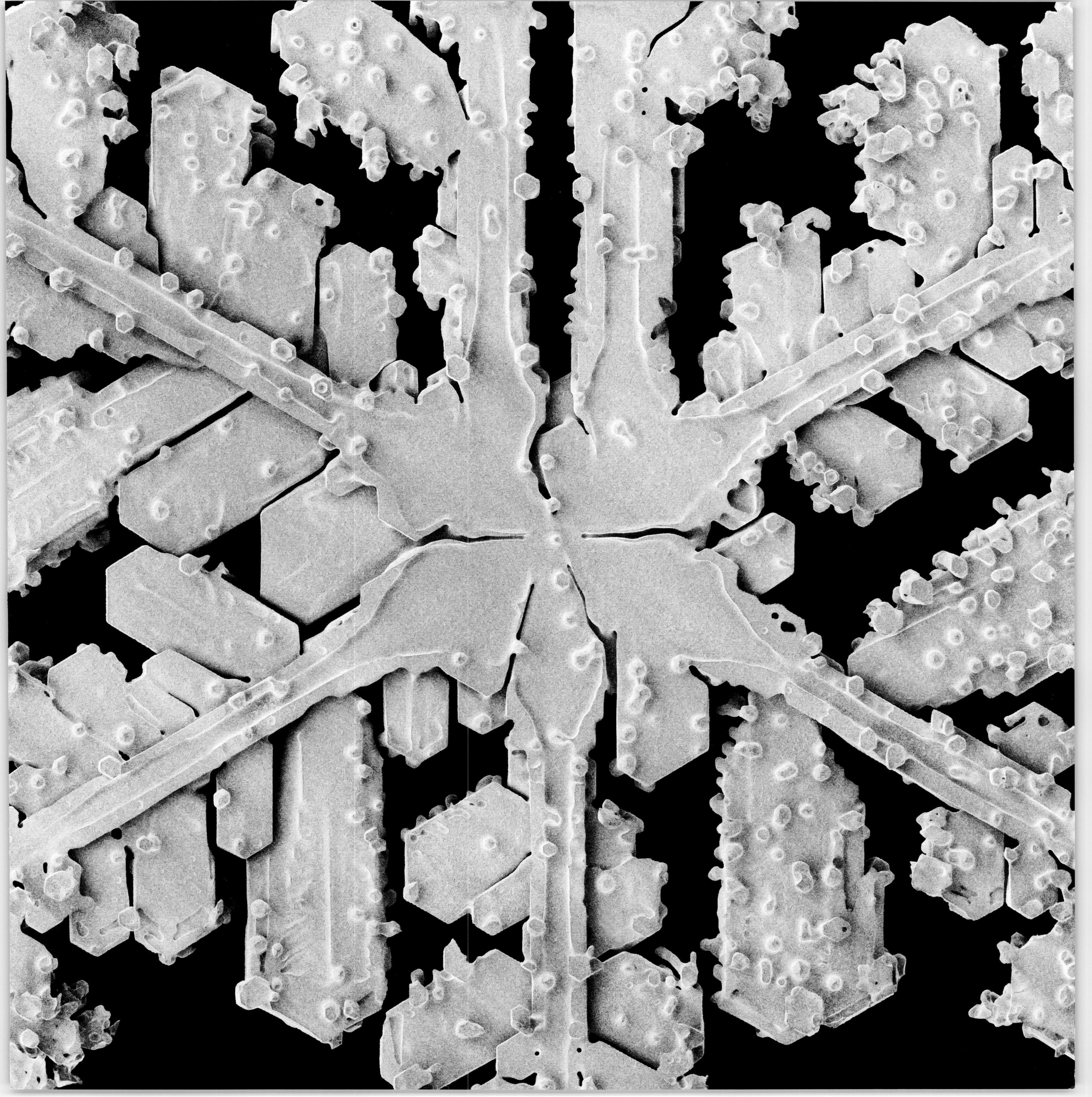

# Acknowledgments

NOBODY SUPPORTED THIS PROJECT MORE than my wife, Melita Gabrič, who has my enduring gratitude and much more. My friend and mentor Eric Himmel of Abrams stuck with me across a number of false starts, and *Nanocosmos* simply wouldn't have been possible without him. Also utterly crucial to the work presented here was the president and CEO of the Canadian Museum of Nature, Margaret Beckel—and subsequently, that outstanding institution writ large. And in 2018, Sir Michael Dixon, then the Director of the Natural History Museum in London, kindly wrote to Margaret on my behalf—an invaluable recommendation.

This is my fifth book with Boris Balant, whose acutely discerning design sensibility is in a class of its own. One of the leading practitioners of literary nonfiction in English, Lawrence Weschler, has taken a gratifying interest in my projects for years, with one result being thoughtful interventions such as the Pascal passage now serving as epigraph to this book.[1] Sarah Lazin, my friend and literary agent since the turn of the century, has supported *Nanocosmos* from the beginning. Danika Goosney, successor to Meg Beckel at the CMN, backed my continuing SEM work with sensitivity and enthusiasm, for which I owe her many thanks.

The opening of my Introduction riffs on Auden's poem *Musée des Beaux Arts*. Its final section quotes Poe's *The Raven*.

My editor at Abrams, Zack Knoll, shepherded *Nanocosmos* to print with sympathetic intelligence throughout. Zack made sure I had the additional time I needed at the very end—a huge bonus. His assistant Ruby Pucillo read every word of this book and offered valuable suggestions. Meanwhile Abrams publisher Michael Sand stuck with me for the duration, for which I'm thankful.

This project had a long and complicated gestation. Back in 2013, Jill Golden, my friend since high school, introduced me to her old Swarthmore pal Neil Gershenfeld, now director of MIT's Center for Bits and Atoms. This led to an invitation to come to the Media Lab, and within a few months I was a visiting scholar using their SEM. I owe Neil and Jill my thanks, as well as then-CBA members Shuguang Zhang and Andreas Mershin, who have been fast friends ever since. Meanwhile CBA postdoc Prashant Patil provided pointers in using the microscope.

Also in Cambridge I met Ben Weiss, the chair of planetary sciences at MIT, who helped me gain access to the lunar samples seen here. I thank him and NASA Lunar Samples Curator Ryan Zeigler, who not only sent me a dozen pieces of the Moon, but also gave me more time with them when I needed it.

Prior to all that, pioneering SEM artist Dee Breger of Columbia University heard of my interest in the field and reached out. Unfortunately, Dee, the founder of Columbia's first electron microscopy facility, passed in September 2016. So while I didn't have the chance to share my results with her, I acknowledge her support and encouragement, as well as her trailblazing status.

In 2012, NYU biology professor Tyler Volkheard about my nascent project and suggested I contact O. Roger Anderson of Columbia University, a leading researcher in radiolarians and other protists. This led to the loan of samples by Dr. Anderson, who also provided much advice on how to prepare them for the SEM. Another leading NYU prof, novelist Larry Lockridge, kindly contacted his friend, Columbia professor and 2008 Nobel laureate in chemistry Martin Chalfie, on my behalf. This led to an introduction to neuroscientist David Hall of the Albert Einstein College of Medicine in the Bronx, who helped arrange my access to the AECM SEM. My visits to their Analytical Imaging Facility were also greatly expedited by senior staff scientist Frank Macaluso.

Just as I ran out of my alotted time on the AECM scope, Beth Jacobs, with typically kindly resourcefulness, helped arrange access to the SEM at CUNY's Advanced Science Research Center.

1 Check out Ren's always intriguing Substack: lawrenceweschler.substack.com.

Unfortunately our untimely return to Europe precluded that opportunity—but almost immediately thereafter Christian Koeberl, the director of Vienna's Natural History Museum, provided access to that institution's SEM. Once again, however, our peripatetic lives took over, and we moved to Canada! In any case Beth and Christian are big friends of *Nanocosmos*, and I am grateful to both.

At the Museum of Nature, the VP for experience and engagement, Ailsa Barry, chief scientist Mark Graham, research and collections VP Jeff Saarela, and distinguished researchers Jean-Marc Gagnon, Aaron Lussier, and entomologists Robert Anderson and Thomas Onuferko, all provided meaningful support, ideas, samples, and more. I am particularly indebted to Bob Anderson and Francois Genier for last minute caption information. As to the technology permitting these images, nobody was more crucially helpful inthe real-time problem-solving of SEM issues than Paul Hamilton and Glenn Poirier—expert microscopists both.

At the Smithsonian, Sue Lutz, manager of the US National Herbarium, kindly agreed to lend me important dinoflagellate samples constituing the life's work of the late Dr. Maria A. Faust, curator of microalgae at the National Museum of Natural History from 1970 to 2010. I gratefully acknowledge their contributions.

I couldn't have captured the snowflakes visible in the final pages of this volume without the help of Roger Bull, head of operations for Canada's National Diversity Cryobank. At Jeff Saarela's suggestion, Roger lent me one of his lab's futuristic "cryopods"—liquid-nitrogen cooled containers usually used to transport DNA samples between Canadian provinces. (Roger also taught me how to charge the pod with N2o.) It served as an invaluable snowflake capture and conveyance tool. No less important was Carlton University Nanoimaging Lab chief Jianqun Wang, who established the pressure and power settings necessary to capture such ephemeral objects, and worked closely with me. Carlton professor Dr. Shelley Hepworth introduced me to Jianqun, who also helped me prepare many other kinds of SEM specimens in his lab.

Eric Himmel, Bob Bowen, my brother Nick, and my wife, Melita, all read drafts of the Introduction and provided actionable suggestions. An accomplished photo-based artist, Bob also served as an invaluable image reaction sounding board. So did Paul Roth, the director of The Image Centre at Toronto Metropolitan University and a leading photography curator and public intellectual. In Ljubljana, photographer Miha Koron assembled numerous initial mosaic composites before sending them on to me to finish. (He also insisted I learn how to use a graphic tablet, and provided a tutorial.) In Ottawa, Don Asquin was always ready with words of encouragment and to loan SEM samples. In South Africa, marine biologist Roksana Majewska arranged the loan of protist samples. In London, Stephen Stukins, Senior Curator in Micropaleontology at the Natural History Museum, took me on a revelatory tour of his department and also kindly arranged a sample loan. In France, Spain and Germany, researchers Fabrice Not, Miguel Mendez, and Johan Renaudie helped me establish credible radiolarian species names at the very last minute.

In Barcelona, Javier Hidalgo Gil and my other friends at the La Caixa Foundation invited me to stage a successful early *Nanocosmos* touring exhibition, for which I am very grateful. Jernej Gregoric, my friend and exhibitions tour manager, has stuck by me and this project for the duration. In Toronto, Dmitri Levanoff made prepress adjustments to my SEM image files. Also in Toronto, my fellow image-maker Elaine Waisglass has proven to be a friend and fellow conspirator about the use of digital tools to fuse photographic and visual-arts techniques. Robert Polidori and Ed Burtynsky have both encouraged me in my work in multiple thoughtful ways. Mark Mayer, Ann Thomas, and David Pantalony have served as valuable sounding boards as well. I'm lucky to call them all friends.

In Cairo, Aliaa Al-Sharif enabled me to see what I described in the Introduction's Egypt section, kindly arranging a guide and car. That 2024 trip wouldn't have happened without the support of my fellow cosmists James Green and Wasim Ahmed. And finally, many others at the CMN have graciously provided much kindness, help and encouragement over the years. You know who you are, and I thank you for it. ■

# Index

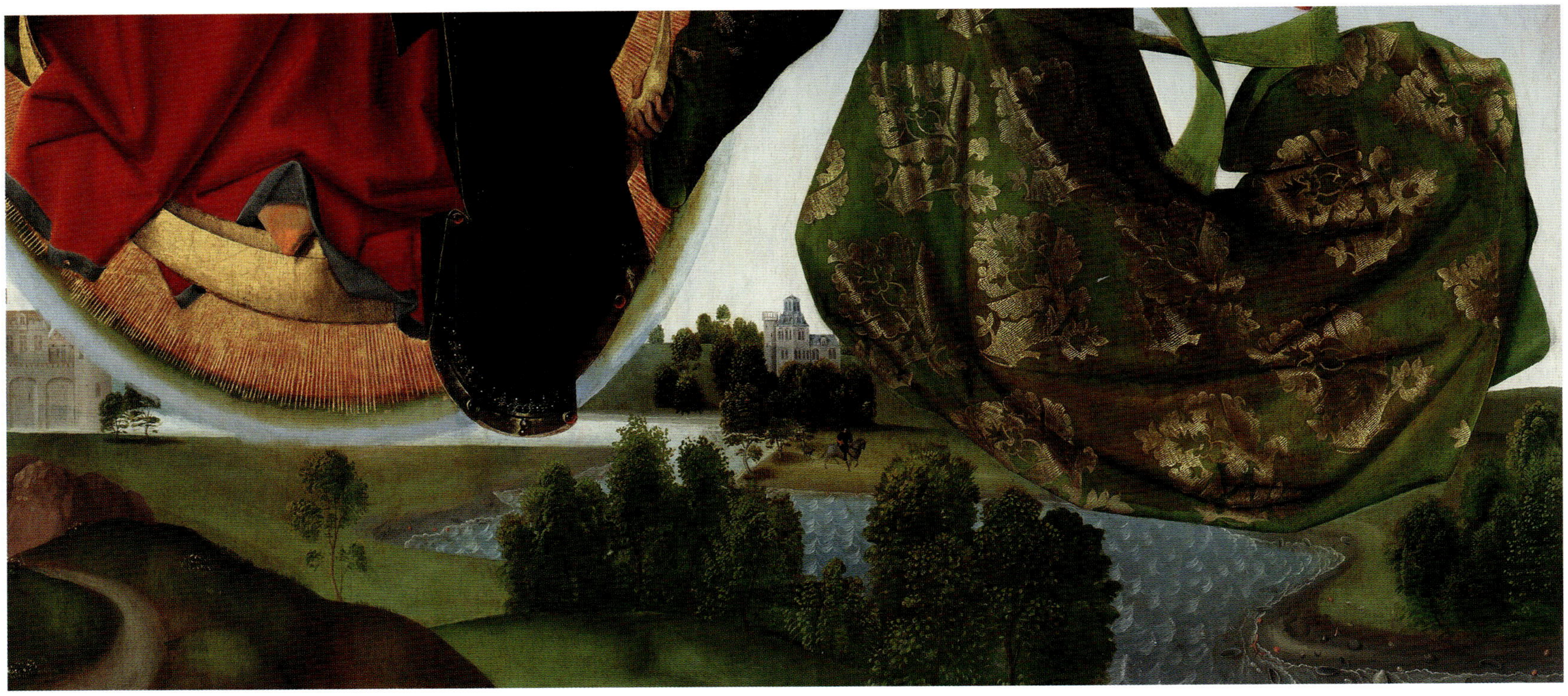



# IMAGE CREDITS

Page 6: National Gallery of Art, Washington, DC; page 9, National Gallery of Art; page 12, Smithsonian Libraries and Archives, Washington; page 16, Rijksmuseum, Amsterdam; page 18, Rijksmuseum; page 19, British Library, London; page 24, Rijksmuseum; page 29, Michael Benson; page 30, NASA/ESA/Hubble; page 31, courtesy Kathryn Finter; page 32-33, M. Benson; page 34 National Gallery of Art; page 36, M. Benson; page 288, Wolbach Library, Harvard; page 292, Smithsonian Libraries and Archives; page 295, Wikimedia Commons; page 296, University of California Libraries; page 299, Rijksmuseum; page 301, Archive.org; page 303, Getty Research Institute, Los Angeles; page 305, US Dept. Agriculture, Washington; page 307, Biodiversity Heritage Library; page 308, Marygrove College Library, Detroit; page 312, Public Domain Review, Manchester; Page 318, National Gallery of Art.

**For additional *Nanocosmos* content, including subjects that didn't make it into the book, size comparisons of images presented here, and multimedia content, visit:**

Master of the Saint Lucy Legend, *Mary, Queen of Heaven*, circa 1485, oil on panel, detail view

Editor: Zack Knoll
Image Processing and Sequencing: Michael Benson
Designers: Boris Balant and Michael Benson
Design Manager: Darilyn Lowe Carnes
Managing Editor: Krista Keplinger
Production Manager: Sarah Masterson Hally

Editor emeritus: Eric Himmel

Library of Congress Control Number: 2025931666

ISBN: 978-0-8109-9797-4
eISBN: 979-8-89684-048-0



Printed and bound in China
10 9 8 7 6 5 4 3 2 1



**ABRAMS** The Art of Books
195 Broadway, New York, NY 10007
abramsbooks.com

ABRAMS is represented in the UK and Europe by Abrams & Chronicle Books, 1 West Smithfield, London EC1A 9JU and Média Participations, 57 rue Gaston Tessier, 75166 Paris, France.
abramsandchronicle.co.uk and media-participations.com
info@abramsandchronicle.co.uk